Intelligent Systems Reference Library

Volume 150

The aim of this series is to publish a Reference Library, including novel advances and developments in all aspects of Intelligent Systems in an easily accessible and well structured form. The series includes reference works, handbooks, compendia, textbooks, well-structured monographs, dictionaries, and encyclopedias. It contains well integrated knowledge and current information in the field of Intelligent Systems. The series covers the theory, applications, and design methods of Intelligent Systems. Virtually all disciplines such as engineering, computer science, avionics, business, e-commerce, environment, healthcare, physics and life science are included. The list of topics spans all the areas of modern intelligent systems such as: Ambient intelligence, Computational intelligence, Social intelligence, Computational neuroscience, Artificial life, Virtual society, Cognitive systems, DNA and immunity-based systems, e-Learning and teaching, Human-centred computing and Machine ethics, Intelligent control, Intelligent data analysis, Knowledge-based paradigms, Knowledge management, Intelligent agents, Intelligent decision making, Intelligent network security, Interactive entertainment, Learning paradigms, Recommender systems, Robotics and Mechatronics including human-machine teaming, Self-organizing and adaptive systems, Soft computing including Neural systems, Fuzzy systems, Evolutionary computing and the Fusion of these paradigms, Perception and Vision, Web intelligence and Multimedia.

More information about this series at http://www.springer.com/series/8578

Jude Hemanth · Valentina Emilia Balas
Editors

Nature Inspired Optimization Techniques for Image Processing Applications

Springer

Editors
Jude Hemanth
Department of Electronics and Communication Engineering
Karunya University
Coimbatore, Tamil Nadu, India

Valentina Emilia Balas
Department of Automation and Applied Informatics
Aurel Vlaicu University of Arad
Arad, Romania

ISSN 1868-4394 ISSN 1868-4408 (electronic)
Intelligent Systems Reference Library
ISBN 978-3-319-96001-2 ISBN 978-3-319-96002-9 (eBook)
https://doi.org/10.1007/978-3-319-96002-9

Library of Congress Control Number: 2018948603

This Springer imprint is published by the registered company Springer Nature Switzerland AG
The registered company address is: Gewerbestrasse 11, 6330 Cham, Switzerland

Foreword

The active quest to endow machines with human abilities has been a feature of modern times. The ultimate goal of creating an artificially intelligent and autonomous entity has been approached through many intermediate steps by providing human-like functionality in a myriad of applications, including industrial automation, health care and security. A chief biological function that has been pursued is that of analysing and understanding visual information. Advances in image processing and computer vision have been adopted in a range of applications and have transformed what is possible to be done automatically and without the need for human visual intervention. In certain applications, machine capabilities have even surpassed what humans can do. However, while in some of these limited cases they have outstripped the human capabilities in terms of scale and speed, there are still areas where humans have the edge and, therefore, the search for better approaches and algorithms for image understanding continues.

At the same time, a better understanding of the emergence of biological systems, including humans, has drawn the designers of machine vision systems to try to learn from Nature. Through a very long process, spanning millennia, the Nature's own search for effective autonomous entities has resulted in efficient and effective mechanisms for understanding and interacting with the world. Scientists and designers are now learning from the fruits of Nature's long labour to expedite the development of artificial systems.

This volume brings together some of these naturally inspired approaches for image understanding in one place and also provides a sample of the vast array of applications to which they can be applied. For the reader new to these approaches, it will provide a good starting point and for the more advanced algorithm designers, it may suggest new ideas that they have not considered before.

The deep and vast experience of Nature is a great resource for engineers and designers in their quest for novel solutions to the current and emerging challenges that face humanity. It is hoped that this book will contribute to this quest and strengthens the case for the continued study of Nature in search of new insights.

Canterbury, UK
August 2018

Farzin Deravi, CEng, FIET
Professor of Information Engineering
Head of School of Engineering & Digital Arts
University of Kent

Preface

This edited book is one of the significant contributions in the field of intelligent systems for practical applications. This book is interdisciplinary with a wide coverage of topics from nature-inspired optimization techniques and image processing applications. The main objective of this book is to highlight the state-of-the-art methods in these interdisciplinary areas to the researchers and academicians. Variety of practical applications are covered in this book which can assist the budding researcher to choose his own area of research. This book also covers in-depth analysis of the methods which will attract high-end researchers to further explore or innovate in these areas. In a nutshell, this book is a complete product for usage by anyone working in the areas of intelligent systems. A brief introduction about each chapter is as follows.

Chapter 1 illustrates the application of firefly optimization algorithm for brain image analysis. Specifically, the methodology of CT and MRI brain image segmentation is analysed in detail. Chapter 2 deals with image compression using bat optimization techniques. An in-depth analysis of codebook generation for image compression is analysed which will attract the readers. Chapter 3 deals with natural language processing using particle swarm optimization methods. Few modified swarm approaches are suggested in this chapter for efficient categorization of alphabets in languages. The proposed approach is tested with Tamil language but can be extended to different languages across the globe.

Chapter 4 covers the application of grey wolf optimization algorithm for image steganography applications. Feature optimization for efficient data hiding is the main objective of the work covered in this chapter. Literature survey is one critical area of research which will attract several readers. With this idea, a detailed survey on nature-inspired techniques for image processing applications is dealt in Chap. 5. The application for ant colony optimization for visual cryptography is discussed in Chap. 6. The primary focus of this work is image enhancement which can assist in developing efficient cryptographic systems. Qualitative and quantitative analyses are covered in this chapter which is more beneficial to the readers.

The necessity of image analysis methods is significantly increasing in the area of agriculture. The application of swarm intelligence techniques for detecting the quality of crops via images is illustrated in Chap. 7. Analysing the quality of different stages of wheat is the main focus of this chapter. Chapter 8 discusses the various concepts of image preprocessing using cuckoo search optimization techniques. Different types of input images are used in this chapter to validate the proposed methodology. Automatic skin disease identification in mango fruits using artificial bee colony algorithm is the focus of Chap. 9. The optimization algorithm is used to select the optimal features for skin classification in this chapter.

Chapter 10 covers the different optimization techniques for fixing the structure of the complex deep convolutional neural networks. An efficient architecture will enhance the performance of any automated system. Chapter 11 deals with the application of differential evolution method for quality enhancement in underwater images. Chapter 12 covers the application of genetic algorithm for biometrics application. Fetal biometrics-based abnormality detection is the prime focus of this chapter.

We are grateful to the authors and reviewers for their excellent contributions for making this book possible.

Our special thanks go to Janus Kacprzyk and Lakhmi C. Jain (Series Editors to Intelligent Systems Reference Library) for the opportunity to organize this guest-edited volume.

We are grateful to Springer, especially to Dr. Thomas Ditzinger (Senior Editor) for the excellent collaboration.

We would like to express our gratitude and thanks to Handling Editor Ms. Rajalakshmi Narayanan, Springer, Chennai and her team for their wholehearted editorial support and assistance while preparing the manuscript.

This edited book covers the fundamental concepts and application areas in detail which is one of the main advantages of this book. Being an interdisciplinary book, we hope it will be useful to a wide variety of readers and will provide useful information to professors, researchers and graduated, and all will find this collection of papers inspiring, informative and useful.

Coimbatore, India
Arad, Romania
August 2018

Jude Hemanth
Valentina Emilia Balas

Contents

Chapter 1
Firefly Optimization Based Improved Fuzzy Clustering for CT/MR Image Segmentation

S. N. Kumar, A. Lenin Fred, H. Ajay Kumar and P. Sebastin Varghese

Abstract The segmentation is the process of extraction of the desired region of interest. In medical images, the anatomical organs and anomalies like a tumor, cysts, etc. are of importance for the diagnosis of diseases by physicians for telemedicine applications. The thresholding, region growing, and edge detection are termed as classical segmentation algorithms. Clustering is an unsupervised learning technique to group similar data points and fuzzy partitioning merges similar pixels based on the fuzzy membership value. The classical FCM algorithm lacks sensitivity in the cluster centroid initialization and often gets trapped in local minima. The optimization algorithm gains its importance in cluster centroids initialization, thereby improving the efficiency of FCM algorithm. In this work, firefly optimization is coupled with FCM algorithm for CT/MR medical image segmentation. Fireflies are insects having a natural capacity to illumine in dark with glowing and flickering lights and firefly optimization algorithm was modeled based on its biological traits. The preprocessing stage comprises of artifacts removal and denoising by Nonlinear Tensor Diffusion (NLTD) filter. The computation time was minimized by reducing the total pixels count for the processing. The Firefly optimization, when coupled with FCM, generates satisfactory results inconsistent with FCM when coupled with Cuckoo, Artificial Bee Colony, and Simulated annealing algorithms. The cluster validity performance metrics are used for the determination of optimum number of clusters. The algorithms are developed in Matlab 2010a and tested on real-time abdomen datasets.

S. N. Kumar (✉)
Sathyabama Institute of Science and Technology, Chennai, India
e-mail: appu123kumar@gmail.com

A. Lenin Fred · H. Ajay Kumar
Mar Ephraem College of Engineering and Technology, Elavuvilai, India
e-mail: leninfred.a@gmail.com

H. Ajay Kumar
e-mail: ajayhakkumar@gmail.com

P. Sebastin Varghese
Metro Scans and Research Laboratory, Trivandrum, India
e-mail: sebastin464@gmail.com

J. Hemanth and V. E. Balas (eds.), *Nature Inspired Optimization Techniques for Image Processing Applications*, Intelligent Systems Reference Library 150,
https://doi.org/10.1007/978-3-319-96002-9_1

Keywords Unsupervised learning · Clustering · Fuzzy C means
FCM-firefly algorithm · FCM-artificial bee colony algorithm · FCM-cuckoo algorithm

1.1 Introduction

Medical image processing refers to the application of computer-aided algorithms for the extraction of anatomical organs and analysis of anomalies like a tumor, cyst, etc. The various steps in image processing are restoration, enhancement, segmentation, classification, and compression. The segmentation can be defined as the extraction of Region of Interest (ROI). The Computer Tomography (CT), Magnetic Resonance Imaging (MRI), Ultrasound and Positron Emission Tomography (PET) are the widely used medical imaging modalities for the disease diagnosis. The choice of segmentation algorithm depends on the medical imaging modality and its characteristics.

The CT images, in general, are corrupted by Gaussian noise and its distribution is as follows.

$$p(z) = \frac{1}{\sqrt{2\pi}\sigma} e^{-(x-\mu)^2/2\sigma^2}$$

where x represents random variable normally distributed with mean μ and standard deviation σ.

The MR images are corrupted by rician noise, artifacts and intensity inhomogeneity due to the non-uniform response of RF coil. The rician noise distribution is as follows

$$p(z) = \frac{z}{\sigma^2} \exp\left(-\frac{z^2 + I^2}{2\sigma^2}\right) B\left(\frac{z\alpha}{\sigma^2}\right)$$

where, I is the true intensity value, σ is the standard deviation of the noise, and B is the modified zeroth order Bessel function.

The Ultrasound images, in general, are corrupted by speckle noise and its distribution is as follows.

$$F(x) = \left(\frac{g^{\gamma-1}}{(\gamma-1)!a^{\gamma}} e^{\frac{g}{a}}\right)$$

where, α is the variance, γ is the shape parameter of gamma distribution and g is the gray level.

Prior to segmentation, the preprocessing was performed by appropriate filtering technique; Filter selection is based on the medical imaging modality and noise characteristics. The role of preprocessing is inevitable in signal and image

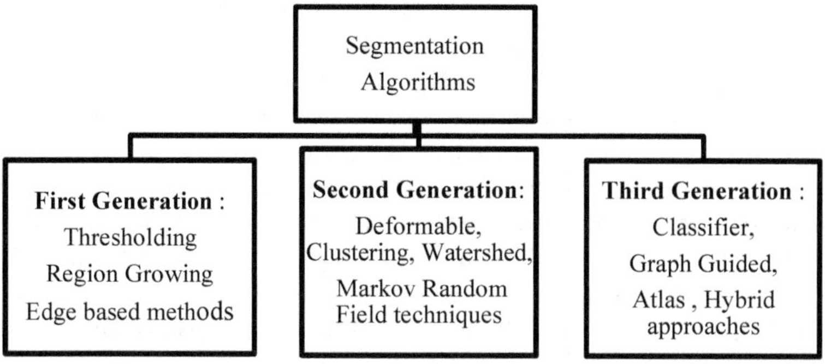

Fig. 1.1 Classification of segmentation algorithms

processing for subsequent operations like segmentation and classification [1, 2]. The segmentation algorithms can be categorized based on the generation of evolution and are depicted in Fig. 1.1.

Image segmentation is the process of grouping the pixels of an image to form meaningful regions. Medical image segmentation is the visualization of the region of interest such as anatomical structures and anomalies like tumor, cyst, etc. for medical applications such as diagnostics, therapeutic planning, and guidance. Lay Khoon Lee et al. performed a review on different types of segmentation algorithms for medical imaging modalities like X-ray, CT, MRI, 3D MRI and Ultrasound [3]. Similarly, S. N. Kumar et al. performed a detailed study on the different generation of the medical image segmentation techniques; qualitative and quantitative analysis was performed for the widely used medical image segmentation algorithms [4]. Neeraj Sharma et al. state the necessity of automated medical image segmentation technique in diagnosis, and radiotherapy planning in medical images and also explained the limitations of the existing segmentation algorithms [5]. The thresholding is a simple and classical technique that separates the foreground and background regions in an image based on the threshold value. The multilevel thresholding eliminates the discrepancy of the bi-level thresholding that uses a single threshold value. The optimization techniques when employed in the multilevel thresholding yield efficient results, since it provides the proper choice of threshold values. The 3D Otsu thresholding was found to be efficient for MR brain images; better results were produced than bi-level and multithresholding techniques [6]. Among the region based approaches, the classical region growing is the semi-automatic segmentation technique that relies on the seed point selection [6]. The multiple-seed point based region growing for brain segmentation was found to be effective on a multi-core CPU computer [7]. The manual seed point selection can be replaced by the deployment of the optimization algorithm for yielding efficient results [8]. The edge detection traces the boundary of objects in an image and among the classical edge detector, canny produces better results [9]. The Markov basics and Laplace filter were coupled to form an edge detection model that gives better results for medical images than the classical techniques [10]. The teaching learning-based optimization was found to be effective for medical image edge detection than the classical edge detectors [11]. The interactive medical image segmentation algorithms are discussed in [12]. J. Senthilnath et al. did a performance study on nature-inspired firefly optimization algorithm in the thirteen benchmark classification datasets [13]. Superior results were produced when compared with classical techniques like Particle Swarm Optimization (PSO), Bayes net, Multilayer Perceptron, Radial Basis Function Neural Network, KStar, Bagging, MultiBoost, Naive Bayes Tree, Ripple Down Rule, Voting Feature Interval.

Iztok Fister et al. made a detailed analysis of the types of firefly algorithm for engineering applications in solving the real world challenges [14]. Hui Wang et al. proposed a modification in the parameter of classical firefly algorithm to reduce the complexity of the algorithm [15]. The proposed adaptive firefly algorithm generates better solution when compared with standard Firefly Algorithm, Variable step size Firefly Algorithm (VSSFA), Wise step strategy Firefly Algorithm (WSSFA), Memetic Firefly Algorithm (MFA), Firefly Algorithm with chaos and Firefly Algorithm with random attraction. Mutasem K et al. proposed a hybrid technique comprising of the Fuzzy C-Means algorithm with Firefly algorithm for the segmentation of brain tumor [16]. The experimental analysis was carried out on 181 brain images obtained from brain-web Simulated Brain Database (SBD) repository; robust results were produced when compared with Dynamic clustering algorithm based on the hybridization of Harmony Search and Fuzzy Variable String Length Genetic Point symmetry techniques. K. Vennila et al. proposed multilevel Otsu image segmentation based on Firefly optimization and good results were obtained in terms of PSNR, computation cost and mean value when compared with Darwinian Particle Swarm Optimization [17]. Cholavendhan Selvaraj et al. made a detailed survey of the bio-inspired optimization algorithms such as Ant Colony Optimization, Particle Swarm optimization, Artificial Bee Colony algorithm and their hybridizations [18].

The summarization of results reflects the status of the optimization techniques in solving the wide range of engineering problems. In the medical image processing, the FCM plays a major role in the clustering and classification of the image for the analysis, diagnosis, and recognition of anomalies [19]. Janmenjoy Nayak et al. performed a survey on major modification and advancement in the classical FCM algorithm and their applications towards the image analysis [20]. Chih Chin Lai et al. proposed a hierarchical evolutionary algorithm based on genetic algorithm for the segmentation of skull images which enhances the diagnostic efficiency than the dynamic thresholding, Competitive Hopfield Neural Networks (CHNN), K-Means and Fuzzy C-Means algorithms [21].

Emrah Hancer et al. proposed a methodology for the segmentation of brain tumor in the MRI images by using artificial bee colony algorithm. Efficient results were produced when compared with K-Means, FCM, and Genetic Algorithm based image segmentation techniques [22]. The FCM, when coupled with PSO was found to be effective for the segmentation of noisy images when compared with K-means, Enhanced FCM, and Fast Global Fuzzy Clustering techniques [23].

The Convolution Neural Network (CNN) was employed for the automatic segmentation of MR brain images, multiple convolution kernels of varying size was used for the generation of accurate results [24]. The CNN with multiple kernels of smaller size was used for the efficient brain tumor segmentation in MR images [25]. The Deep Learning Neural Network (DLNN) gains its importance in attenuation correction of PET/MR images [26]. The DLNN along with deformable model was proposed for the automatic segmentation of left ventricle in cardiac MR images [27]. The Deep Convolution Neural Network (DCNN) along with the 3D deformable model generates good segmentation results for the extraction of tissues in musculoskeletal MR images [28]. Vijay Badrinarayanan et al. proposed SegNet, a novel DCNN architecture for semantic pixel-wise segmentation [29]. In this chapter, firefly optimization algorithm was coupled with FCM for CT/MR medical image segmentation. The preprocessing stage comprises of artifacts removal and denoising by Non-Linear Tensor Diffusion (NLTD) filter. The computation complexity of the algorithm was minimized by sampling the total pixel count for manipulation. The Cluster Validity Indexes (CVI's) are used for the validation of results to determine the optimum number of clusters.

1.2 Materials and Methods

1.2.1 Data Acquisition

The real-time abdomen CT data sets are used in this work for the analysis of algorithms. The images are acquired from Optima CT machine with a slice thickness of 3 mm. The images in DICOM format with a size of 512 × 512 are used in this work. The Metro Scans and Research Laboratory approved the study of human datasets for research purpose. The five abdomen CT data sets, each comprising of 200 slices are used in this work. The results of typical slice from each dataset are depicted here.

1.2.2 Fuzzy C Means Clustering

In this chapter, the Fuzzy c-means Clustering algorithm coupled with optimization technique was proposed for the segmentation of medical images. In the perspective of image processing, clustering is defined as the grouping of pixels into a cluster which is similar between them, while dissimilar pixels belong to other clusters. The concept of clustering is depicted below in Fig. 1.2. The clustering algorithms can be classified into two groups; Supervised and Unsupervised. The requirement of prior knowledge termed as training samples is the key concept of the supervised classifier. Artificial Neural Network (ANN), Naive Bayes Classifier, and Support

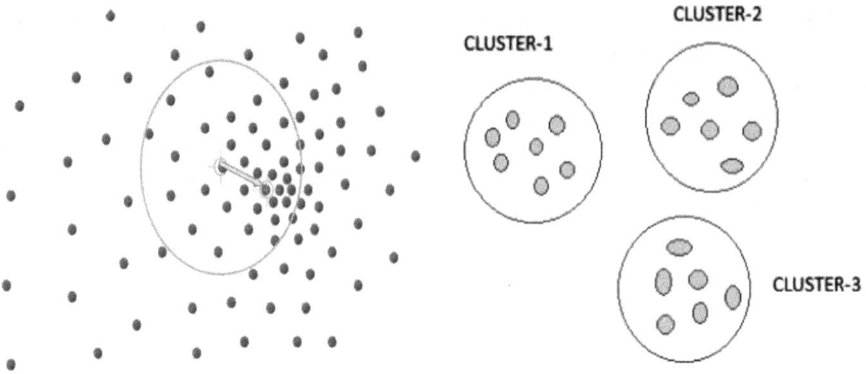

Fig. 1.2 Principle of clustering

Vector Machine are some of the widely used supervised algorithms. The unsupervised technique doesn't need any prior information and is particularly well suited for huge unlabeled datasets. The unsupervised clustering techniques can be further classified into two categories; hierarchical and partitional. The role of partitional clustering is prominent in image analysis and pattern recognition. The K-means and Fuzzy c-means (FCM) are well-known partition clustering algorithms. The K- means clustering is termed as Crisp (hard) since the objects are assigned to only one cluster. The FCM clustering is termed as soft (Fuzzy) since an object can be accommodated in more than one cluster based on the fuzzy membership value.

The FCM overcomes the issues of classical K-means clustering; since the data can belongs to more than one cluster. The FCM was developed by Dunn [30] and modified by Hathaway and Bezdek [31] which was widely used for pattern classification. FCM is an unsupervised algorithm based on the minimization of the objective function.

$$J_m = \sum_{i=1}^{N} \sum_{j=1}^{C} U_{ij}^{m} \left\| y_i - c_j \right\|^2, \quad 1 \leq f < \infty$$

The pixels are grouped into clusters in such a manner that, the intracluster similarity is maximized and the intracluster similarity is minimized.

The fuzzy partition represents the fuzzy membership matrix of the pixel in the cluster. The parameter U_{ij} represents the fuzzy membership of the ith object (pixel) in the jth cluster. The parameter 'f' depicts weighting exponent that determines the degree of fuzziness for the fuzzy membership function. The fuzzy classification is based on the iterative optimization of the objective function depicted above with the updation of membership function u_{ij} and the cluster center c_j as follows.

$$U_{ij} = \frac{1}{\sum_{K=1}^{C} \left(\frac{\|y_i - c_j\|}{\|y_j - c_k\|} \right)^{\frac{2}{f-1}}}$$

$$c_j = \frac{\sum_{i=1}^{N} U_{ij}^{f} \cdot y_i}{\sum_{i=1}^{N} U_{ij}^{f}}$$

The iterative calculation is terminated, when $max_{ij}\left\{ \left| u_{ij}^{(k+1)} - u_{ij}^{(k)} \right| \right\} < \delta$, where δ is a termination criterion between 0 and 1, and k represents the iteration count. The convergence of the algorithm occurs when the objective function (J_m) attains local minima or saddle point.

The steps in FCM clustering algorithm are summarized as follows

1. *Initialise* $U = [U_{ij}]$ *matrix*, $U^{(0)}$
2. *At* k^{th} *step: Calculate the cluster center vector* $C^{(k)} = [c_j]$ *with* $U^{(k)}$

$$c_j = \frac{\sum_{i=1}^{M} U_{ij}^{m} \cdot x_i}{\sum_{i=1}^{M} U_{ij}^{m}}$$

3. *Update* $U^{(k)}, U^{(k+1)}$

$$U_{ij} = \frac{1}{\sum_{K=1}^{C} \left(\frac{\|x_i - c_j\|}{\|x_j - c_k\|} \right)^{\frac{2}{m-1}}}$$

4. *If* $\left\| U^{(k+1)} - U^{(k)} \right\| < \delta$, *then Stop; otherwise return to step* 2.

The operating principle of FCM is based on the fact that, the minimization of the objective function ends up with the solution. In many real-time cases, classical FCM stuck into local minima. The optimization algorithm can be employed to achieve global minima. The parameter selection is vital for optimization algorithms and it influences the performance of the algorithm to maximize or minimize the objective function subjecting to certain constraints. The cluster centers are randomly initialized by classical FCM, hence the optimization based clustering solves this problem. The cluster centers generated by the optimization technique is utilized by the FCM for image segmentation. The pixels in the image are mapped into the particular cluster based on similarity and distance. The initialization of the cluster centers by optimization improves the performance in terms of the convergence rate, computation complexity, and segmentation accuracy.

1.2.3 Firefly Optimization Algorithm

In this chapter, the performance of firefly optimization in the FCM algorithm was analyzed for the estimation of optimal cluster center values for image segmentation. The biological trait of the firefly is the motivation for Yang [32] to propose an optimization technique. The rhythmic flashes generated by the firefly was used as a mode of communication between them to search for prey and for mating. More than 2000 species of fireflies are there in the world and they have natural characteristics to create illumination in the dark with flickering and glowing lights. Fister et al. found that the attraction capacity of the fireflies is proportional to the brightness [14]. The fireflies tend to move towards ones which emits a brighter light.

The population-based firefly algorithm was found to generate a global optimal solution for many engineering problems. The biological chemical substance luciferin present in the body of the fireflies was responsible for flashing the light. The intensity of light emitted is directly proportional to the discharge of luciferin. The degree of attraction tends to decrease as the distance between the fireflies increases. If any firefly fails to discover another firefly which is brighter than itself, it will travel arbitrarily. The optimization algorithm when employed for clustering applications, cluster centers are the decision variables and the objective function is associated with the euclidean distance. Based on the objective function, initially, all the fireflies will be spread randomly over the search space.

The two stages of firefly algorithm are summarized as follows:

The first stage is based on the difference in the intensity that is associated with the objective function values. Depending on the nature of the problem that requires maximization/minimization, a firefly with higher/lower intensity will entice another individual with higher/lower intensity.

Consider that there are n swarms (fireflies), where Y_i signifies the solution of a firefly i. The fitness value is expressed by $f(Y_i)$ moreover the current position I of the fitness value $f(y)$ is estimated by the brightness of a firefly [32].

$$I_i = f(y_i), \quad 1 \leq i \leq n$$

The second stage is the movement towards the firefly with high brightness intensity. The attraction factor of the firefly is represented by β that indicates the attraction power of firefly in the swarm and it changes with distance (R_{ij}) between two fireflies i and j at positions Y_i and Y_j respectively.

$$R_{ij} = \|Y_i - Y_j\| = \sqrt{\sum_{k=1}^{d} (Y_{ik} - Y_{jk})^2}$$

The attraction function $\beta(R)$ of the firefly is expressed as follows.

$$\beta(R) = \beta_0 e^{-\gamma R^2}$$

where β_0 is the attraction function value for $R = 0$, γ is the coefficient of ingestion of light.

The pseudo code for firefly optimization algorithm is as follows

```
Define objective function f(Y), Y=[Y1,Y2,Y3,--------Yd]

Generate initial population of fireflies Yi =[1,2,3-----n]

Estimate the light intensity of firefly Ii using the objective function f(Y)

 Define light absorption coefficient(γ)

While (t<max generation )

        for i=1:n   //all n fireflies

            for j=1: n  // all n fireflies

                    if (Ij > Ii)

                            Move firefly i towards j in d dimensions.

        end if

                                // Attraction capacity changes with distance

                                //Validate new solutions and update light intensity

    end for j

            end for i

Estimate the current best by ranking the fireflies

end while
```

The motion of a firefly 'i' from the position Y_i which is attracted towards another brighter firefly 'j' at position Y_j is expressed as follows

$$Y_i(t+1) = Y_i(t) + \beta(R)\left(Y_i - Y_j\right) + \alpha\left(rand - \frac{1}{2}\right)$$

$$Y_i(t+1) = Y_i(t) + \beta_0 e^{-\gamma R^2}\left(Y_i - Y_j\right) + \alpha\left(rand - \frac{1}{2}\right)$$

where α depicts the maximum radius of the random step. The term rand represents randomization parameter uniformly distributed between 0 and 1.

There are two special cases

Case i: $r = 0$, then $\beta = \beta_0 e^0 = \beta_0$, The air is absolutely clear with no light dispersion. The fireflies can see each other; exploration and exploitation is out of balance.
Case ii: $r = \infty$, then $\beta = I_0 e^{-\infty d^2} = 0$, The air is foggy with extreme light dispersion. The fireflies can't see each other; exploration and exploitation is out of balance.

1.2.4 Improved FCM-Firefly Optimization Segmentation Algorithm

The FCM clustering algorithm proposed here comprises of two stages. In the first stage, firefly optimization is employed to determine the near-optimal cluster centers. In the second stage, the cluster centers are used for the initialization of FCM algorithm. The firefly optimization algorithm makes the clustering an effective tool for medical image segmentation by eliminating the problem of stucking at local minima. The firefly optimization is a swarm intelligence based algorithm and hence it mimics its advantages.

The solution vector is expressed as follows

$$S = \begin{pmatrix} V_1 & V_2 & V_3 \\ S_1, S_2, \ldots, S_i \ldots S_d & S_1, S_2, \ldots, S_i \ldots S_d & S_1, S_2, \ldots, S_i \ldots S_d \end{pmatrix}$$

where S_i represents characteristics in numerical form such that S_i € S. The 'S' depicts the array representing pixel attribute. Each cluster center V_i is represented by d numerical features $(S_1, S_2, \ldots S_d)$. Each solution vector is of the size (c $\times$ d), where c indicates given number of clusters and d represents the features of the dataset.

For the delineation of anomalies like tumor or cyst or anatomical organs, each pixel in the image is mapped into the clustering sector. The cluster centers are randomly initialized from the image pixel gray values with the randomly initialized solution vector, the fitness value is determined by the objective function. The solution vector is then rearranged based on the decreasing order of the objective function value. The firefly optimization determines near optimal cluster centers thereby ensuring global minima for FCM algorithm and hence eliminates the trapping at local minima. The improved FCM based on firefly optimization replaces the classical techniques of random initialization.

Prior to filtering, the medical image film artifacts are eliminated by a statistical technique coupled with convex hull computational geometry [33]. The threshold value determined by standard deviation technique was used for the binarization of input image. The binarized image was then subjected to connected component labeling for the elimination of patient details and technical information. The convex hull of the resultant image was multiplied with the original image for the generation of artifacts removed image. The preprocessing of input image was performed by Non-Linear Tensor Diffusion (NLTD) filter prior to segmentation [34]. The NLTD ensures good edge preservation since the smoothing is heterogeneous and non-noisy pixels are not disturbed.

The computation complexity was minimized by reducing the pixel count for the processing by segmentation algorithm.

$$Rp = randperm\,(L)$$

The parameter S_p represents the pixels taken for optimization, here in this work 50% of the total pixels are taken. The L represents the total pixel count of the image to be segmented and randperm function returns a row vector depicting a random permutation of the integers from 1 to n.

$$S_n = ceil\left(L * S_p\right)$$

The S_n represents the number of pixels selected for optimization and the function below represents the subset of pixels chosen for optimization process

$$X_2 = X(Rp(1 : S_n), :)$$

The optimization of the objective function relies on the brightness and movements of the firefly. The firefly algorithm starts by initializing the population of fireflies. The intensity of light emitted by the firefly estimates the movement of the fireflies. The algorithm works in the iterative fashion. The intensity of ith firefly is compared with the jth firefly as follows

if $\beta(i) > \beta(j)$
firefly j move towards firefly i
else
firefly i move towards firefly j

The incorporation of firefly algorithm has significantly improved the segmentation results. There were four stages of Improved FCM-Firefly segmentation algorithm.

i. Initialization phase
ii. Intensity calculation phase
iii. Movement calculation phase
iv. FCM algorithm phase.

The goal of incorporating firefly optimization in FCM is to minimize the objective function with a global minima value. The cluster centers represent the decision parameters to minimize the objective function. The initialization of the firefly population is as follows

$$y_{if} = \left[y_{i1}, y_{i2}, \ldots, y_{ij}, \ldots y_{id}\right] \quad 2 < j < C$$

Each firefly in the population is represented by using the above equation. Where y_{ij} represents the *j*th cluster centre.

The population of the fireflies are initialized and randomly distributed in search space. The position of firefly depicts the possible solution (centroids) for the clustering problem. In this phase, the parameter like β_0, γ, α and maximization iteration are also initialized. Once the initialization process is over, the intensity of each firefly is determined by estimating the distance between the position of the firefly and the entire data in the dataset. The minimum distance value among the population with respect to data from the dataset is considered. The intensity value of each firefly is determined based on the sum of minimum distance with respect to the data from the dataset.

The expression for determination of intensity is as follows

$$\beta(FF_j) = \sum_{i=1}^{n} d_i$$

where FF represents firefly, d_i represents the minimum distance value for a particular firefly.

The brightness of the fireflies indicates the movement of the fireflies in the search space. The intensity of fireflies is compared to determine the new position. The difference in the brightness triggers the movement. The firefly optimization is employed in the FCM algorithm to enhance the clustering operation. The new position of the entire swarm of the fireflies is determined by the FCM operator based on the current intensity value.

The FCM-Firefly algorithm is carried out through the updation of the membership value u_{ij} and position of the firefly y_j using the below equations

$$U_{ij} = \frac{1}{\sum_{k=1}^{s}\left(\frac{\|y_i - f_j\|}{\|y_i - f_k\|}\right)^{2/f-1}}, \quad 1 \le i \le N$$

where U_{ij} depicts the degree of membership of y_i in the firefly j, degree of fuzziness f = 2 and y_i is the data associated with the firefly under study.

$$F_j = \frac{\sum_{i=1}^{N} u_{ij}^m \cdot x_i}{\sum_{i=1}^{N} u_{ij}^m}$$

The F_j represents the solution after applying FCM in the firefly j.

The new position of Firefly is determined and the intensity value is updated. The fixed number of iterations will be provided and at the end of the iteration, the best solution was determined.

1.3 Results and Discussion

The algorithms are developed in Matlab 2010a and tested on CT abdomen data sets. The system specifications are as follows; Intel Core i3 processor of 3.30 GHz with 4 GB RAM. In the scenario of medical image segmentation, fixing the number of clusters is cumbersome, since it cannot be initialized roughly by viewing the image. The validation metrics are employed for the optimal cluster selection.

This is performed in 3 steps

i. The parameters of clustering algorithm except the cluster number is fixed.
ii. The cluster number is varied from an initial value of 2 to an upper limit (max). The data partition is carried out for each cluster number.
iii. The cluster validity indexes are applied on the data partition obtained from the previous stage for evaluation. Based on the values of CVI's, the cluster number selection is done.

The terminologies used in the formulation of cluster validity function are as follows

N: the count of data objects for clustering
f: the fuzzifier factor that represents the level of cluster fuzziness
u: the ith data object, $1 \le i \le N$
P: the number of clusters
C_p: the pth cluster, $1 \le p \le P$
$|C_p|$: the count of data objects in the p-th cluster
V_p: the centroid of the p-th cluster
$\|u - v\|$: the distance between a pair of data objects
μ_{ip}: the membership degree of u_i corresponding to C_p.

The FCM algorithm is an iterative technique in which the pixels are grouped into a cluster based on the membership degree through the minimization of the objective function.

$$\sum_{i=1}^{M}\sum_{p=1}^{P}\mu_{ip}^{f}\left\|u_i - V_p\right\|^2, \quad f \geq 1$$

The number of the cluster is taken initially P and randomly P centroids are selected. The objective function represented above is optimized in an iterative fashion by the updation of μ_{ip} and V_p as follows

$$\mu_{ip} = \frac{1}{\sum_{j=1}^{P}\left(\frac{\left\|u_i - v_p\right\|^2}{\left\|u_j - v_c\right\|^2}\right)^{\frac{2}{f-1}}}$$

$$V_p = \frac{\sum_{i=1}^{N}\mu_{ip}^{f}u_i}{\sum_{i=1}^{N}\mu_{ip}^{f}}$$

The iteration terminates when, $\left\|U_{T+1} - U_T\right\| < \in$, where $U_T = \left[\mu_{ip}\right]$ represents the matrix comprising of all $\mu_{ik}'s$. T is the number of iteration and $\in$ is a threshold specified by the user. The clustering validity metrics are used to estimate the quality of clustering result. The partition coefficient (PC) and partition entropy (PE) is based only on the membership values of fuzzy partition dataset. The criteria for optimum cluster number selection is the maximization of (PC) or minimum of PE. The issues in the performance metrics, PC or PE is that they do not consider the geometrical properties of the dataset.

Xie Beni's index (XBI) and Fukuyama's and Sugeno's index (FSI) are also widely used classical CVI's. XBI and FSI focus on the characteristics, compactness, and separation. The numerical part of the expression XBI in Table 1.1 represents the compactness of fuzzy partition, the denominator part represents the strength of separation between the cluster for optimal clustering. The value of XBI should be minimized for optimum cluster number selection. The expression for FSI in Table 1.1 comprises of two terms. The first term represents the compactness measure and the second term represents separation measure. Though FSI and XBI consider the inter-cluster information, geometrical properties are not considered.

The DB index was obtained by the mean of cluster similarities. For each cluster P, the similarity between P and all other clusters are determined. The term *Sp* is represented as follows

$$Sp = \frac{1}{|C_P|}\sum_{U_i \in C_p}\left\|U_i - V_p\right\|^2$$

Table 1.1 Classical clustering validation metrics

Cluster validity index	Formula
Partition coefficient (PC) [35, 36]	$PC = \frac{1}{N}\sum_{p=1}^{P}\sum_{i=1}^{N}\mu_{ip}^{2}$
Partition entropy (PE) [35, 36]	$PE(K) = \frac{1}{N}\sum_{p=1}^{P}\sum_{i=1}^{N}\mu_{ip}\log_2(\mu_{ip})$
Xie and Beni index (XBI) [37]	$XBI(K) = \frac{\sum_{p=1}^{P}\sum_{i=1}^{N}\mu_{ip}^{2}\|\|u_i - v_p\|\|^2}{N \cdot \min_{i \neq j}\|\|v_i - v_j\|\|^2}$
The Fukuyama and Sugeno Index (FSI) [37]	$FSI(K) = \sum_{p=1}^{P}\sum_{i=1}^{N}\mu_{ip}^{f}\|\|u_i - v_p\|\|^2 - \sum_{p=1}^{P}\sum_{i=1}^{N}\mu_{ip}^{f}\|\|v_p - \hat{v}\|\|^2$

Table 1.2 Clustering validation metrics based on compactness and separation ratio

Cluster validity index	Formula
Calinski-Harabasz index (CHI) [38]	$CHI(K) = \frac{B_P}{P-1} / \frac{W_p}{N-P}$
Silhouette coefficient index (SCI) [38]	$SCI(P) = SC_1(P) - SC_2(P)$
Centroid similarity index (CSI) [38]	$CSI(K) = \frac{\sum_{p=1}^{P}\left(\frac{1}{\|C_P\|}\left(\sum_{u_j \in C_P} \max_{u_j \in C_p}\|\|u_j - u_i\|\|\right)\right)}{\sum_{j=1}^{P}\min_{i \neq j}\|\|v_i - v_j\|\|}$
Davies Bouldin index (DBI) [38]	$DBI = \frac{1}{P}\sum_{P=1}^{P}\max\frac{S_j + S_p}{\|\|V_j - V_p\|\|}$
Partition coefficient and exponential separation index (PCAESI) [38]	$PCAESI = \sum_{P=1}^{P}\sum_{i=1}^{N}\frac{\mu_{iP}^{2}}{\mu_M} - \exp\left(\frac{-\min_{h \neq p}\|\|V_p - V_h\|\|^2}{\beta_T}\right)$
Pakhira-Bandyopadhyay-Maulik index (PBMFI) [39]	$PBMFI = \frac{\max_{j \neq lp}\{\|\|V_j - V_p\|\|\} \times \sum_{i=1}^{N}\mu_{il}\|\|U_i - V_l\|\|}{P\sum_{P=1}^{P}\sum_{i=1}^{N}\mu_{iK}^{f}\|\|U_i - V_K\|\|}$
WL index (WLI) [38]	$WLI = \sum_{P=1}^{P}\sum_{i=1}^{N}\frac{\mu_{ilP}^{2}\|\|U_i - V_P\|\|^2}{\sum_{i=1}^{N}\mu_{iP}}$

Table 1.2 represents the clustering validity metrics based on compactness and separation ratio.

The shortcoming of the traditional CVI's is that they are focusing only on the distance between the cluster centroids. The classical clustering validity indexes were not found to be good for large cluster numbers.

The CS index is a function of the cluster diameter and the mean distance between the cluster centers. The PCAES index is a function of exponential separation component, and normalized partition coefficient.

The CH index is based on the mean between and within the sum of squares. The terms in the CH index are represented as follows

$$B_P = \sum_{p=1}^{P} |C_P| \, \|v_P - \overline{v}\|^2$$

$$W_P = \sum_{p=1}^{P} \sum_{U_i \in C_P} \|u_i - v_p\|^2$$

The SC index is based on the combination of two functions and evaluates the compactness-separation ratio. The terms in the SC index are represented as follows

$$SC_1(P) = \frac{\frac{1}{P}\sum_{p=1}^{P} \|v_p - \overline{v}\|^2}{\sum_{p=1}^{P}\left(\sum_{i=1}^{N} \mu_{ip}^{m} \|u_i - v_p\|^2 / \sum_{i=1}^{N} \mu_{ip}\right)}$$

$$SC_2(P) = \frac{\sum_{p=1}^{P-1}\sum_{j=p+1}^{P}\left(\sum_{i=1}^{N}\left(\min(\mu_{ip}, \mu_{ij})^2\right)/n_{ij}\right)}{\left(\sum_{i=1}^{N} \max_{1 \le p \le P} \mu_{ip}^2\right) / \left(\sum_{i=1}^{N} \max_{1 \le p \le P} \mu_{ip}\right)}$$

where SC_1 is related with the geometric properties of data; SC_2 is related with the membership degree properties.

The PBMF index is based on the compactness within clusters and a large separation between clusters. The WL index estimates the compactness of clusters by taking into account fuzzy weighted distance and the fuzzy cardinality of clusters. The five abdomen medical data sets are used for the analysis of algorithms. The cluster number was changed from P = 2 to 6 and for each cluster number, 10 times the executions are done and the performance metrics are validated.

The expression for μ_M and β_T in PCAES index are as follows

$$\mu_M = \min_{1 \le p \le P} \left\{ \sum_{i=1}^{N} \mu_{iP}^2 \right\}$$

$$\beta_T = \frac{1}{P}\sum_{P=1}^{P} \|V_p - \overline{V}\|^2$$

The performance metrics of the first run for the data set (ID1) is represented below in Table 1.3. Each cluster validity metric was represented with ± sign, the "+" indicates that the CVI value should be high and "−" sign indicates that the CVI value should be low. The representative input images corresponding to data sets (ID1 to ID5) after the removal of artifacts are depicted in Fig. 1.3. Figure 1.4 represents the NLTD filtering result. Compared with classical filters like median filter, Gaussian filter, and bilateral filter, the NLTD filter generates efficient result. In the median filter, the noise-free pixels are also affected. The edge preservation is poor in Gaussian and bilateral filter. The performance of Anisotropic Diffusion Filter (ADF) was clearly stated in [40]. The NLTD filter is an improved version of ADF, thereby providing promising restoration results. The FCM results when

Table 1.3 Cluster validity performance metrics values of ID1 in the first run

Cluster validity index	P = 2	P = 3	P = 4	P = 5	P = 6
PC+	0.9119	0.9130	**0.9160**	0.8940	0.8946
PE−	0.2164	0.2240	**0.1989**	0.2703	0.2692
CHI+	8512.6035	8541.22	**8652.3935**	7300.4416	7520.1463
DBI+	0.2027	**0.2107**	0.1619	0.1870	0.1793
XBI+	0.0157	0.0227	**0.0321**	0.0148	0.0125
FSI+	**−108.7419**	−216.2876	−249.4163	−262.5782	−266.6921
SCI+	1.3704	2.6804	**4.8570**	3.7541	4.7213
CSI−	**3.5026**	6.7195	11.7389	16.2443	22.5481
PCAES+	15,264.6105	15,633.7410	**15,846.0836**	15,739.2628	15,684.7342
PBMF+	0.1536	0.1220	**0.1719**	0.1597	0.1393
WLI−	0.0240	0.0154	0.0134	**0.0117**	0.0098

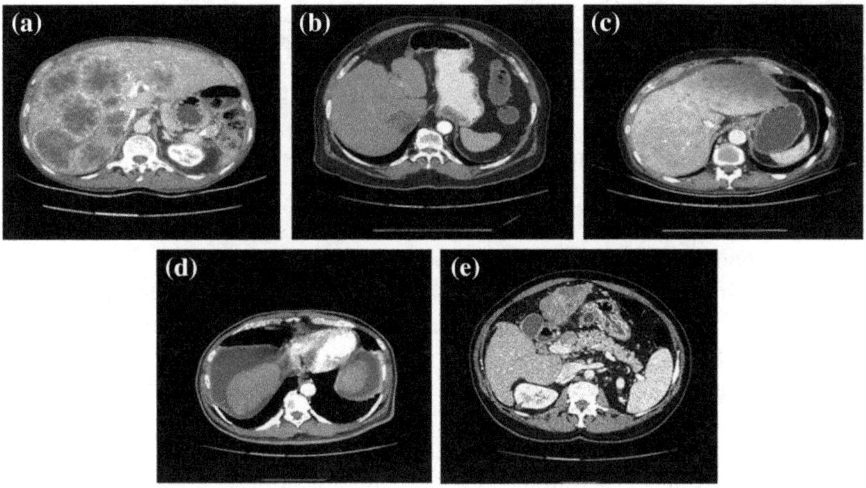

Fig. 1.3 Input images corresponding to five data sets (ID1, ID2, ID3, ID4, and ID5) after the removal of artifacts

coupled with ABC, Cuckoo, SA and Firefly algorithms for a typical slice from the dataset (ID1) are depicted in Fig. 1.5. Figure 1.6 represents the firefly algorithm results for the typical slice from datasets (ID2, ID3, ID4, and ID5). The cluster number selection for the input images is determined from the analysis of CVI's values.

The artifacts removed image was subjected to NLTD filter. The parameters of NLTD filter are step size (λ = 0.24), diffusion constant (K = 0.1) and number of iterations (iter = 10). The Gaussian smoothed image was used for the estimation of

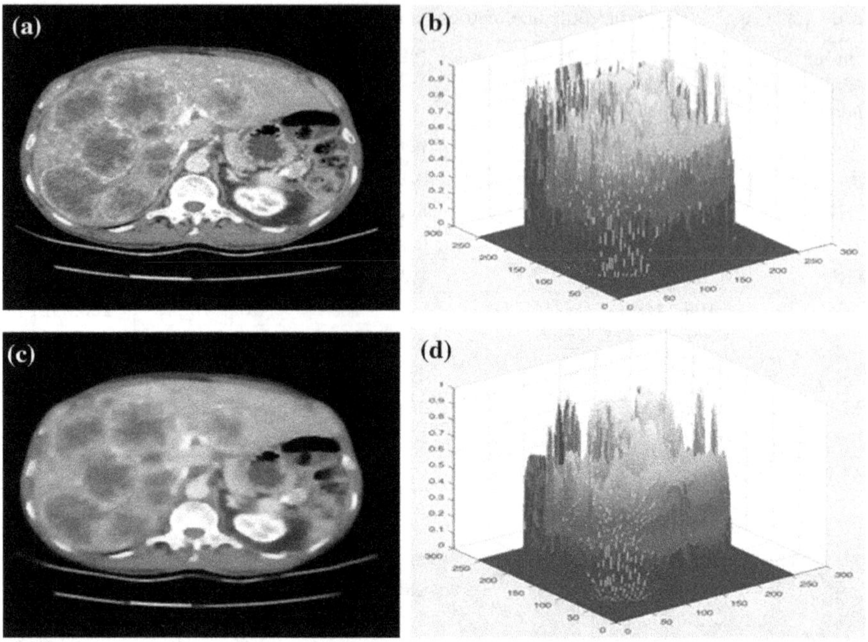

Fig. 1.4 **a**, **b** Artifacts removed image and its color map. **c**, **d** NLTD filter output and its color map corresponding to an input image from dataset (ID1)

conduction coefficient and the elements of the tensor matrix are functions of Gaussian smoothed image components.

The performance metrics are calculated by executing the algorithms 10 times for the images from each data set and the cluster number count are determined. Table 1.8 represents the parameters of optimization algorithms used in this work.

The bold values in Tables 1.3, 1.4, 1.5, 1.6 and 1.7 represents the appropriate value of CVI's in the first run. In Cuckoo, Firefly and ABC optimization, the population (N_p) was set to 20 and the number of iterations was also initialized to 20. Table 1.9 depicts the cluster number count by improved FCM firefly algorithm for 10 runs. For each metric, the cluster count was determined and by the majority voting system, for each metric the optimal cluster number was determined and tabulated in Table 1.10.

The Simulated Annealing was found to be efficient than GA when coupled with FCM, however, the setting of initial temperature is crucial in SA algorithm since it will generate erroneous results [41]. The ABC–FCM yield efficient results than GA-FCM and PSO-FCM in terms of computation time, convergence rate and accuracy [42]. The number of parameters to be tuned is less in cuckoo optimization when compared with the ABC, PSO and GA algorithms. The number of parameters to be tuned for GA and PSO were 4, 6 respectively [43]. The firefly algorithm was found to be efficient for image enhancement when compared with cuckoo optimization in terms of robustness, fitness function and convergence rate [44]. In [45],

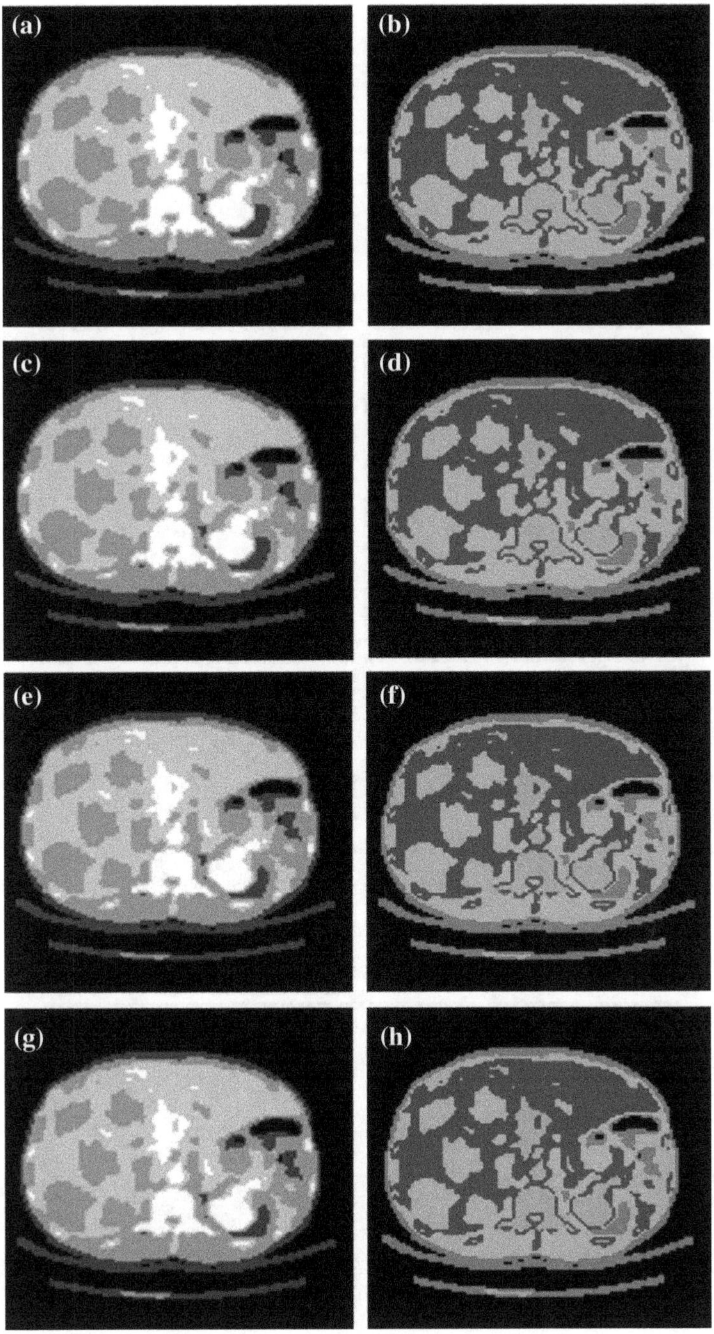

Fig. 1.5 Segmentation results of ID1 **a**, **b** ABC-FCM, **b**, **c** cuckoo-FCM, **d**, **e** simulated annealing-FCM, and **f**, **g** firefly-FCM corresponding to cluster number value of 4

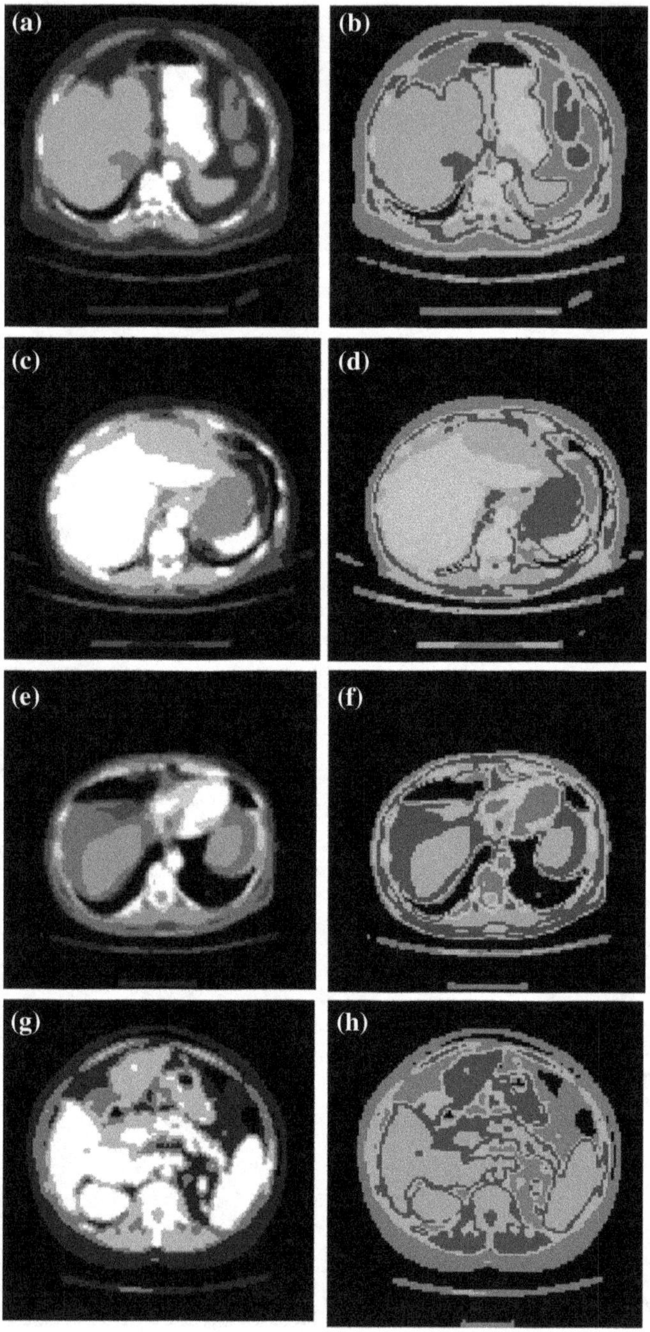

Fig. 1.6 Firefly-FCM segmentation results of ID2 (**a**, **b**), ID3 (**c**, **d**), ID4 (**e**, **f**), ID5 (**g**, **h**) for cluster values 5, 5, 6, 4

Table 1.4 Cluster validity performance metrics values of ID2 in the first run

Cluster validity index	C = 2	C = 3	C = 4	C = 5	C = 6
PC+	0.9218	0.9285	0.9462	**0.9491**	0.9192
PE−	0.1353	0.1997	0.1794	**0.1239**	0.2062
CHI+	**20,575.164**	14,329.983	13,451.709	10,958.149	11,552.368
DBI+	0.1131	0.1379	0.1577	**0.1673**	0.1595
XBI+	0.0122	0.0130	0.0152	**0.0269**	0.0094
FSI+	−416.1045	−406.7087	−402.6203	−364.536	**−254.7033**
SCI+	1.5489	5.7416	6.4141	8.2199	**8.8638**
CSI−	**2.7142**	8.8507	12.2696	25.4724	28.7029
PCAES+	16,991.165	17,193.821	17,115.650	**17,373.889**	17,076.531
PBMF+	0.2229	**0.3725**	0.3053	0.3252	0.2768
WLI−	0.0237	0.0144	0.0106	0.0092	**0.0075**

Table 1.5 Cluster validity index's values of ID3 in the first run

Cluster validity index	C = 2	C = 3	C = 4	C = 5	C = 6
PC+	0.9032	0.9296	0.9339	**0.9350**	0.8975
PE−	0.1952	0.1815	0.1652	**0.1570**	0.2600
CHI+	9406.4576	9366.5511	11,069.808	11,442.9662	**15,577.7010**
DBI+	0.1419	0.1354	0.1551	**0.2354**	0.2278
XBI+	**0.0280**	0.0165	0.0148	0.0203	0.0132
FSI+	−374.5702	−397.8660	**−265.7651**	−392.2532	−395.9922
SCI+	1.8105	3.6604	4.2628	5.0546	**6.8118**
CSI−	**2.2733**	4.6216	8.0308	13.2241	18.5982
PCAES+	13,458.899	13,645.782	13,693.402	**13,832.283**	13,766.516
PBMF+	0.2256	**0.2590**	0.2010	0.2456	0.1286
WLI−	0.0260	0.0144	0.0109	0.0100	**0.0080**

a comparison of Firefly, Bat and Cuckoo optimization was performed; experimental results reveal that firefly outperforms bat and cuckoo algorithm. The tuning of parameters is simple in firefly optimization.

The parameter study reveals that $\beta_0 = 1$ can be used for most of the applications [46]. The studies indicate that the Accelerated PSO [46] is a special case of firefly with $\gamma = 0$. The merits of firefly optimization algorithm are automatic subcategory of population and the capacity to deal with multimodality (Fig. 1.7).

Table 1.6 Cluster validity index's values of ID4 in the first run

Cluster validity index	C = 2	C = 3	C = 4	C = 5	C = 6
PC +	0.9120	**0.9152**	0.9149	0.8909	0.9027
PE−	0.2169	**0.2140**	0.2142	0.2781	0.2473
CHI +	**8509.6424**	5277.0495	6468.0569	5454.4758	5443.1152
DBI +	0.1634	0.1636	0.1533	0.1904	**0.2633**
XBI +	0.0232	0.0198	0.0173	**0.0430**	0.0111
FSI +	−237.1674	−232.2133	−232.2133	−197.3121	**−87.2592**
SCI +	0.8770	2.9675	3.9067	3.6105	**6.8062**
CSI−	20.0393	12.2458	8.9944	5.9863	**2.9196**
PCAES+	10,066.8628	10,350.1937	10,614.0788	**10,714.7197**	10,412.3122
PBMF+	0.1487	0.1843	0.1843	0.2248	**0.2550**
WLI−	0.0354	0.0214	0.0150	0.0135	**0.0098**

Table 1.7 Cluster validity index's values of ID5 in the first run

Cluster validity index	C = 2	C = 3	C = 4	C = 5	C = 6
PC+	0.9297	0.9128	**0.9397**	0.9155	0.9177
PE−	0.2213	0.1760	**0.1440**	0.2131	0.2094
CHI+	12,878.1624	**18,625.8092**	11,641.9111	11,420.381	11,654.719
DBI+	0.1344	0.1612	**0.1749**	0.1507	0.1474
XBI+	0.0168	0.0200	**0.0292**	0.0123	0.0097
FSI+	−571.2952	−561.8331	−532.6463	−506.1773	**−378.2250**
SCI+	1.8957	3.2453	3.6688	6.5740	**7.9765**
CSI−	7.8721	4.6173	**2.1393**	13.1834	17.4610
PCAES+	15,227.6157	15,314.8496	15,075.6335	15,246.377	**15,491.691**
PBMF+	0.2759	0.2510	0.2255	**0.2925**	0.1876
WLI−	0.0300	0.0190	0.0152	0.0109	**0.0085**

Table 1.10 depicts the cluster count manipulated from the CVI's. Table 1.11 represents the optimal cluster value for each dataset determined from the analysis of CVI's.

Table 1.8 Parameters of optimization algorithm

Sl. No.	Optimization algorithm	Parameters
1	Simulated annealing	Initial temperature (T_o): 500 Number of iterations: 250 Cooling schedule (α): 0.4
2	Artificial bee colony	Pheromone evaporation parameter (ρ) [−1 1]: 1 Stagnation limit (L): 10 No. of employed bees [10–30% of total population]: 0.20 * N_p
3	Cuckoo	Step size (α): 1 Levy distribution coefficient(λ) [$1 \leq \lambda \leq 3$]: 1.5 Discovery rate of alien eggs (P_α): 0.25
4	Firefly	Attractiveness coefficient (β_0): 1 Coefficient of ingestion of light (γ): 1 Step size (α): 0.25

Table 1.9 Cluster number count by the improved FCM-firefly algorithm for 10 runs

Data set ID	PC^+	PE^-	CHI^+	DBI^-	XBI^-	FSI^-	SCI^+	CSI^-	$PCAES^+$	$PBMF^+$	WLI^-
1	4^6 3^4	4^6 3^4	4^7 3^3	3^8 4^2	4^8 3^2	2^6 3^4	4^8 3^2	2^8 3^2	4^7 3^3	4^8 3^2	5^5 4^3 3^2
2	5^6 4^4	5^6 4^4	2^3 4^3 5^4	5^8 4^2	5^6 4^4	6^4 5^6	6^8 5^2	2^6 3^4	5^6 4^2 3^2	3^6 5^4	6^4 5^6
3	5^6 6^4	5^6 6^4	6^4 5^3 4^3	5^6 6^3 4^1	2^2 5^3 6^5	5^5 4^3 3^2	6^4 5^3 3^3	2^4 5^6	5^8 4^2	5^6 3^4	6^8 5^2
4	6^6 2^2 5^2	3^2 6^6 5^2	2^2 3^6 6^2	6^8 5^2	5^8 6^2	6^8 5^2	6^8 5^2	6^8 4^1 5^1	5^6 6^4	6^6 4^1 5^3	6^7 5^3
5	4^8 3^2	4^8 3^2	3^6 4^4	4^6 3^4	4^7 3^3	6^6 4^4	6^5 4^4 5^1	4^7 3^3	6^3 5^6 4^1	5^8 4^2	6^6 4^4

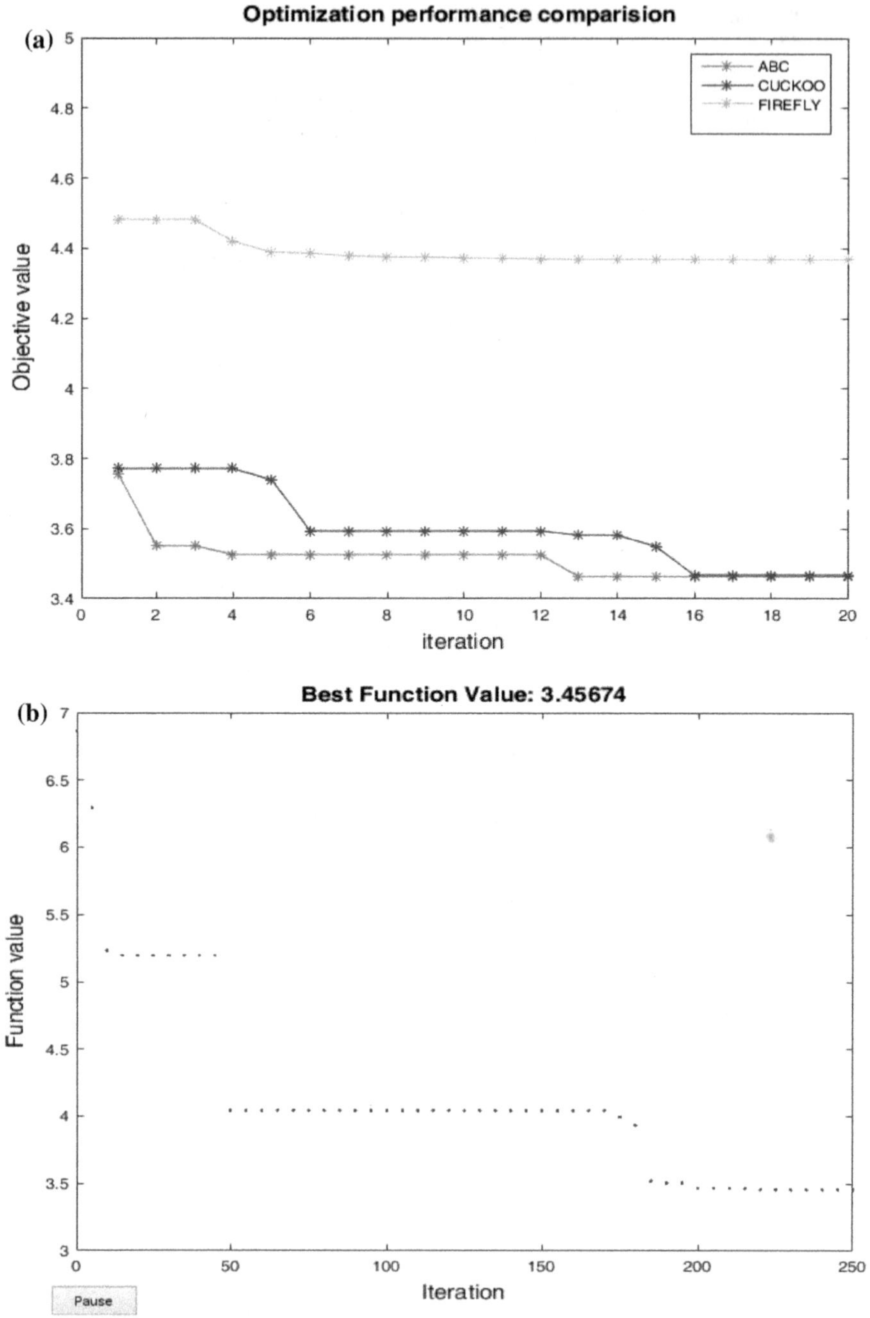

Fig. 1.7 Optimization algorithms performance of ID1: **a** ABC-FCM, cuckoo-FCM and firefly-FCM, **b** SA-FCM

Table 1.10 Possible cluster number decided by CVI's

Data set ID	PC^+	PE^-	CHI^+	DBI^-	XBI^-	FSI^-	SCI^+	CSI^-	$PCAES^+$	$PBMF^+$	WLI^-
1	4	4	4	3	4	2	4	2	4	4	5
2	5	5	5	5	5	5	6	2	5	3	5
3	5	5	6	5	6	5	6	5	5	5	6
4	6	6	3	6	5	6	6	6	5	6	6
5	4	4	3	4	4	6	6	4	6	5	6

Table 1.11 Actual cluster number decided by CVI's

Data set ID	1	2	3	4	5
Possible cluster values	2^2 3 5 4^7	2^2 3 5^7 6	5^7 6^4	3 5^2 6^8	3 4^5 5 6^4
Chosen cluster value	4	5	5	6	4

1.4 Conclusion

The FCM algorithm when coupled with the firefly optimization generates promising results for medical image segmentation. After the removal of medical image film artifacts, denoising of input image was performed by NLTD filtering approach prior to segmentation and the computation complexity was minimized by the optimal pixel count selection. The Firefly optimization based improved FCM generates satisfactory and consistent results with FCM-Cuckoo, FCM-ABC and FCM-SA segmentation techniques. The few parameter tuning of firefly optimization makes it a robust choice for image processing applications. A detailed analysis of cluster validity performance metrics were also done for the appropriate determination of number of clusters.

This work highlights the importance of optimization algorithms in image segmentation. The FCM is a widely used clustering segmentation technique, however to solve the issues in FCM, numerous algorithms like SFCM, ARKFCM, FLICM, T2FCM etc. have been proposed. But the above-mentioned techniques are concentrating mainly on the performance of FCM when the input images are noisy. The optimization techniques are then employed in FCM for solving the problem of random centroid initialization, but the tuning of parameters and computation complexity generate issues. The novelty of the proposed work is as follows; the filtering was accomplished by NLTD filter which has good edge preservation, firefly optimization was employed which has fewer parameters to be tuned and has quick convergence rate. The outcome of this work will provide a path for the researchers to develop novel optimization algorithm for solving the problems in image processing.

Acknowledgements The authors would like to acknowledge the support provided by DST under IDP scheme (No: IDP/MED/03/2015). We thank Dr. Sebastian Varghese (Consultant Radiologist, Metro Scans & Laboratory, Trivandrum) for providing the medical CT images and supporting us in the preparation of the manuscript.

References

1. Fida, B., Bernabucci, I., Bibbo, D., Conforto, S., Schmid, M.: Pre-processing effect on the accuracy of event-based activity segmentation and classification through inertial sensors. Sensors **15**(9), 23095–23109 (2015)
2. Hemanth, D.J., Anitha, J.: Image pre-processing and feature extraction techniques for magnetic resonance brain image analysis. In: Computer Applications for Communication, Networking, and Digital Contents, pp. 349–356. Springer, Berlin (2012)
3. Lee, L.K., Liew, S.C., Thong, W.J.: A review of image segmentation methodologies in medical image. In: Advanced Computer and Communication Engineering Technology, pp. 1069–1080. Springer, Cham (2015)
4. Kumar, S.N., Muthukumar, S., Kumar, A., Varghese, S.: A voyage on medical image segmentation algorithms. Biomed. Res. (2018)
5. Sharma, N., Aggarwal, L.M.: Automated medical image segmentation techniques. J. Med. Phys. Assoc. Med. Physicists India **35**(1), 3 (2010)
6. Banerjee, S., Mitra, S., Shankar, B.U.: Single seed delineation of brain tumor using multi-thresholding. Inf. Sci. **10**(330), 88–103 (2016)
7. Smistad, E., Elster, A.C., Lindseth, F.: GPU accelerated segmentation and centerline extraction of tubular structures from medical images. Int. J. Comput. Assist. Radiol. Surg. **9** (4), 561–575 (2014)
8. Pal, S.K., Rosenfeld, A.: Image enhancement, and thresholding by optimization of fuzzy compactness. Pattern Recogn. Lett. **7**(2), 77–86 (1988)
9. Li, M., Yan, J.H., Li, G., Zhao, J.: Self-adaptive Canny operator edge detection technique. J. Harbin Eng. Univ. **9**(28) (2007)
10. Günsel, B., Jain, A.K., Panayirci, E.: Reconstruction and boundary detection of range and intensity images using multiscale MRF representations. Comput. Vis. Image Underst. **63**(2), 353–366 (1996)
11. Thirumavalavan, S., Jayaraman, S.: An improved teaching-learning based robust edge detection algorithm for noisy images. J. Adv. Res. **7**(6), 979–989 (2016)
12. Zhao, F., Xie, X.: An overview of interactive medical image segmentation. Ann. BMVA **2013** (7), 1–22 (2013)
13. Senthilnath, J., Omkar, S.N., Mani, V.: Clustering using firefly algorithm: performance study. Swarm Evol. Comput. **1**(3), 164–171 (2011)
14. Fister, I., Fister Jr., I., Yang, X.S., Brest, J.: A comprehensive review of firefly algorithms. Swarm Evol. Comput. **1**(13), 34–46 (2013)
15. Wang, H., Zhou, X., Sun, H., Yu, X., Zhao, J., Zhang, H., Cui, L.: Firefly algorithm with adaptive control parameters. Soft. Comput. **21**(17), 5091–5102 (2017)
16. Alsmadi, M.K.: A hybrid firefly algorithm with Fuzzy-C mean algorithm for MRI brain segmentation. Am. J. Appl. Sci. **11**(9), 1676–1691 (2014)
17. Vennila, K., Thamizhmaran, K.: Multilevel image segmentation based on firefly algorithm. Biometrics Bioinform. **9**(3), 57–60 (2017)
18. Selvaraj, C., Kumar, R.S., Karnan, M.: A survey on application of bio-inspired algorithms. Int. J. Comput. Sci. Inf. Technol. **5**(1), 366–370 (2014)
19. Nayak, J., Naik, B., Behera, H.S.: Fuzzy C-means (FCM) clustering algorithm: a decade review from 2000 to 2014. In: Computational Intelligence in Data Mining, vol. 2, pp. 133–149. Springer, New Delhi (2015)

20. Nayak, J., Nanda, M., Nayak, K., Naik, B., Behera, H.S.: An improved firefly fuzzy c-means (FAFCM) algorithm for clustering real-world data sets. In: Advanced Computing, Networking and Informatics, vol. 1, pp. 339–348. Springer, Cham (2014)
21. Lai, C.C., Chang, C.Y.: A hierarchical evolutionary algorithm for automatic medical image segmentation. Expert Syst. Appl. **36**(1), 248–259 (2009)
22. Hancer, E., Ozturk, C., Karaboga, D.: Extraction of brain tumors from MRI images with artificial bee colony based segmentation methodology. In: 2013 8th International Conference on Electrical and Electronics Engineering (ELECO), 28 Nov 2013, pp. 516–520. IEEE (2013)
23. Mirghasemi, S., Rayudu, R., Zhang, M.: A heuristic solution for noisy image segmentation using particle swarm optimization and fuzzy clustering. In: 2015 7th International Joint Conference on Computational Intelligence (IJCCI), 12 Nov 2015, vol. 1, pp. 17–27. IEEE (2015)
24. Moeskops, P., Viergever, M.A., Mendrik, A.M., de Vries, L.S., Benders, M.J., Išgum, I.: Automatic segmentation of MR brain images with a convolutional neural network. IEEE Trans. Med. Imaging **35**(5), 1252–1261 (2016)
25. Pereira, S., Pinto, A., Alves, V., Silva, C.A.: Brain tumor segmentation using convolutional neural networks in MRI images. IEEE Trans. Med. Imaging **35**(5), 1240–1251 (2016)
26. Liu, F., Jang, H., Kijowski, R., Bradshaw, T., McMillan, A.B.: Deep learning MR imaging-based attenuation correction for PET/MR imaging. Radiology **286**(2), 676–684 (2017)
27. Avendi, M.R., Kheradvar, A., Jafarkhani, H.: A combined deep-learning and deformable model approach to fully automatic segmentation of the left ventricle in cardiac MRI. Med. Image Anal. **30**, 108–119 (2016)
28. Liu, F., Zhou, Z., Jang, H., Samsonov, A., Zhao, G., Kijowski, R.: Deep convolutional neural network and 3D deformable approach for tissue segmentation in musculoskeletal magnetic resonance imaging. Magn. Reson. Med. **79**, 2379 (2017)
29. Badrinarayanan, V., Kendall, A., Cipolla, R.: Segnet: a deep convolutional encoder-decoder architecture for image segmentation. IEEE Trans. Pattern Anal. Mach. Intell. **39**(12), 2481–2495 (2017)
30. Dunn, J.C.: A fuzzy relative of the ISODATA process and its use in detecting compact well-separated clusters. J. Cybern. **3**, 32–57 (1973)
31. Hathaway, R.J., Bezdek, J.C.: Recent convergence results for the fuzzy c-means clustering algorithms. J. Classif. **5**(2), 237–247 (1988)
32. Yang, X.S.: Firefly algorithm, Levy flights and global optimization. In: Research and Development in Intelligent Systems, vol. XXVI, pp. 209–218. Springer, London (2010)
33. Roy, S., Bandyopadhyay, S.K.: Classification of brain disorder using medical imaging. Int. J. Curr. Med. Pharm. Res. **2**(11), 989–997 (2016)
34. Kumar, S.N., Fred, A.L., Kumar, H.A., Varghese, P.S.: Nonlinear tensor diffusion filter based marker-controlled watershed segmentation for CT/MR images. In: Proceedings of International Conference on Computational Intelligence and Data Engineering, pp. 317–331. Springer, Singapore (2018)
35. Wu, K.L., Yang, M.S.: A cluster validity index for fuzzy clustering. Pattern Recogn. Lett. **26** (9), 1275–1291 (2005)
36. Pal, N.R., Bezdek, J.C.: On cluster validity for the fuzzy c-means model. IEEE Trans. Fuzzy Syst. **3**(3), 370–379 (1995)
37. Kim, D.W., Lee, K.H., Lee, D.: Fuzzy cluster validation index based on inter-cluster proximity. Pattern Recogn. Lett. **24**(15), 2561–2574 (2003)
38. Wu, C.H., Ouyang, C.S., Chen, L.W., Lu, L.W.: A new fuzzy clustering validity index with a median factor for centroid-based clustering. IEEE Trans. Fuzzy Syst. **23**(3), 701–718 (2015)
39. Pakhira, M.K., Bandyopadhyay, S., Maulik, U.: Validity index for crisp and fuzzy clusters. Pattern Recogn. **37**(3), 487–501 (2004)
40. Chao, S.M., Tsai, D.M., Chiu, W.Y., Li, W.C.: Anisotropic diffusion-based detail-preserving smoothing for image restoration. In: 2010 17th IEEE International Conference on Image Processing (ICIP), 26 Sept 2010, pp. 4145–4148 (2010)

41. Ghazanfari, M., Alizadeh, S.: Learning FCM with Simulated Annealing. INTECH Open Access Publisher (2008)
42. Bose, A., Mali, K.: Fuzzy-based artificial bee colony optimization for gray image segmentation. SIViP **10**(6), 1089–1096 (2016)
43. Bhandari, A.K., Soni, V., Kumar, A., Singh, G.K.: Cuckoo search algorithm based satellite image contrast and brightness enhancement using DWT–SVD. ISA Trans. **53**(4), 1286–1296 (2014)
44. Katiyar, S., Patel, R., Arora, K.: Comparison and analysis of cuckoo search and firefly algorithm for image enhancement. In: International Conference on Smart Trends for Information Technology and Computer Communications, 6 Aug 2016, pp. 62–68. Springer, Singapore (2016)
45. Arora, S., Singh, S.: A conceptual comparison of firefly algorithm, bat algorithm, and cuckoo search. In: 2013 International Conference on Control Computing Communication & Materials (ICCCCM), 3 Aug 2013, pp. 1–4. IEEE (2013)
46. Yang, X.S., He, X.: Firefly algorithm: recent advances and applications. Int. J. Swarm Intell. **1** (1), 36–50 (2013)

Chapter 2
Bat Optimization Based Vector Quantization Algorithm for Medical Image Compression

A. Lenin Fred, S. N. Kumar, H. Ajay Kumar and W. Abisha

Abstract Image compression plays a significant role in medical data storage and transmission. The lossless compression algorithms are generally preferred for medical images. The variants of lossy vector quantization algorithm are also used in many cases, where the reconstructed image quality is fairly good with optimum compression ratio. Bat optimization algorithm is formulated based on the biological trait of bats to detect prey and avoid obstacles by using echolocation. In this chapter, the application of bat optimization algorithm in medical image compression is highlighted. The bat optimization algorithm is employed here for the optimum codebook design in Vector Quantization (VQ) algorithm. The performance of the BAT-VQ compression scheme was compared with the Classical VQ, Contextual Vector Quantization (CVQ) and JPEG lossless schemes for the abdomen CT images. Satisfactory results were obtained by BAT-VQ in terms of picture quality measures.

Keywords Segmentation · Bat optimization algorithm · Compression
Vector quantization · Contextual vector quantization

A. L. Fred (✉) · H. Ajay Kumar · W. Abisha
Mar Ephraem College of Engineering and Technology, Elavuvilai, India
e-mail: leninfredip@gmail.com

H. Ajay Kumar
e-mail: ajayhakkumar@gmail.com

W. Abisha
e-mail: abishawilson01@gmail.com

S. N. Kumar
Sathyabama Institute of Science and Technology, Chennai, India
e-mail: appu123kumar@gmail.com

J. Hemanth and V. E. Balas (eds.), *Nature Inspired Optimization Techniques for Image Processing Applications*, Intelligent Systems Reference Library 150,
https://doi.org/10.1007/978-3-319-96002-9_2

2.1 Introduction

The image segmentation and compression plays vital role in telemedicine for data storage and transfer. The objective of image segmentation is to partition an image into meaningful regions. In the perspective of medical imaging, segmentation refers to the extraction of anatomical organs or anomalies like tumor, cyst. The compression is defined as a process that minimizes the overall size of data by encoding and representing the data through the suitable algorithm. The role of compression is vital in data storage and communication due to the enormous amount of medical data. The compression algorithms can be classified into lossy and lossless techniques. More precisely, the compression can be defined as the process of reducing number of bits required to represent an image by eliminating redundancies in pixel data.

The objectives of image compression are as follows

- Elimination of redundancy
- Minimize data storage and hence the cost
- Minimized bandwidth
- Less time to retrieve and transmit data.

The mechanism of image compression can be explained with a simple example as follows; Consider a Lena image of size 512 × 512 (8 bits), the uncompressed file requires 286,740 bytes of disk space. When the image is represented in TIFF (Lempel-Ziv-Welch (LZW) coding employed) format, the resultant compressed file is of 224,420 bytes.

The compression ratio (CR) is defined as follows

$$CR = \frac{Original\,data\,size}{Compressed\,data\,size}$$

Hence, with respect to the above discussion, the compression ratio is 1.277.

$$CR = (286{,}740/224{,}420) = 1.277$$

The vector quantization was initially proposed for the speech compression; Linear Predicted Coded (LPC) speech systems compress the data size of 6000–1400 bits/s [1]. In [2], the set partitioning in hierarchical trees (SPIHT) algorithm was used; better compression ratio and low computation complexity was obtained, when compared with Embedded Zero-tree Wavelet (EZW) [3] and modified EZW [4] algorithm. The JPEG2000 employs wavelet transform and relies mainly on block-based coder with truncation [5, 6]. The scaling of wavelet coefficient gives rise to Region of Interest (ROI) based coding [7, 8]. Xiong et al. [9] proposed a lossless compression scheme based on 3D Integer Wavelet Transform. The Contextual Set Partitioning in Hierarchical Trees (CSPIHT), algorithm yields robust results for context-based ultrasound medical image compression, when compared with scaling, max-shift, implicit and Embedded Block Coding with Optimal Truncation (EBCOT) methods [10]. The generation of the efficient codebook is vital in vector quantization and different techniques of codebook generation are discussed in [11]. The contextual vector quantization [12] proposed for the compression of ultrasound images generates efficient results when compared with the classical methods such JPEG and JPEG 2000, and ROI based techniques such as EBCOT and CSPIHT.

In image processing, the solution space will be complex for many problems and an optimum solution is required. The optimization is a technique in which, iteratively the execution takes place till an optimum solution is found. The evolutionary optimization algorithm relies on nature and hence they are termed as bio-inspired optimization technique. The optimization algorithms play a vital role in various aspects of image processing such as restoration, segmentation, classification, and compression. The bat optimization algorithm [13] when coupled with vector quantization was proven to generate efficient results for compression of standard test images; the computation complexity was minimized and high PSNR was produced, when LBG was coupled with Particle Swarm Optimization (PSO) [14], Firefly [15], Honey Bee Mate [16], Quantum behaved Particle Swarm Optimization (QPSO) [17] algorithms. The optimization algorithms can be employed for optimum codebook design in VQ algorithm [18–20]. The adaptive LBG algorithm was proposed in [21], good edge preservation with high compression ratio, when compared with classical LBG and modified LZ77 compression techniques. A novel compression scheme based on LBG and Singular Value Decomposition (SVD) was proposed in [22] for the compression of ECG data. The vector quantization [23] when coupled with hybrid wavelet transform was found to be efficient for color image compression. The hybrid optimization scheme [24] consisting of GA and ABC was incorporated in VQ algorithm for codebook design; superior results were produced for standard test images in terms of PSNR and compression ratio. In [25], gravitation search algorithm is employed for the codebook design in VQ algorithm, the results outperform LBG-PSO and LBG-Firefly algorithms. The Learning Vector Quantization (LVQ) [26] gain its importance in classification problems and was found to be a good alternative for support vector machine and deep learning techniques. The vector quantization [27] was also found to be robust in speech processing for automatic text-independent speaker identification when coupled with Gaussian mixture model-universal background model (GMM-UBM). In [28], the types of modified transform based compression schemes are discussed. The modified DWT and hybrid compression technique yield efficient results when compared with the JPEG and JPEG2000 model. The Discrete Wavelet Transform (DWT) [29] based compressed sensing algorithm generates efficient results when compared with Discrete Cosine Transform (DCT) based scheme for the compression of CT/MR images. In [30], a comparative analysis of various optimization algorithms when employed in LBG algorithm was discussed. The Bat optimization technique was found to be efficient when compared with Firefly, cuckoo and hybrid cuckoo search algorithms. The nonlinear transform coding [31] based compression scheme was found to be robust when compared with the JPEG and JPEG2000 algorithms.

A hybrid optimization model [32] comprising of simulated annealing and genetic algorithm was employed for codebook design in kernel VQ. The region of interest based medical image compression schemes is highlighted in [33]. The SPIHT HAAR algorithm was found to be proficient when compared with the EZW HAAR and HAAR based global thresholding. The VQ algorithm [34] gains its importance in the invisible watermarking scheme for image authentication. The search order coding scheme [35] was employed in VQ algorithm, computation complexity was minimized and efficient for hardware implementation. In this work, BAT optimization was used for the optimum codebook design in VQ algorithm. The performance of BAT-VQ compression scheme was analyzed in terms of picture quality measures and the results outperform the Classical VQ, CVQ, and JPEG lossless techniques.

2.2 Materials and Methods

2.2.1 Data Acquisition

The real-time abdomen CT data sets are used in this work for the analysis of algorithms. The images are acquired from Optima CT machine with a slice thickness of 3 mm. The images in DICOM format with a size of 512 × 512 are used in this work. The Metro Scans and Research Laboratory approved the study of human datasets for research purpose. The four abdomen CT data sets, each comprising of 200 slices are used in this work and the results of typical slices are depicted here.

2.2.2 Overview of Image Compression

The field of compression has undergone significant growth through the practical application of the theoretic work that began in the 1940s when C. E. Shannon and others first formulated the probabilistic view of information and its representation, transmission and compression [36, 37]. The digital image compression can be broadly classified into two categories: lossy and lossless technique. In lossless compression, the reconstructed image is identical to the original image with no loss of data. The lossy compression technique involves loss of data and the reconstructed image is not identical with the original image. The storage and transmission are the vital features for telemedicine applications; hence the lossy compression technique can be employed for medical images while preserving the reconstructed image quality [38]. The lossy compression techniques can yield high compression ratio thereby improving the efficiency of computing system [39].

Fig. 2.1 Generic classification of image compression algorithms

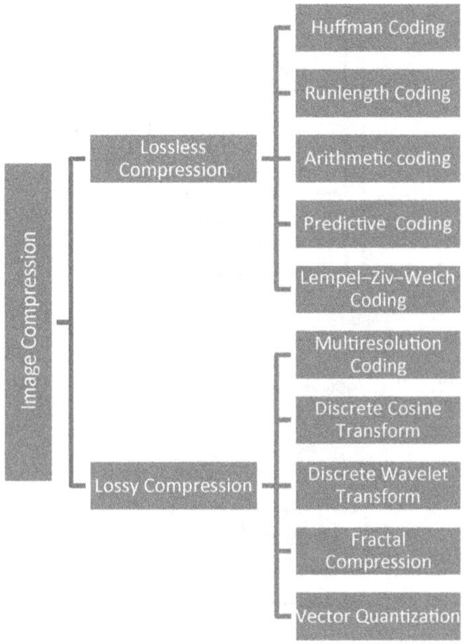

The general classification of compression algorithms is depicted above in Fig. 2.1. The Huffman coding is an entropy coding invented by a person named Huffman and it is now widely used as the backend of many hybrid compression models [40, 41]. The Run Length Coding (RLE) replaces data with (L, V) pair. The 'V' represents the repeated value and 'L' is the count of repeated values [42]. The Arithmetic Coding is based on the probabilistic occurrence of symbols and depicts a message in some finite interval between 0 and 1 [43]. The predictive coding relies on the prediction error based on the neighborhood pixels of the current pixel under investigation [44, 45]. Lempel-Ziv–Welch (LZW) coding is called as dictionary coding and is appropriate for encoding text. The LZW coding can be static or dynamic and it depends on the application [46]. The LZW compression is used in the GIF and TIFF image format. Ahmed et al. [47] proposed that, the Discrete Cosine Transform (DCT) has a vital role in frequency domain transformation with its applications in signal and image processing. The classical JPEG compression is based on DCT transform The blocking effect of DCT was overcome by Discrete Wavelet Transform (DWT), JPEG2000 uses DWT [48].

The fractal image compression was introduced by Barnsley [49, 50] and the compressed image was represented by contractive transforms and functions. It relies on the Collage theorem and is applicable for reconstruction of damaged images [51]. The fundamental concept of Vector Quantization relies on the generation of the codebook and the modifications of classical VQ gain its importance in medical image compression [52, 53].

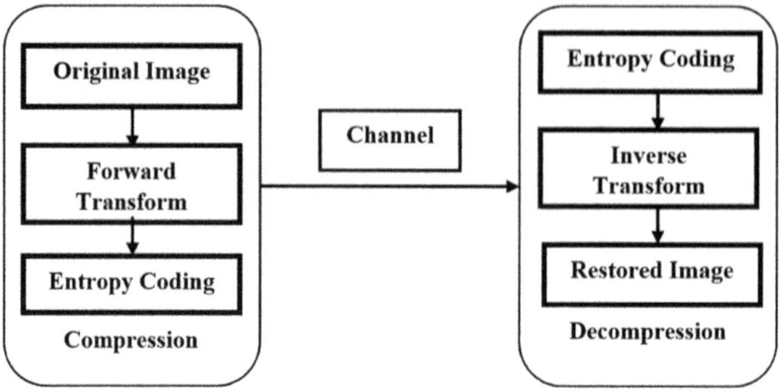

Fig. 2.2 Block diagram of lossless compression

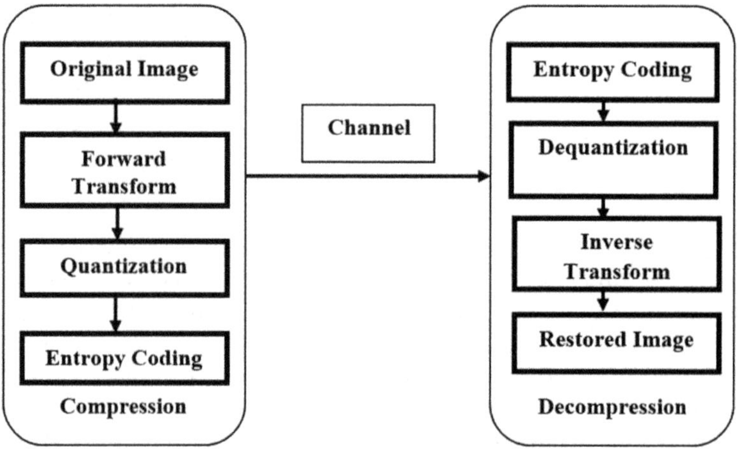

Fig. 2.3 Block diagram of lossy compression

The generic block diagram of the lossless compression is depicted in Fig. 2.2. The first stage involves the removal of inter-pixel redundancy by the application of the mathematical transform. The second stage involves the removal of coding redundancy by entropy coding technique.

The decoder section performs the reverse operation for the image reconstruction.

The lossy compression involves three stages as depicted in Fig. 2.3; the coefficients obtained from transform coding are quantized for the removal of psycho-visual redundancy.

The key principle behind the data reduction is the removal of redundancy. The image is a 2D array of pixels and mathematically compression is a process of transforming 2D pixel cluster into statistically uncorrelated data for transmission. In the receiver side, the compressed image is reconstructed for the original image. The basic characteristics of most images are that adjacent pixels are correlated and redundant information exists. The two basic components of compression are redundancy and irrelevant data. Redundancy minimization aims at removal of duplication from the signal source (Image or Video).

In the perspective of digital image compression, three types of redundancy exist; coding redundancy, inter-pixel redundancy, and psycho-visual redundancy. The data compression objective is to remove or eliminate one or more of the redundancy. The inter-pixel dependency relies on the statistical relationship between pixels, especially between neighboring pixels. The inter-pixel redundancy also occurs between the images, when the co-relation represents the structural or geometrical relationship between the objects in the image. Consider two images in Fig. 2.4 as below. The histogram of the two images looks similar, but the structure and geometry of the objects in the image are different. The variable length coding can be employed for minimizing the coding redundancy; however, it cannot alter the correlation between the pixels within the images.

The autocorrelation is determined as follows

$$f_1(x,y)^\circ f_2(x,y) = \frac{1}{PQ}\sum_{m=0}^{M-1}\sum_{n=0}^{N-1} f^*(m,n)h(x+m,y+n)$$

where

$f_1(x,y)\, and f_2(x,y)$ are the two functions whose auto co-relation coefficient to be computed,
f^* denote the complex conjugate of f.

The normalized form of the above equation is as follows

$$\gamma(\Delta n) = \frac{A(\Delta n)}{A(O)}$$

where,

$$A(\Delta n) = \frac{1}{N-\Delta n}\sum_{y=0}^{N-1-\Delta n} f(x,y)f(x,y+\Delta n)$$

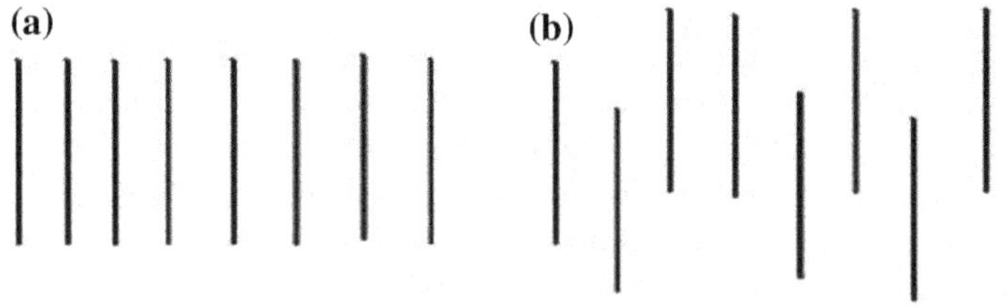

Fig. 2.4 Images representing interpixel redundancy

The scaling factor in the equation accounts for varying number of sum terms that arise for each integer value of Δn. The value Δn must be strictly less than N. The dramatic difference between the shapes of the functions can easily be identified in their respective plots of Δn versus γ. The histogram gives the pixels gray level profile and the entropy coding technique does not minimize correlation between the pixels in the image. The interpixel correlation can be determined from the auto-correlation plot. The interpixel redundancy exists, since the objects are regularly and closely arranged. For the minimization of interpixel redundancy, the image is transformed into an efficient format.

The technique of removal of the interpixel redundancy by transformation is called mapping. The mapping is said to be reversible if the original image is obtained from the transformed dataset.

The histogram gives the gray level profile of the image. It specifies the number of pixels in each gray value. Each gray value is represented by using a constant number of bits say 8 bits. For an image of size $p \times q$, number of bits of memory required to store the image is $p \times q \times 8$.

The average number of bits required to represent mth gray value is as follows.

$$n_{b(avg)} = \sum_{m=0}^{L-1} b(g_m)p(g_m)$$

where $b(g_m)$ represents number of bits needed to represent mth gray level, $p(g_m)$ represents the probability associated with the mth gray level.

If we represent the image using k-bits, then $b(g_m) = k$ bits (since sum of $p(g_m) = 1$). The average word length needed to represent an image can be minimized, if the gray levels of the image are represented using varying number of bits based on the probability associated with the gray levels. The variable length coding was found to efficient than fixed length coding, since it requires a less average number of bits for representation. The low number of bits are used to represents the gray levels whose probability is high and high number of bits are used to represent the gray levels whose probability is low. The variable length coding is employed to minimize the coding redundancy.

The psycho-visual redundancy refers to those features that have relatively less importance in visual processing. This redundancy can be eliminated without significantly degenerating the quality of the image. When an image is perceived, the notable features like edges, texture, and color will be distinguished. The above-said features are coupled into recognizable grouping, the brain then correlates these features with the prior knowledge for image interpretation. It is associated with the real or quantifiable information and this information can be removed since it is not needed for visual processing. The elimination of psycho-visual redundancy is termed as quantization and it is an irreversible process. Figure 2.5a depicts an image with 256 possible gray levels of 8-bit image and Fig. 2.5b depicts the same image after uniform quantization of 1 bit or 128 possible gray levels.

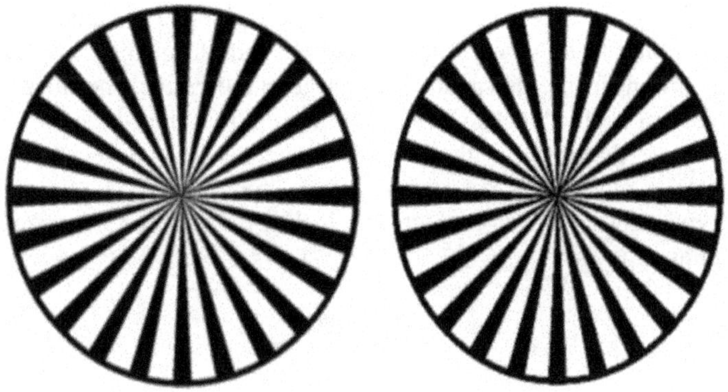

Fig. 2.5 Images depicting the psycho-visual redundancy

The two images look identical to a human observer, however, quantization is performed on the second figure. The above same thing can be done by bit plane slicing; each pixel of the image was represented by 8-bit planes.

2.2.3 *Vector Quantization Scheme*

The vector quantizer is termed as a system which maps a sequence of continuous or discrete vectors into a digital sequence appropriate for communication over a channel. The vector quantization initially finds its role in speech processing.

Mathematically, the process of vector quantization comprises of two mappings; encoding and decoding. The encoding (α) is termed as process in which each input vector $X = (X_0, X_1, \ldots X_{k-1})$ of the channel is assigned with a channel symbol $\alpha(x)$ and decoding (β) is termed as process in which encoded sample gives a reproduction value. The basic block diagram of data compression by vector quantization is depicted below in Fig. 2.6.

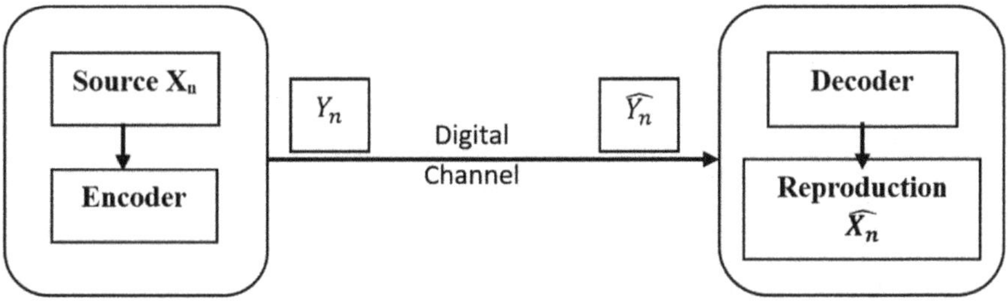

Fig. 2.6 Data compression using vector quantization

Vector quantization is an effective and simple technique for data compression. The VQ was found to be efficient in pattern recognition, image compression, and speech recognition. In the perspective of image compression, the image is subdivided into several vector blocks and the codewords from the codebook are used to map each vector. Compared with the scalar quantization, the vector quantization allows fractional rates in bits per sample.

The squared error distortion measure is widely used in vector quantization and it is expressed as follows.

$$d(x,\hat{x}) = \|x-\hat{x}\|^2 = \sum_{i=0}^{k-1} (x_i - \hat{x}_J)^2$$

The Quantization Error (QE) is represented by D(x, q(x)), where D is the distance vector.

The mean quantization error (MQE) is defined as follows

$$MQE = \frac{1}{N_n} \sum_{i=1}^{N_n} d(X_n, q(X_n))$$

where N_n is the number of elements in the input datasets.

For distortion measures, many metrics are there, however, squared Euclidean distance is widely used.

$$D(X,X') = \sum_{i=1}^{k} (X_i - X'_i)^2$$

The mean quantization error is also called as squared root of MSE (Root mean square error (RMSE)). In some cases, Normalized mean square error (NMSE) is used. NMSE is defined as the MSE divided by the MSE from the codebook of a codeword placed at the centroid of the entire dataset.

$$NMSE = \frac{MSE}{\frac{1}{N_n}\sum_{i=1}^{N_n} D(X_i - C)}$$

The quantizer is said to be optimum when a higher MQE is obtained for each other quantizer with a same number of codewords.

Mathematically q* is optimum if, for each other q.

$$D(q^*) \leq D(q)$$

The two conditions usually employed for the optimum quantizer design are Nearest Neighbor Condition (NNC) and Centroid condition (CC). The NNC is a process of assigning to each input vector the nearest code word.

The input dataset is expressed as follows.

$$\overline{I}_i = \{x \in X; D(x, y_j) \leq D(x, y_j), j = 1, 2, \ldots, N_c, j \neq 1\}$$
$$i = 1, 2, 3, \ldots N_c$$

The set $\overline{I}_i$ constitute a partition of the input dataset termed as Voronoi partition. The three stages in image compression using vector quantization are codebook generation, encoding and decoding. The foremost step is the code book generation; images are split into n dimension training vectors and each vector is encoded by the index of codeword by table lookup method. The process of encoding and decoding of an image by vector quantization is depicted in Fig. 2.7.

For example, consider an image f(x, y) comprising of 16-pixel values. The image is split into four sub-blocks of size 2 × 2. A typical codebook comprising of code words with the index is depicted here. The next step is the mapping of vectors using the index from the codebook.

The above example comprises of four vectors (X_1, X_2, X_3, X_4). The vectors $X_1(7, 9, 15, 9)$ is mapped to 7th code word of the codebook. Similarly the vector $X_2(7, 10, 14, 6)$ is mapped to the 9th code word, $X_3(10, 37, 60, 20)$ is mapped to the 9th code word, $X_4(22, 55, 14, 75)$ is mapped to the 10th codeword. The vital feature of vector quantization is that for efficient codebook generation, distortion between original image and reconstructed image should be minimized. The computation time involved in the generation of codebook is high; the widely used technique for the codebook generation is Generalized Lloyd algorithm (GLA) which is also called Linde Buzo Gray (LBG) algorithm. The codebook is created by clustering technique and euclidean distance is the widely used distortion criterion. The minimum distortion indicates smallest euclidean distance between the input vector and the reconstructed vector. The code vector with smallest distortion is stored or transmitted through the communication channel. In the decoding section, the index of the codeword vector is searched in the codebook for the image reconstruction.

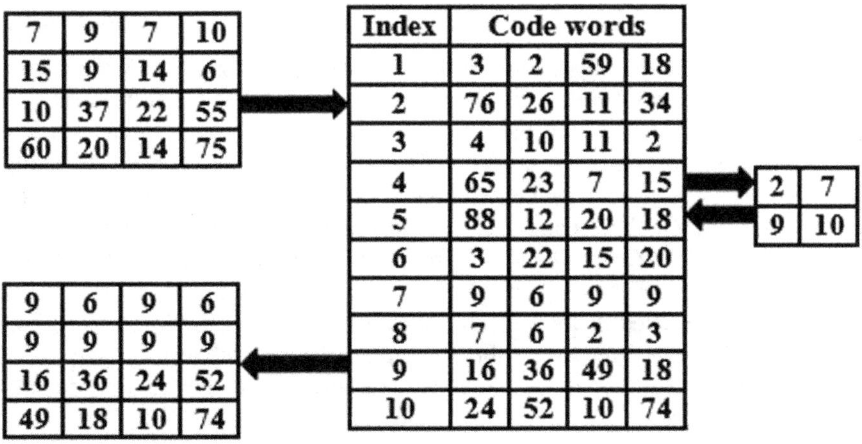

Fig. 2.7 Encoding and decoding process by VQ

2.2.4 Linde Buzo Gray Algorithm

Linde et al. put forward Generalized Lloyd algorithm (GLA) which is also termed as Linde Buzo Gray (LBG) algorithm. A mapping function is used to partition training vectors into C clusters.

The mapping function is defined as follows

$$F^n \rightarrow CB$$

Let $X = (x_1, x, \ldots x_n)$ be a training vector and $D(x, y)$ represents the euclidean distance between any two vectors.

The steps in LBG algorithm for the codebook generation are represented as follows.

Step 1: The initial codebook (CB_0) is generated randomly.
Step 2: $i = 0$
Step 3: For each training vectors, the following steps are performed
Determine the euclidean distance between the training vector and the codeword in CB_i
The euclidean distance is expressed as follows

$$D(x, c) = \sqrt{\sum_{t=1}^{k} (x_t - c_t)^2}$$

Track the closest codeword among CB_i
Step 4: Split the codebook into N cells
Step 5: Determine the centroid of each cell to determine the new codebook CB_{i+1}
Step 6: Determine the average distortion for CB_{i+1}
Step 7: This iterative procedure terminates, when the distortion value change is small enough for the last iteration, otherwise $i = i + 1$ and go to step 3.

Consider an example that comprises of five training vectors, N = 3 be the total number of the codewords in the codebook. The initial codebook (CB_0) is generated randomly and is depicted in Table 2.1. The distance between the training vectors and codewords among CB_0 is computed. The training vectors that have the similar closest codeword are partitioned into the same cell. The vectors X_1 and X_4 have the closest codeword C_3. The centroid of the two vectors is the 3rd code word of new generated codebook (CB_1). The above procedure is repeated until the convergence of codebook occurs. The computation of euclidean distance by Generalized Lloyd algorithm for the determination of codebook is depicted in Table 2.2. The bold values in the Table 2.2 represents the minimum value of euclidean distance.

The C1, C2, C3 represents the Euclidean distance computed from the training vectors and codebooks are depicted in Table 2.3.

Table 2.1 Example depicting five training vectors

X_i	x_1	x_2	x_3	x_4
X_1	240	190	21	154
X_2	212	76	123	36
X_3	10	219	108	232
X_4	160	108	155	41
X_5	109	52	19	247

Table 2.2 Example depicting GLA algorithm for the computation of codebook

i	X	Nearest codeword	C_1	C_2	C_3	CB_i
0	X_1	C_3	247.89	220.41	**215.61**	CB_0
	X_2	C_2	270.69	**99.64**	126.83	
	X_3	C_1	**168.63**	350.58	254.55	
	X_4	C_3	223.27	171.39	**87.93**	
	X_5	C_1	**195.70**	255.58	282.29	
1	X_1	C_3	211.42	131.747	**3**	CB_1
	X_2	C_2	268.35	**68.922**	197.74	
	X_3	C_1	**106.96**	266.613	259.682	
	X_4	C_2	242.60	**82.591**	211.655	
	X_5	C_1	**107.4**	223.534	212.859	
2	X_1	C_3	211.42	195.21	**102.65**	CB_2
	X_2	C_2	268.35	**0**	103.38	
	X_3	C_1	**106.96**	316.06	245.42	
	X_4	C_2	242.60	**69.115**	106.80	
	X_5	C_1	**107.4**	257.91	212.72	

Table 2.3 Example depicting the codebooks for the computation of euclidean distance

CB_i	CB_0				CB_1				CB_2			
C_1	32	177	143	210	59.5	136	63.5	240	59.5	136	63.5	240
C_2	196	16	46	24	212	76	123	36	206	125.3	99.6	77.6
C_3	180	130	212	101	203	150	88	98.5	241	192	21	156

Linde et al. extended Lloyd's results from mono to n-dimensional cases and the algorithm is termed as Generalized Lloyd algorithm (GLA). The block diagram of the LBG algorithm is depicted in Fig. 2.8. It comprises of two main stages; codebook initialization and optimization. The codebook initialization can be random or by splitting procedure and the generation of optimum codebook is depicted in Fig. 2.9.

In the Fig. 2.9, m represents the iterative number, f_m is the mth codebook and d_m represents the quantization error (MQE). The term 'ε' represents the precision of the optimization process. If $\left(\frac{d_{m-1}-d_m}{d_m}\right) < \varepsilon$. then the code book optimization process terminates.

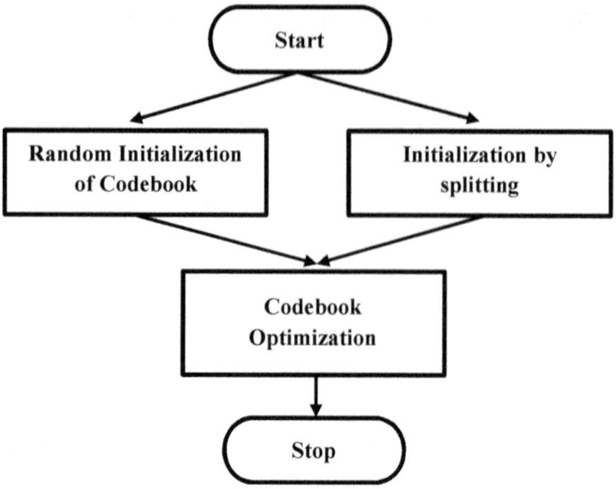

Fig. 2.8 LBG codebook optimization

2.2.5 *Bat Optimization Algorithm*

Yang [54] proposed the meta-heuristic bat algorithm based on echolocation of bats. The echolocation principle was used to detect prey and obstacle avoidance. The minimum parameters requirement is the key feature of bat optimization algorithm. The bat optimization was found to be efficient in a wide range of applications; image processing [55], Scheduling problems [56], data mining [57], and global optimization [58].

The bat optimization is based on the following three rules.

i. The loudness changes from the maximum value (A_{max}) to the minimum value (A_{min}).
ii. The bats use echolocation to sense distance.
iii. The bats can automatically adjust the wavelength of the emitted pulses.

The steps in bat optimization algorithm are summarized as follows

Step 1: Initialize the population of bats. The initial population is randomly generated $(i = 1, 2, \ldots N)$, where each solution has K dimensions. The matrix representation of the solution is as follows

$$Y = \begin{bmatrix} y_{1,1} & y_{1,2} & y_{1,3} & \cdots & y_{1,K} \\ y_{2,1} & y_{2,2} & y_{2,3} & \cdots & y_{2,K} \\ \vdots & \vdots & \vdots & \ddots & \vdots \\ y_{N,1} & y_{N,2} & y_{N,3} & \cdots & y_{N,K} \end{bmatrix}$$

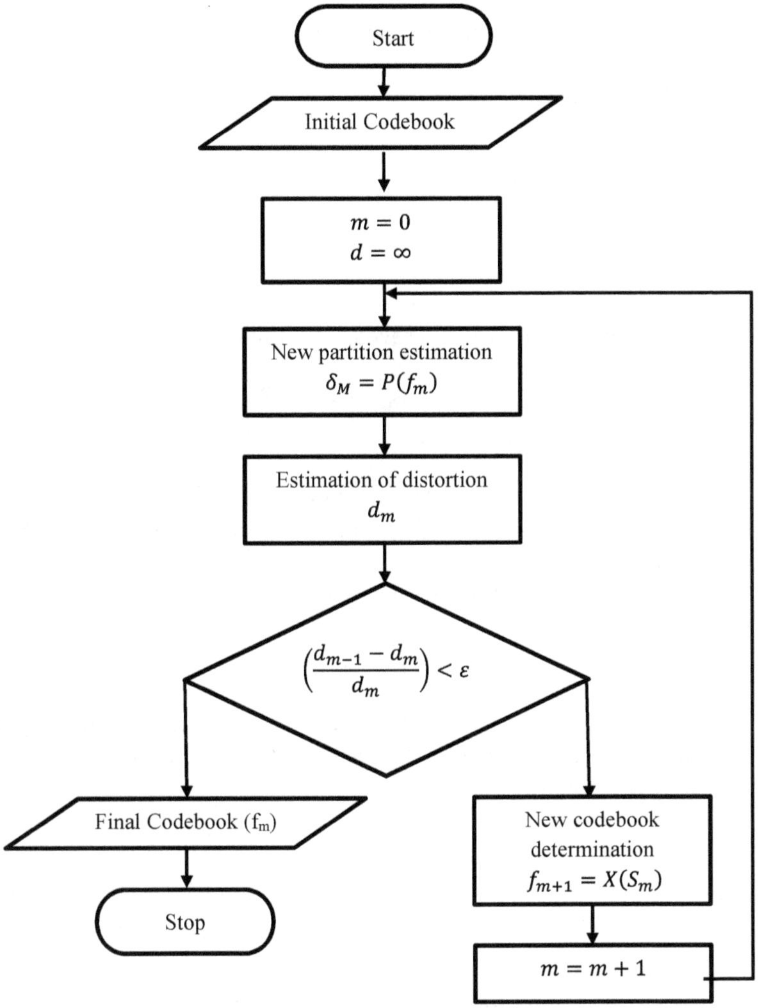

Fig. 2.9 Flowchart depicting codebook optimization

Step 2: Determine the new solution by the updation of position, velocity, and frequency.

Frequency update

$$f_i(t+1) = f_{min} + (f_{max} - f_{min}) * \beta$$

where β is a random variable with uniform distribution [0, 1], and $f_{min} = 0, f_{max} = 2$.
Distance update

$$U_i(t+1) = U_i(t) + (Y_i - Y_{best}) \cdot f_i(t+1)$$

Position Update

$$Y_i(t+1) = Y_i(t) + V_i(t+1)$$

Step 3: The current solution is improved as follows

$$Y_{new} = \begin{cases} y_{best} + \varepsilon L_t & if\ rand_1 > P_i^t \\ y_i^t & otherwise \end{cases}$$

where $rand_1$ is a random number in range [0, 1], ε is a scaling factor in the range [−1, 1], $L_t = \langle L_i^t \rangle$ is the mean loudness of all bats, P_i^t is the pulse rate function.
The pulse rate function is expressed as follows.

$$P_i^t = P_i^0 \left(1 - e^{-\beta t}\right)$$

where β is a constant and P_i^0 is the initial pulse rate in the range [0, 1]

Step 4: The solution obtained in the above step is accepted based on the expression below

$$\left(y_i^t, fitness\left(y_i^t\right)\right) = \begin{cases} \left(y_{new}, f\left(y_{new}^t\right)\right), & if\left(\left(rand_2 < L_i^t\right) and \left(f\left(y_{new}^t\right) > f\left(y_i^{t-1}\right)\right)\right) \\ \left(y_i^{t-1}, f\left(y_i^{t-1}\right)\right) & otherwise \end{cases}$$

where $rand_2$ is uniform random number in range [0, 1].
The loudness function $\left(L_i^t\right)$ is expressed as follows

$$L_i^t = \alpha L_i^{t-1}$$

where α is a constant.

Step 5: The current best solution is recorded, the solution with highest objective function value is considered

Step 6: The number of iterations will be predefined and the stopping criteria is checked.

2.2.6 Bat-VQ Image Compression Algorithm

The steps of BAT-VQ compression are summarized here. The BAT optimization was incorporated in VQ for the efficient codebook design. The less parameter tuning makes it an attractive choice for selection in this work for medical image compression

Step 1: Initialize the parameters of the bat optimization algorithm and choose the size of the codebook. Each codebook is assumed as a bat.
Step 2: Determine the fitness of all codebooks and the codebook with the best fitness is set as y_{best}.
Step 3: The codebooks are zoomed towards y_{best} by the adjustment of frequency, velocity and position.
Step 4: The codebooks are moved around the y_{best}, when the generated random number (step size for random walk) is greater than the pulse rate (P).
Step 5: A new codebook is chosen by random number selection and when the random number is greater than loudness and the newly selected codebook fitness is better than the previous codebook, the new one is considered.
Steps 6: The bats are ranked and determine the current best y_{best}.
Steps 7: The steps (2)–(6) are repeated until a stopping criterion is reached.

2.3 Results and Discussion

The algorithms have been developed in Matlab 2010a and tested on real-time medical images. The system specifications are as follows; Intel Core i3 processor of 3.30 GHz with 4 GB RAM. The analysis of algorithms has been carried out on four DICOM abdomen data sets. The results of the typical image from each dataset are depicted here. The proposed BAT-VQ algorithm results are compared with classical LBG-VQ, Contextual Vector Quantization (CVQ) and Classical JPEG techniques. The efficiency of bat optimization for the codebook design was proved in [13], superior results were produced when compared with LBG-VQ coupled with optimization techniques like Particle Swarm Optimization (PSO), Quantum Particle Swarm Optimization (QPSO), Honey Bee Mating Optimization (HBMO), and Firefly algorithm. For the evaluation of compression algorithms, performance metrics like Peak to Signal Noise Ratio (PSNR), Mean Square Error (MSE) and Compression Ratio (CR) are used. The reconstructed image quality was also validated in terms of picture quality metrics like Structural Content (SC), Normalized Cross Correlation (NCC). The proposed BAT-VQ compression scheme was compared with the classical vector quantization using LBG [59], Contextual Vector Quantization (CVQ) [60], JPEG lossy scheme [61].

The parameters of BAT optimization algorithm incorporated in the VQ for codebook design are as follows; BAT population = 20, Maximum iteration = 50, Loudness of BAT = 0.2, Pulse rate = 0.2. The codebook size of 16 is used for the analysis of all the data sets. In CVQ scheme, region growing algorithm was used to separate foreground and background. The foreground region was encoded using high bit rate since it comprises of vital information (Contextual region). The background region was encoded using low bit rate. The foreground and background regions are encoded by separate codebooks and the merged codebook is transmitted. The receiver side decodes the codebooks and reconstruction of the image take place. In this work for CVQ, the foreground ground and background codebook size were chosen 16 and 8 respectively for all datasets.

Compression ratio is defined as the ratio between uncompressed image file size and compressed image file size.

$$CR = \frac{S_{UC}}{S_C}$$

where S_{UC} represents the file size of input image and S_C represents the file size of the compressed image.

The PSNR and MSE are expressed as follows

$$MSE = \frac{1}{N}\sum_{x=1}^{M}\sum_{y=1}^{N}\left(I(x,y) - \hat{I}(x,y)\right)^2$$

$$PSNR = 10\log\frac{255^2}{MSE}$$

The $I(x,y)$ represents the pixel value of input image and $\hat{I}(x,y)$ represents the pixel value of reconstructed image.

The PSNR is used to determine the quality of the reconstructed image in compression. A higher value of PSNR and lower value of MSE indicates the efficiency of the compression algorithm. The Normalized Cross Correlation (NCC) measures the resemblance between two digital images. The ideal value of correlation measure is 1, which indicates the perfect matching of two images. The quality of the reconstructed image is good as the value of NCC approaches 1.

$$NCC = \frac{\sum_{x=1}^{M}\sum_{y=1}^{N} I(x,y)\hat{I}(x,y)}{\sum_{x=1}^{M}\sum_{y=1}^{N} I(x,y)^2}$$

The PSNR and MSE plot of compression schemes are represented in Figs. 2.10 and 2.11.

The PSNR, MSE, NCC and SC plot reveals that BAT-VQ is better than classical VQ based on LBG, CVQ, and JPEG lossy techniques. The Structural Content (SC) is also termed as Structural Correlation and it also measures the degree of

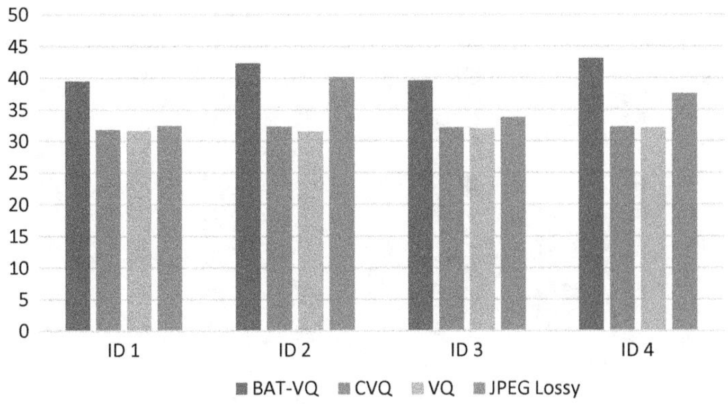

Fig. 2.10 PSNR plot of compression schemes

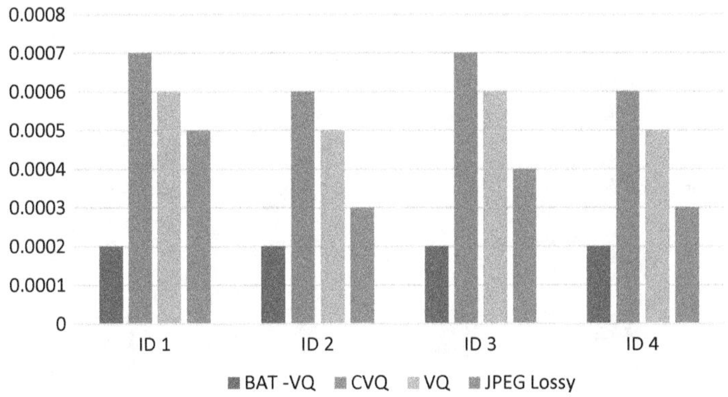

Fig. 2.11 MSE plot of compression schemes

closeness between two images. If the value of SC is spread at 1, the quality of the reconstructed image is good. A higher value of SC indicates the poor quality of the reconstructed image.

$$SC = \frac{\sum_{x=1}^{M} \sum_{y=1}^{N} I(x,y)^2}{\sum_{x=1}^{M} \sum_{y=1}^{N} \hat{I}(x,y)^2}$$

The NCC and SC plot of compression schemes are represented in Figs. 2.12 and 2.13.

The performance table of CVQ and BAT-VQ shows that in terms of elapsed time, BAT-VQ is superior. The compression ratio of BAT-VQ is less when compared with the CVQ approach. The performance of the CVQ is represented in Table 2.4.

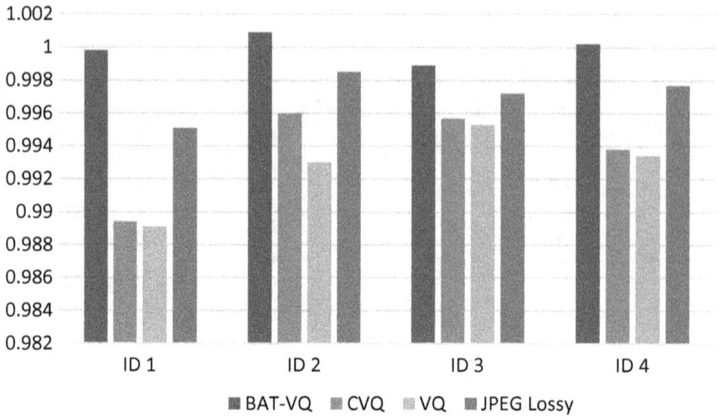

Fig. 2.12 NCC plot of compression schemes

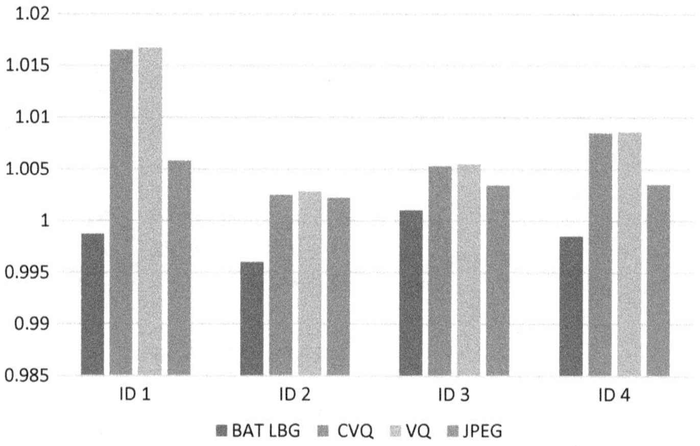

Fig. 2.13 SC plot of compression schemes

Table 2.4 Contextual vector quantization compression performance

Dataset	CB	S_{uc}	S_C	CR	Space saving	bpp	Elapsed time (s)
ID1	16	557,056.00	92,033.25	6.05	0.83	7	70.261368
ID2	16	2,228,224.00	445,991.00	5.00	0.80	7	325.560597
ID3	16	557,056.00	95,636.25	5.82	0.83	7	56.253661
ID4	16	2,228,224.00	351,068.75	6.35	0.84	7	845.045550

Table 2.5 BAT-vector quantization compression performance

Dataset	CB	S_{uc}	S_C	CR	Space saving	bpp	Elapsed time
ID1	16	557,056.00	163,840.00	3.40	0.71	5.00	2.660046
ID2	16	2,228,224.00	655,360.00	3.40	0.71	5.00	5.037085
ID3	16	557,056.00	163,840.00	3.40	0.71	5.00	2.647725
ID4	16	2,228,224.00	655,360.00	3.40	0.71	5.00	5.452841

The performance of the BAT-VQ compression is represented in Table 2.5. The reconstructed image quality plays a vital role, hence BAT-VQ is considered as an efficient technique because the metrics like PSNR, MSE, NCC, and SC reflects the image quality.

The optimized codebook values by BAT optimization algorithm is depicted in Table 2.6. The bits per pixel (bpp) represents the average number of bits required to encode each pixel in the image. For a grayscale image, bpp = 8 and for a color image, it relies on the color model. In the scenario of image compression, bpp is defined as the ratio of the size of compressed file to the number of pixels in the image. The BAT-VQ has low bpp when compared with CVQ indicating the proficiency of the proposed compression technique.

The fitness function plot of bat optimization for the datasets ID1–ID4 are depicted in Fig. 2.14. The maximum number of iteration was set to 50 and a codebook size of 16 was used. The proposed BAT-VQ algorithm in this work was compared with CVQ, Classical VQ, and JPEG lossy techniques. The classical VQ algorithm cannot ensure that the resultant codebook will be optimum. In CVQ, the

Table 2.6 Optimized codebook values of BAT optimization algorithm

Index	ID1	ID2	ID3	ID4
1	0.0000	0.5671	0.2064	0.7070
2	0.1937	0.2933	0.3629	0.9997
3	0.6494	0.8284	0.5332	0.5184
4	0.2417	0.8762	0.4250	0.0675
5	0.9995	0.0574	0.1267	0.8395
6	0.1569	0.3914	0.2702	0.2509
7	0.0902	0.4881	0.0000	0.3135
8	0.5643	0.6837	0.8736	0.3992
9	0.7700	0.4442	0.6434	0.7870
10	0.0264	0.9992	0.4862	0.7394
11	0.2885	0.4739	0.0284	0.6098
12	0.3479	0.0000	0.9980	0.1245
13	0.4411	0.2188	0.0540	0.4597
14	0.7399	0.3659	0.0892	0.1866
15	0.8684	0.1020	0.5779	0.0000
16	0.1032	0.7336	0.1715	0.6605

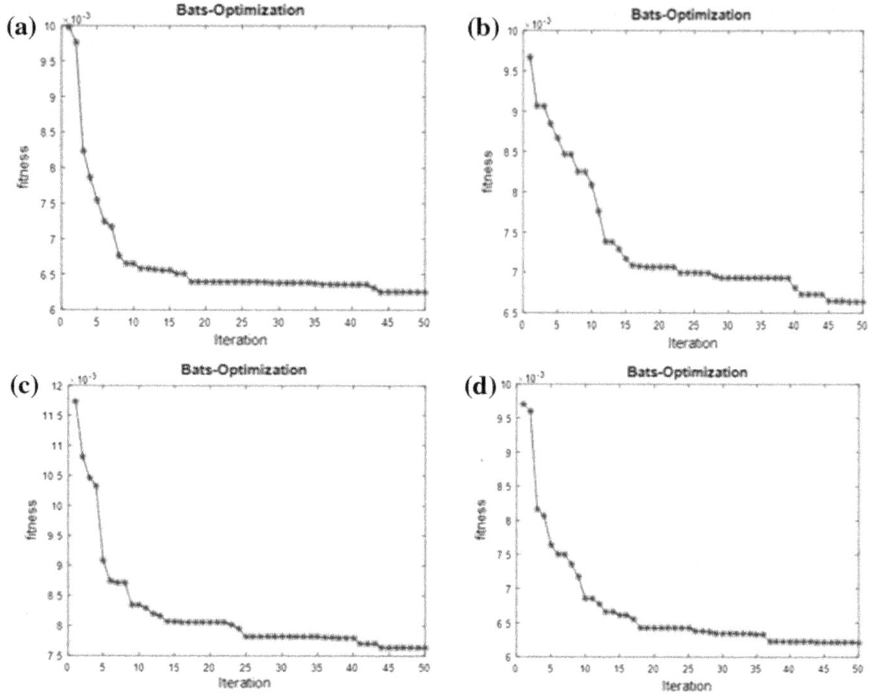

Fig. 2.14 BAT optimization algorithm performance for the datasets (ID1, ID2, ID3, and ID4)

codebook size has to be properly tuned for the foreground and background. Initially, the performance was analyzed by classical metrics like PSNR and MSE. The SC and NCC picture quality metrics also reveals the efficiency of the proposed BAT-VQ approach. The algorithm was found to yield efficient results for all images in for data sets. The results of typical images from each dataset are depicted here. The parameter tuning will be crucial in many optimization algorithms; however, the bat optimization needs only less number of parameters to be tuned and kept constant for all datasets. The computation time has also to be considered for real-time applications and hence the proposed BAT-VQ was found to be proficient. The computation time changes for each dataset since the uncompressed image file size is different. The BAT-VQ compression scheme can be employed for real-time data transfer for telemedicine applications and can be efficiently implemented in hardware. In the BAT-VQ algorithm, the codebook size was manually set to 16. In future, the optimization algorithm can be employed for choosing the appropriate size of the codebook. The performance analysis reveals that BAT optimization was found to be efficient in the codebook design for VQ algorithm. The promising results are obtained for compression of medical images, when compared with the CVQ, Classical VQ and JPEG lossy algorithms. The BAT-VQ compression results are depicted in Fig. 2.15.

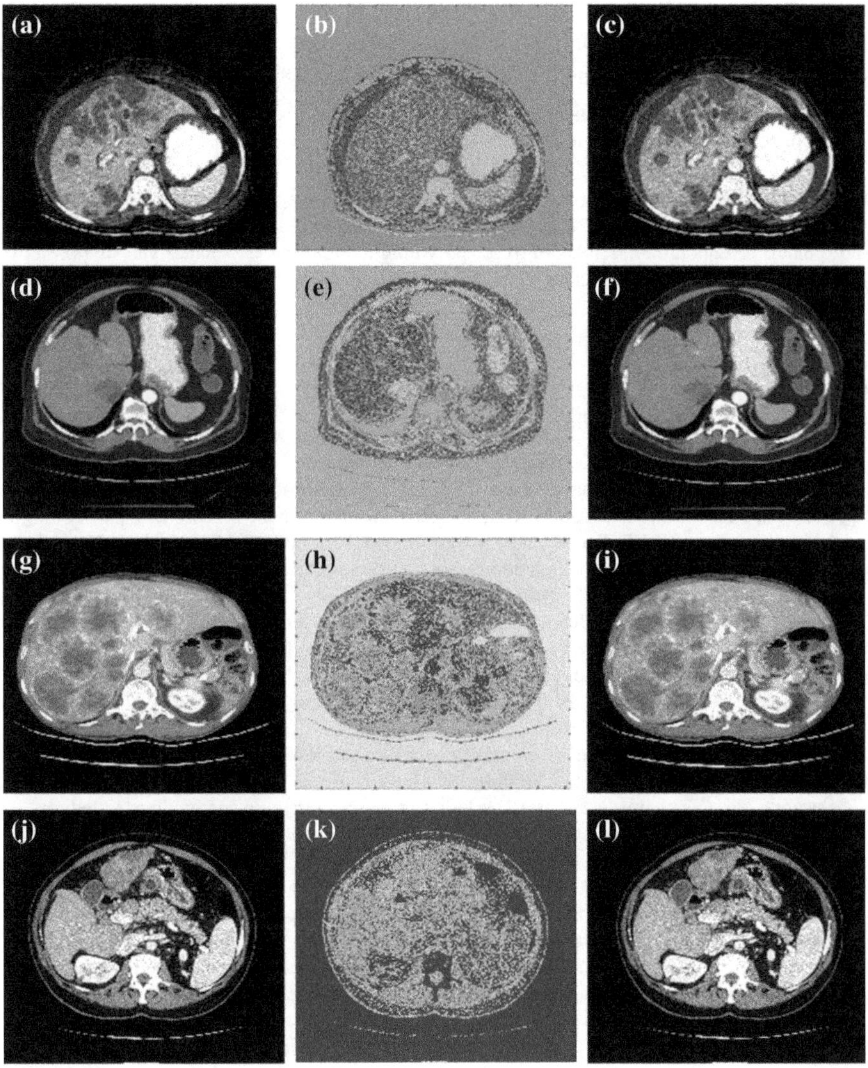

Fig. 2.15 BAT-VQ compression results; first column represents the input image, second column represents the compressed image and the third column represents the decompressed image

2.4 Conclusion

In this work, BAT-VQ compression method is proposed and experimental analysis was done for the compression of abdomen CT images. The key idea is the incorporation of BAT optimization in the codebook design for VQ algorithm. The BAT-VQ compression scheme generates efficient results in comparison with classical VQ, CVQ, and JPEG lossy techniques. The compression ratio of BAT-VQ was less when compared with CVQ, however, in terms of picture quality metrics like PSNR, MSE, NCC, and SC, the BAT-VQ outperforms other compression techniques. The reconstructed image quality is vital since medical images play important role in disease diagnosis and hence BAT-VQ compression scheme is a good choice for telemedicine applications.

Acknowledgements The authors would like to acknowledge the support provided by DST under IDP scheme (No: IDP/MED/03/2015). We thank Dr. Sebastian Varghese (Consultant Radiologist, Metro Scans & Laboratory, Trivandrum) for providing the medical CT images and supporting us in the preparation of the manuscript.

References

1. Buzo, A., Gray, A., Gray, R., Markel, J.: Speech coding based upon vector quantization. IEEE Trans. Acoust. Speech Signal Process. **28**(5), 562–574 (1980)
2. Said, A., Pearlman, W.A.: A new, fast, and efficient image codec based on set partitioning in hierarchical trees. IEEE Trans. Circuits Syst. Video Technol. **6**(3), 243–250 (1996)
3. Shapiro, J.M.: Embedded image coding using zerotrees of wavelet coefficients. IEEE Trans. Signal Process. **41**(12), 3445–3462 (1993)
4. Ouafi, A., Ahmed, A.T., Baarir, Z., Doghmane, N., Zitouni, A.: Color image coding by modified embedded zerotree wavelet (EZW) algorithm. In: Information and Communication Technologies. ICTTA'06, vol. 1, 2nd edn, pp. 1451–1456. IEEE (2006)
5. Taubman, D., Zakhor, A.: Multirate 3-D subband coding of video. IEEE Trans. Image Process. **3**(5), 572–588 (1994)
6. Taubman, D.: High performance scalable image compression with EBCOT. IEEE Trans. Image Process. **9**(7), 1158–1170 (2000)
7. Atsumi, E., Farvardin, N.: Lossy/lossless region-of-interest image coding based on set partitioning in hierarchical trees. In: Proceedings of the 1998 International Conference on Image Processing. ICIP 98, vol. 1, pp. 87–91. IEEE, 4 Oct 1998
8. Nister, D., Christopoulos, C.: Lossless region of interest with a naturally progressive still image coding algorithm. In: Proceedings of the 1998 International Conference on Image Processing. ICIP 98, pp. 856–860. IEEE, 4 Oct 1998
9. Xiong, Z., Wu, X., Cheng, S., Hua, J.: Lossy-to-lossless compression of medical volumetric data using three-dimensional integer wavelet transforms. IEEE Trans. Med. Imaging **22**(3), 459–470 (2003)
10. Ansari, M.A., Anand, R.S.: Context based medical image compression for ultrasound images with contextual set partitioning in hierarchical trees algorithm. Adv. Eng. Softw. **40**(7), 487–496 (2009)
11. Lu, T.C., Chang, C.Y.: A survey of VQ codebook generation. J. Inf. Hiding Multimedia Signal Process. **1**(3), 190–203 (2010)
12. Hosseini, S.M., Naghsh-Nilchi, A.R.: Medical ultrasound image compression using contextual vector quantization. Comput. Biol. Med. **42**(7), 743–750 (2012)
13. Karri, C., Jena, U.: Fast vector quantization using a bat algorithm for image compression. Eng. Sci. Technol. Int. J. **19**(2), 769–781 (2016)
14. Eberhart, R., Kennedy, J.: A new optimizer using particle swarm theory. In: Proceedings of the Sixth International Symposium on Micro Machine and Human Science. MHS'95, pp. 39–43. IEEE, 4 Oct 1995
15. Yang, X.S.: Nature-Inspired Metaheuristic Algorithms. Luniver Press (2010)

16. Karaboga, D., Basturk, B.: On the performance of artificial bee colony (ABC) algorithm. Appl. Soft Comput. **8**(1), 687–697 (2008)
17. Rini, D.P., Shamsuddin, S.M., Yuhaniz, S.S.: Particle swarm optimization: technique, system, and challenges. Int. J. Comput. Appl. **14**(1), 19–26 (2011)
18. Horng, M.H., Jiang, T.W.: Image vector quantization algorithm via honey bee mating optimization. Exp. Syst. Appl. **38**(3), 1382–1392 (2011)
19. Horng, M.H.: Vector quantization using the firefly algorithm for image compression. Exp. Syst. Appl. **39**(1), 1078–1091 (2012)
20. Chang, C.C., Li, Y.C., Yeh, J.B.: Fast codebook search algorithms based on tree-structured vector quantization. Pattern Recogn. Lett. **27**(10), 1077–1086 (2006)
21. Abouali, A.H.: Object-based VQ for image compression. Ain Shams Eng. J. **6**(1), 211–216 (2015)
22. Soussi, I., Ouslim, M.: A new compression scheme based on adaptive vector quantization and singular value decomposition. Int. Rev. Comput. Softw. (IRECOS) **11**(5), 445–455 (2016)
23. Kekre, H.B., Natu, P., Sarode, T.: Color image compression using vector quantization and hybrid wavelet transform. Proc. Comput. Sci. **1**(89), 778–784 (2016)
24. Zhao, M., Yin, X., Yue, H.: Genetic simulated annealing-based kernel vector quantization algorithm. Int. J. Pattern Recognit. Artif. Intell. **31**(05), 1758002 (2017)
25. Tripathi, D.P., Jena, U.R.: Vector codebook design using gravitational search algorithm. In: 2016 International Conference on Signal Processing, Communication, Power and Embedded System (SCOPES), pp. 553–558. IEEE, 3 Oct 2016
26. Villmann, T., Bohnsack, A., Kaden, M.: Can learning vector quantization be an alternative to SVM and deep learning?-recent trends and advanced variants of learning vector quantization for classification learning. J. Artif. Intell. Softw. Comput. Res. **7**(1), 65–81 (2017)
27. Trabelsi, I., Bouhlel, M.S.: Learning vector quantization for adapted gaussian mixture models in automatic speaker identification. J. Eng. Sci. Technol. **12**(5), 1153–1164 (2017)
28. Al-Fayadh, A., Abdulkareem, M.: Improved transform based image compression methods. Appl. Math. Sci. **11**(47), 2305–2314 (2017)
29. Kher, R., Patel, Y.: Medical image compression framework based on compressive sensing, DCT and DWT. Biol. **2**(2), 1–4 (2017)
30. Chiranjeevi, K., Jena, U., Dash, S.: Comparative performance analysis of optimization techniques on vector quantization for image compression. Int. J. Comput. Vis. Image Process. (IJCVIP) **7**(1), 19–43 (2017)
31. Ballé, J., Laparra, V., Simoncelli, E.P.: End-to-end optimized image compression. arXiv preprint arXiv:1611.01704, 5 Nov 2016
32. Zhao, M., Yin, X., Yue, H.: Genetic simulated annealing-based kernel vector quantization algorithm. Int. J. Pattern Recognit. Artif. Intell. **31**(05), 1758002 (2017)
33. Vallabhaneni, R.B., Rajesh, V.: On the performance characteristics of embedded techniques for medical image compression. J. Sci. Ind. Res. **76**, 662–666 (2017)
34. Tiwari, A., Sharma, M.: Novel watermarking scheme for image authentication using vector quantization approach. Radioelectron. Commun. Syst. **60**(4), 161–172 (2017)
35. Shah, P.K., Pandey, R.P., Kumar, R.: Vector quantization with codebook and index compression. In: International Conference on System Modeling & Advancement in Research Trends (SMART), pp. 49–52. IEEE, 25 Nov 2016
36. Das, S., Sethy, R.R. Image compression using discrete cosine transform & discrete wavelet transform. Doctoral dissertation (2009)
37. Shannon, C.E.: Communication in the presence of noise. Proc. IRE **37**(1), 10–21 (1949)
38. Dennison, D., Ho, K.: Informatics challenges—lossy compression in medical imaging. J. Digit. Imaging **27**(3), 287–291 (2014)
39. Moura, L., Furuie, S.S., Gutierrez, M.A., Tachinardi, U., Rebelo, M.S., Alcocer, P., Melo, C. P.: Lossy compression techniques, medical images, and the clinician. MD Comput. Comput. Med. Pract. **13**(2), 155–159 (1996)

40. Raeiatibanadkooki, M., Quchani, S.R., Khalil Zade, M., Bahaadinbeigy, K.: Compression and encryption of ECG signal using wavelet and chaotically Huffman code in telemedicine application. J. Med. Syst. **40**(3), 73 (2016)
41. Han, S., Mao, H., Dally, W.J. Deep compression: compressing deep neural networks with pruning, trained quantization, and Huffman coding. arXiv preprint arXiv:1510.00149, 1 Oct 2015
42. Subramanya, A.: Image compression technique. IEEE Potentials **20**(1), 19–23 (2001)
43. Howard, P.G., Vitter, J.S.: Practical implementations of arithmetic coding. In: Image and Text Compression, pp. 85–112. Springer, Boston (1992)
44. Said, A., Pearlman, W.A.: Reversible image compression via multiresolution representation and predictive coding. In: Visual Communications and Image Processing'93, vol. 2094, pp. 664–675. International Society for Optics and Photonics, 22 Oct 1993
45. Robinson, J.A.: Efficient general-purpose image compression with binary tree predictive coding. IEEE Trans. Image Process. **6**(4), 601–608 (1997)
46. Badshah, G., Liew, S.C., Zain, J.M., Ali, M.: Watermark compression in medical image watermarking using Lempel-Ziv-Welch (LZW) lossless compression technique. J. Digit. Imaging **29**(2), 216–225 (2016)
47. Ahmed, N., Natarajan, T., Rao, K.R.: Discrete cosine transform. IEEE Trans. Comput. **100**(1), 90–93 (1974)
48. Skodras, A., Christopoulos, C., Ebrahimi, T.: The JPEG 2000 still image compression standard. IEEE Signal Process. Mag. **18**(5), 36–58 (2001)
49. Barnsley, M.F.: Fractal Image Compression. AK Peters (1993)
50. Barnsley, M.F.: Fractal modeling of real world images. In: The Science of Fractal Images, pp. 219–242. Springer, New York, NY (1988)
51. Li, H., Liu, K.R., Lo, S.C.: Fractal modeling and segmentation for the enhancement of microcalcifications in digital mammograms. IEEE Trans. Med. Imaging **16**(6), 785–798 (1997)
52. Oehler, K.L., Gray, R.M.: Combining image compression and classification using vector quantization. IEEE Trans. Pattern Anal. Mach. Intell. **17**(5), 461–473 (1995)
53. Riskin, E.A., Lookabaugh, T., Chou, P.A., Gray, R.M.: Variable rate vector quantization for medical image compression. IEEE Trans. Med. Imaging **9**(3), 290–298 (1990)
54. Yang, X.S.: A new metaheuristic bat-inspired algorithm. In: Nature inspired cooperative strategies for optimization (NICSO 2010), pp. 65–74. Springer, Berlin (2010)
55. Alihodzic, A., Tuba, M.: Improved bat algorithm applied to multilevel image thresholding. Sci. World J. (2014)
56. Sagnika, S., Bilgaiyan, S., Mishra, B.S. Workflow scheduling in cloud computing environment using bat algorithm. In: Proceedings of First International Conference on Smart System, Innovations and Computing, pp. 149–163. Springer, Singapore (2018)
57. Chandrasekar, C.: An optimized approach of modified bat algorithm to record deduplication. Int. J. Comput. Appl. **62**(1) (2013)
58. Yılmaz, S., Küçüksille, E.U.: A new modification approach on bat algorithm for solving optimization problems. Appl. Soft Comput. **1**(28), 259–275 (2015)
59. Shabanifard, M., Shayesteh, M.G.: A new image compression method based on LBG algorithm in DCT domain. In: 2011 7th Iranian on Machine Vision and Image Processing (MVIP), pp. 1–5. IEEE, 16 Nov 2011
60. Hosseini, S.M., Naghsh-Nilchi, A.R.: Medical ultrasound image compression using contextual vector quantization. Comput. Biol. Med. **42**(7), 743–750 (2012)
61. Skodras, A., Christopoulos, C., Ebrahimi, T.: The JPEG 2000 still image compression standard. IEEE Signal Process. Mag. **18**(5), 36–58 (2001)

Chapter 3
An Assertive Framework for Automatic Tamil Sign Language Recognition System Using Computational Intelligence

M. Krishnaveni, P. Subashini and T. T. Dhivyaprabha

Abstract The sign language that comprises of various sign patterns is an effective communication medium to convey message, disseminate knowledge and transfer ideas among the deaf people. Understanding of such sign signals and positive responses for the benefit deaf people using intelligent machine learning strategy is essential in the technological era. The objective of this chapter to addresses an optimized Automatic Tamil Sign Language Recognition (ATSLR) framework with natural inspired computing paradigm employing image processing technique for recognition of TSL patterns in computer vision application. The algorithm has been validated with local regional signs of Tamil Language. Algorithm aftermaths and comparison analysis in the context of state-of-art methods reveals the effectiveness of the proposed method. The real time Tamil Sign Language (TSL) images have taken for experimentation include 12 vowels, 1 Aayutha Ezhuthu and 18 consonants representation from 110 different signers and the experimental outcomes demonstrated its effectiveness.

Keywords Tamil Sign Language (TSL) · Particle swarm optimization (PSO)
Motility factor based cellular particle swarm optimization (m-CPSO)
Synergistic fibroblast optimization (SFO) · Neural network · Self organizing map (SOM)

M. Krishnaveni (✉) · P. Subashini · T. T. Dhivyaprabha
Department of Computer Science, Avinashilingam Institute for Home Science and Higher Education for Women, Coimbatore, India
e-mail: krishnaveni.rd@gmail.com

P. Subashini
e-mail: mail.p.subashini@gmail.com

T. T. Dhivyaprabha
e-mail: ttdhivyaprabha@gmail.com

J. Hemanth and V. E. Balas (eds.), *Nature Inspired Optimization Techniques for Image Processing Applications*, Intelligent Systems Reference Library 150,
https://doi.org/10.1007/978-3-319-96002-9_3

3.1 Introduction

Sign language (SL) communication is considered as the most prominent skill for deaf as well as the hearing impaired people. SL has finite set of well-structured gesture codes, which are usually used for sharing information among such communities. With development of technologies and machine learning techniques, various SL recognition tools have been developed over recent past. Many SL include British sign language, American sign language, Japanese sign language, Chinese sign language, Indian sign language. Various methods employ SL in order to convert them into voice or text for learning applications. One of such common language in India is Tamil language which is also necessary to study in an effective manner to improve the communication skill of the Tamil SL. Many SL include British Sign Language (BSL), American Sign Language (ASL), Japanese Sign Language (JSL), Chinese Sign Language (CSL) and so on. Various methods employ SL in order to convert them into voice or text for learning applications. One of such common language in India is Tamil language which is also necessary to study in an effective manner to improve the communication skill of the Tamil SL. Indian Sign Language (ISL) based recognition tool is a broad research field, because, there are totally 22 languages such as Hindi, Tamil, Bengali, Devnagiri, Marathi, Telugu, etc. considered as official languages in India. For the past two decades, ISL based research works are continuously evolved, but very few contributions have been done so far in the area of Tamil Sign Language. Tamil is an ancient language and also, it is a regional language in Tamil Nadu state, India. It is used by most of the people in the country.

A study on a statistical report of physically challenged children, during the past decade, reveals that there is a steady increase in the number of neonates born with defect of hearing impairment in Tamil Nadu. Indian Sign Language (ISL) has a broad spectrum that covers diverse sort of regional languages, signs and communication approach. A statistical profile for disabled person in 2016 report given by Government of India showed that as per the census 2011, In India, out of 121 crore people, 2.68 crore people are disabled. Among these, 19% people have hearing impairment. A study carried out by a medical team of the Madras ENT Research Foundation (MERF) reported that 6 out of every 1000 children are profound to be severe deafness. It is three times higher than national average and six times higher than international average of deaf community. Medical science stated that the brain development of every child is strongly builds in the early age of 0–5 years. During this preliminary stage, child can easily acquire knowledge, infer thoughts, observe ideas and learn language and skills. Henceforth, dissemination of sign language is needful to develop the linguistic skills of hearing impaired people. Automated computational systems were developed to recognize gestures posted by humans through the implementation of image processing and computer vision applications, such as, clipping and boundary tracing, finger tip detection, handshapes and movements, contour recognition and contour matching. But, the limitations identified in the non-selection of generalization and comprehensibility in feature extraction, feature analysis and selection, learning rules and recognition in the computational model leads to deliver poor performance. Despite potential improvements, many research challenges are still exist which are necessary to explore. In this work, an optimized sign language recognition model is developed using computational intelligence techniques for noise removal, edge detection, classification and recognition in which the goal is to motivate the regional based hearing impaired people.

In this research a novel computational intelligence based ATSLR framework is proposed. Digital image processing techniques are employed to recognize TSL characters using nature inspired computing paradigm which is tremendously enhanced the performance efficiency of the conventional system. This computer-based assistive technology offer system which is greatly beneficial to the hearing impaired to correct, enable, maintain and communicate with the normal people and improved the functional capabilities of individuals with disabilities. The main contributions of this chapter are:

1. An Efficient Removal of Impulse Noise from Tamil Sign Language Digital Images using PSO and SFO Based Weighted Median Filter will be explained in the preprocessing phase.
2. A development of an optimized edge detector that increases the localization accuracy of edge detection by introducing a hybrid optimization technique to find optimal threshold values based on the combination of PSO with cellular organism inspired from fibroblast will be demonstrated as the next phase.
3. Region-based analysis on both boundary and interior pixels of an image done to extract the structural features of the sign in the digital images will be explored in the third segment.
4. In classification phase, the inaccuracy in classifying the Tamil Sign Language digital images will be clearly addressed by evident experimentation using neural networks and optimization techniques and
5. The patterns identified are categorized into distinct class labels using SOM network, where the recognition accuracy of conventional SOM is improved, using computational methods illustrated through experimental evaluation.

3.2 Literature Survey

To overcome the communication barrier between hearing impaired and normal hearing community, an automatic machine translation will be very useful. There are various machine interpreter systems being developed to interpret sign language, but some of the systems are especially developed to be used in particular domains such as banking hall, weather forecasting and post offices. A complete and full fledge system that interprets Tamil sign language is not yet developed and only little work is carried out in this area. Detection of object from video file has several limitations in the real time environment. For instance, hand skin colour is approximately formed as homogeneous representation of pixels and hence colour-based hand detection is possible. But, object detection based on colour representation is not a reliable approach. Because, detection of hands from the multiple modality sources involve poor light, low resolution camera becomes a very tedious task and highly non-reliable. Development of the hand detection model using statistical methods is effective and but, it needs specific application training.

Singha and Das introduced Indian sign language recognition system, based on Eigen value weighted Euclidean distance, for the recognition and classification of Indian sign language images. The outcome of the results revealed that the proposed method gives 97% classification accuracy than conventional method [1]. Ravikiran et al. proposed a novel algorithm that improves the boundary detection and finger tip detection for recognition of American Sign Language (ASL) alphabet. It gives better results for the recognition of ASL letters in computer vision application [2]. Jadhav et al. implemented a contour matching algorithm for recognition of alphabet representing Devnagari Sign Language. The results demonstrated that the proposed system achieved good performance on the evaluation of both static and dynamic sign hand gestures [3]. Pathak et al. developed a Marathi sign language recognition system to translate the sign hand gestures into the corresponding textual and vocal format using cost-effective sign language interpreter. The proposed work gives better performance in the recognition of sign language characters that can be used in computer vision application [4]. Gao et al. proposed a Chinese sign language recognition system based on self-organizing feature maps (SOFM) and Hidden Markov models (HMM). The obtained results illustrated that the introduced method achieved greater accuracy of 82.9% for word recognition of about 5113 sign vocabulary and 86.3% for signer-independent continuous Sign Language Recognition (SLR) [5]. Subha Rajam and Balakrishnan had implemented canny edge detection method and Euclidean distance measure with binary-decimal conversion algorithm for recognition of Tamil sign language alphabet which includes 12 Vowels, 18 Consonants and 1 Aayutha Ezhuthu [6].

Caridakis et al. proposed automatic sign language recognition architecture based on novel classification scheme incorporating Self-organizing maps (SOMs), Markov chains and Hidden Morkov Models (HMMs). Extracted hand shape features such as area, fourier, moments and curvature were utilized to train HMM classifier for recognition of sign images [7]. Krishnaveni et al. [8] proposed a hybrid BackPropagation Neural network (BPN) with Particle Swarm Optimization (PSO) algorithm for classification of Tamil Sign Language (TSL) digital images. A number of features extracted from TSL images which include bounding box, centroid, area, perimeter, equidistance, roundness, number of boundaries, angles and distance are used to train the classification model. The proposed work was evaluated and compared with other exiting classification techniques such as Support Vector Machine (SVM) and Probabilistic Neural Network (PNN). The experimental results revealed that the novel optimized classifier achieved 92% classification accuracy rate than conventional methods. Lungociu introduced real time sign language recognition system using Artificial Neural Networks (ANN). Non-manual features such as body position, facial expression and gloves wear hand shape extracted from English alphabet images were utilized for verification and validating the recognition model and it attained 80% character recognition accuracy [9]. Lang developed a framework for sign language recognition using Kinect platform. Manual feature involves hand shape, palm orientation, location and movement extracted from American Sign Language digital images used for training and testing the recognition system based on Hidden Markov Model [HMM] and compared with K-means algorithm. The examined results demonstrated that the proposed work gives 97.7% recognition accuracy than K-means algorithm [10]. Taunk et al. introduced Feed forward neural network classifier for static gesture recognition of Devnagari Sign Language (DSL) image dataset. Feature extraction involves area of hand region, correlation, solidity, mean, entropy, maximum area of cropped region and energy utilized to train the classification model and the evaluation results show 60% recognition accuracy of DSL [11]. Khanduja et al. proposed a hybrid approach integrating the structural features of the character image and a mathematical curve fitting model to find the optimal feature vector space. Neural network classifier was constructed, using the fittest features for recognition of Devnagari scripts, and the novel algorithm gives average recognition accuracy of 93.4% [12].

Survey was also carried out in image processing domain where optimization technique is applied for effective performance. Lee proposed geometric optimization algorithm to optimize 2D and 3D pose estimation for object recognition and classification in computer vision. The experimental study reveals that the geometric optimization method is utilized to design computer vision algorithms that resolve complex problems effectively [13]. Dasgupta et al. developed dictionary tool for multilingual multimedia Indian Sign Language. It was utilized to association of signs based on given text. The system constructed the phonological annotation of Indian signs based on HamNoSys structure. The manually generated HamNoSys string was given as input to develop avatar module in order to produce an animated sign representation [14]. Johnson developed swedish sign language recognition system based on visual information. It focused on the segmentation of hand portion of visual processing area with complex backgrounds as well as poor light conditions. The skin colour model was constructed to extract features which were applied to recognize the gesture movement with highest accuracy [15].

3.3 Proposed Methodology

One of the desired properties of the system is that, the work belongs to signer independent. The design and development of proposed system has emphasized on the extraction and recognition of patterns from signer irrespective of the signers involved in the machine vision application. Figure 3.1 depicts the methodology for the ATSLR system. A framework of Tamil Sign Language recognition system using computational intelligence techniques is proposed based on the vision based technology which is greatly beneficial for the deaf people towards new e-services. The hand shape data are gathered from different volunteers rather than just a single

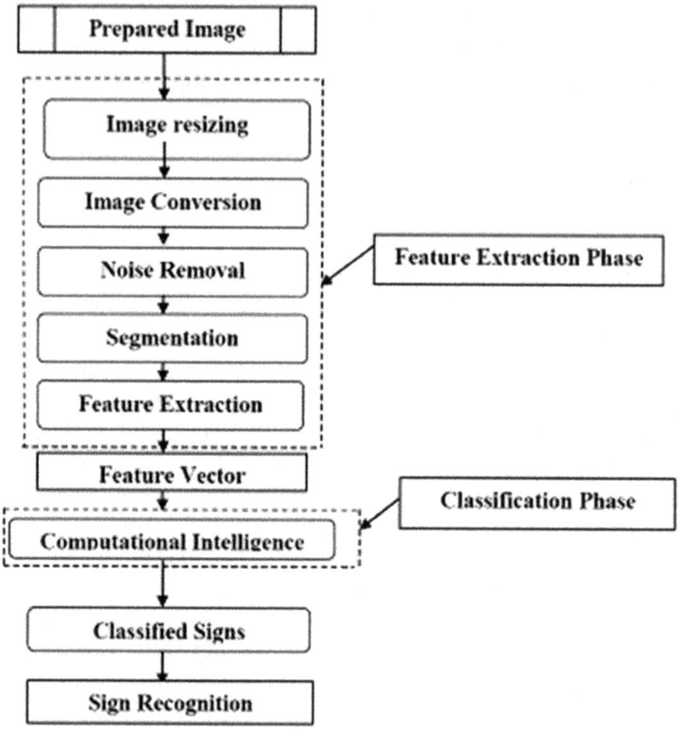

Fig. 3.1 Steps involved in ATSLR framework

signer. The real time Tamil Sign Language hand images which includes 12 vowels, 1 Aayutha Ezhuthu and 18 consonants are taken from 10 different people. Totally, 130 images for vowels and 180 images for consonants are used [6]. Figures 3.2 and 3.3 illustrates the few sample images of the TSL hand gesture images from the database.

3.3.1 Preprocessing

Pre-processing of input information or signals helps to extract relevant sets of features by suppressing or removing the superfluous contents or noise from the original input space. A review on significant literature works stated that noise introduced into digital images during image acquisition and transmission are generally impulse noise. Images are frequently corrupted with impulse noise due to errors generated in sensors and communication channels. The impulse noise should be eliminated to improve the visual quality of images for further analysis such as edge detection, image segmentation, object detection and recognition [16].

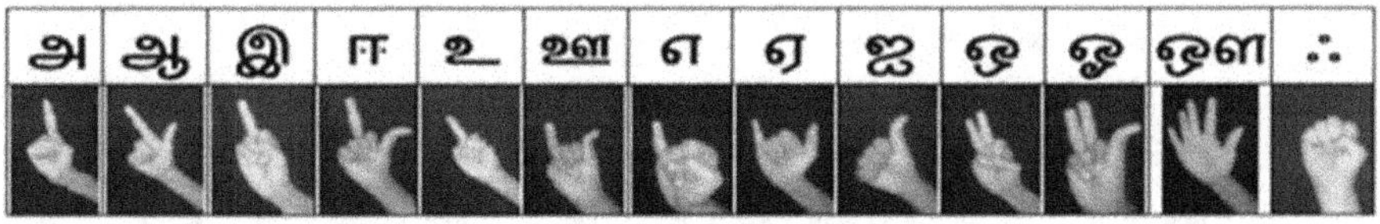

Fig. 3.2 Manually generated Tamil vowels dataset (12 Uyir and 1 Aayutha Ezhuthukal)

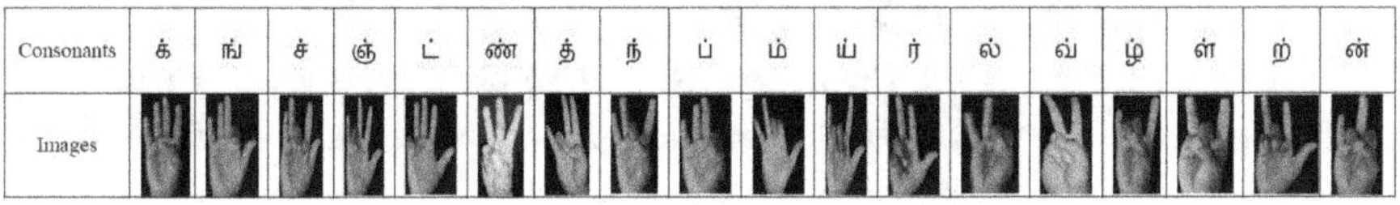

Fig. 3.3 Manually generated Tamil consonants sign language dataset

In this work, the digital images of TSL dataset are corrupted with impulse noise that is reduced at the maximum by using different variants of median filtering techniques which are analyzed and compared with proposed PSO developed on Optimized Weighted Median Filter (OWMF) method. The characteristics of standard benchmark functions are generally categorized into unimodal or multimodal, continuous or discontinuous, differentiable or non-differentiable, separable or non-separable and scalable or non-scalable [17]. A study on the benchmark test function states that the properties of sphere, rotated ellipse2 and schwefel are well suited with univariate characteristic of TSL dataset. Therefore, PSO integrates with OWMF is evaluated with aforementioned standard benchmark functions to achieve better results. The primary objective of the proposed preprocessing method is the selection of weights for the median filter technique which is considered to be a non-linear optimization problem in image localization. PSO algorithm of Nature-inspired computing is implemented to optimize the median weights which are generated in the weighted matrix. Then, the process of convolution of weights is associated with the pixels filters where impulse noise is removed without loss of lines and edges present in the original image. Therefore, an extensive loss of image is reduced without reducing the image fine details using optimized Weighted Median Filter (OWMF). For the improvisation of the obtained results, a newly developed Synergistic Fibroblast Optimization (SFO) algorithm is also applied on the adopted filtering techniques to remove dense noise, especially in digital images (i.e., TSL). SFO is a bio-inspired computing algorithm, which has been developed by the inspiration obtained from the intellectual behaviour of fibroblast cell in the dermal wound healing process [18]. The improvisation is obtained and the obtained results also confirms that SFO based weighted median filtering technique produces promising results than conventional PSO based WMF.

The highest PSNR and lowest MAE values across a wide range of noise densities including the visual clarity shows the significance of the developed preprocessing algorithm. Experimental results are validated for its potential elimination of corrupted noise along with preserving the fine details of the image. The conceptual

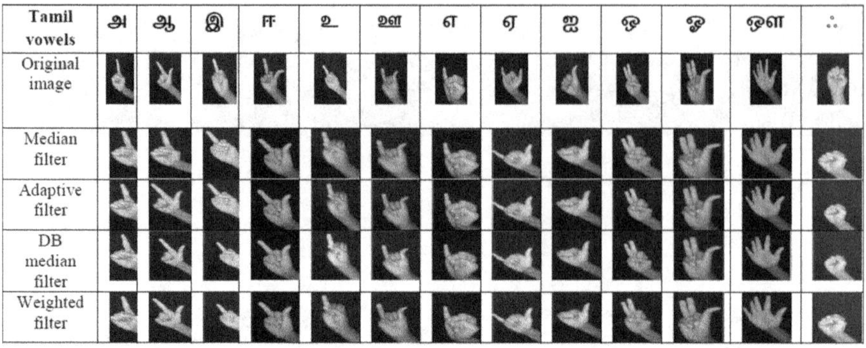

Fig. 3.4 Subjective assessment of the standard filters (a) Tamil vowel signs, (b) original image, (c) median filter, (d) adaptive median filter, (e) decision based median filter, (f) weighted median filter

descriptions of conventional filters are beyond the scope of this chapter. Experiments are conducted by using four filter approaches-median filter, adaptive median filter, decision based median filter and PSO based weighted filter and the performance has been examined in Fig. 3.4.

3.3.2 Optimization Algorithms for Noise Removal

The PSO optimization based weighted median filtering technique optimizes the weights given in the weighted matrix, in order to retrieve the high quality filtered image using standard benchmark functions. PSO is a swarm intelligence algorithm used for optimizing the distributed weights W_{ij}, randomly generated in a weighted median filter technique [4].

Algorithmic steps: Optimized WMF filter using PSO

Step 1: A sign language input image I (u, v) for removing the impulse noise contains in it.

Step 2: A population of weighted particles says p, of size m ($0 \leq m \leq 1$) are generated and initialized with random position x_p and velocity v_p.

Step 3: For each particle, evaluate the objective (fitness) function using multiple benchmark functions such as Rotated Ellipse2, Schwefel and Sphere.

Step 4: Based on the evaluation of objective function (maxima or minima), an individual best (Cognitive) p_{best} is chosen for each iteration.

Step 5: Suppose the current particle p_i value is better than previous value $p_{besti-1}$, set the current particle p_i as the p_{best} value. Otherwise set previous value $p_{besti-1}$ as individual best p_{best}.

Step 6: Particles in the neighborhood with the best success so far g_{best} are identified and assigned to the index variable g.

Step 7: The velocity and position of a particle are updated at time t using the following Eqs. 3.1 and 3.2: The parameters values are referred in [19].

$$V_{ij}^{(t+1)} = \omega * V_{ij} + (c_1 * r_1 * (P_{best} - x_p)) + (c_2 * r_2 * (G_{best} - x_p)) \quad (3.1)$$

$$X_{ij}^{(t+1)} = X_{ij} + V_{ij} \quad (3.2)$$

where

V_{ij} velocity of jth particle of ith iteration
X_{ij} Position of jth particle of ith iteration
ω inertia weight acts an external force, a particle moves in the problem space; set ($\omega = 0.4$)
c_1, c_2 acceleration coefficients. It enables the convergence of P_{best} and G_{best} particles in an stability condition; ($c_1 = c_2 = 2.0$)
r_1, r_2 range of random numbers falls within range of 0 and 1
P_{best} Personal (individual) particle best p
G_{best} Social (neighborhood) particle best g

Step 8: Repeat the steps 3–7 either for maximum number of iterations or criteria (determination) is to be met.

Step 9: PSO based weight values are randomly generated for a weight matrix W_{ij} and assigned to individual pixels of an image.

Step 10: The 3 * 3 resultant vector matrix (weight values) of an input image are sorted, and then a centroid element in a matrix are chosen as a median value which applies to an original image for denoising process.

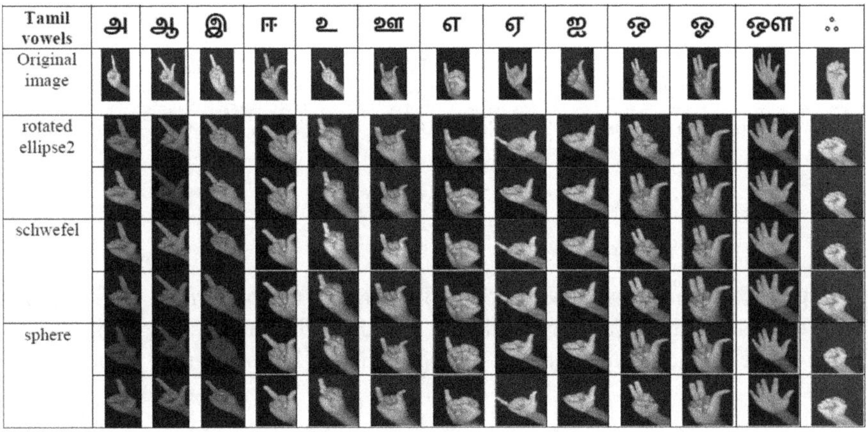

Fig. 3.5 Visual assessment of proposed OWMF using PSO

Step 11: The filtered output image are obtained and used for further analysis. Figure 3.5 portrays the subjective assessment of proposed OWMF using PSO algorithm on TSL datasets.

Though the results based on PSO is significant, Synergistic Fibroblast Optimization (SFO) algorithm is also implemented to select optimal weights for weighted median in order to suppress the intensive noise that can improve the quality of TSL images which are described below [20].

Step 1: Tamil Sign Language images are given as input data.

Step 2: Convert the original image into gray scale image.

Step 3: Initialization—A population of fibroblast cells (fi), i = {1,2,…,10} with arbitrary generation of position (x_i) = {1.0, 3.0, 8.0, 6.0, 2.0, 5.0, 7.0, 0.0, 9.0, 4.0} and velocity (v_i) = {0.7770004, 0.5985996, 0.61042523, 0.43213195, 0.30436742, 0.868674, 0.5339259, 0.44541568, 0.9589586, 3.2305717E−5}. The finite amount of collagen deposition, say c, of size m are ($0.0 \leq m \leq 1.0$) are generated in the extracellular matrix (ecm) and two parameters such as cell speed (s) = 15 μm h^{-1}, L = 10 and diffusion coefficient (ρ) = 0.5 are initialized.

Step 4: Fitness evaluation—Evaluate the individual cell with randomly chosen collagen particles found in ecm using benchmark function such as rotated ellipse, schwefel and sphere. The mathematical representations of test functions are given below.

Rotated Ellipse function (Continuous, Differentiable, Non-Separable, Non-Scalable, Unimodal)

$$f(x) = 7x_1^2 - 6\sqrt{3}x_1x_2 + 13x_2^2 \tag{3.3}$$

Schwefel function (Continuous, Differentiable, Non-Separable, Scalable, Unimodal)

$$f(x) = \sum_{i=1}^{D} x_i^{10} \tag{3.4}$$

Sphere function (Continuous, Differentiable, Separable, Scalable, Multimodal)

$$f(x) = \sum_{i=1}^{D} x_i^2 \tag{3.5}$$

Step 5: Reorientation—The reorientation of cell are performed in the search space to yield optimal (minima) solution based on the fittest collagen it choose. Compare the previous value ($c_{besti-1}$)of the particle with the current particle (c_{best}) value.

if ($c_{best-1} < c_{best}$)

set $C_{best} = c_{best}$;

else

set $C_{best.} = c_{best-1}$;

Step 6: Update the velocity and position of cell—The migration of a cell are performed in the evolutionary region by updating the following Eqs. 3.6 and 3.7.

$$v^{i(t+1)} = v^i(t) + (1-\rho)c\big(f^i(t),t\big) + \rho * \frac{f^i(t-\tau)}{\|f^i(t-\tau)\|} \tag{3.6}$$

$$x^{i(t+1)} = x^i(t) + s * \frac{v^i(t+1)}{\|v^i(t+1)\|} \tag{3.7}$$

Step 7: Remodeling—Synthesis of collagen (c_i) can be performed in the extracellular matrix.

Step 8: Repeat the steps from 3 to 8 until the maximum iterations have attained 10,000.

Step 9: Continuous evolution of swarm in the problem space—Synergistic Fibroblast Optimization algorithm offer best solution (C_{best}) after the fitness evaluation of 10,000 runs.

Step 10: The resultant fittest solutions (C_{best}) are applied to weight matrix given in weighted median filtering technique for removal of impulse noise present in TSL image dataset.

Step 11: The final filtered images are obtained as output. Figure 3.6 and Table 3.1 depict the comparison results of the Optimized Noise filter using PSO and SFO.

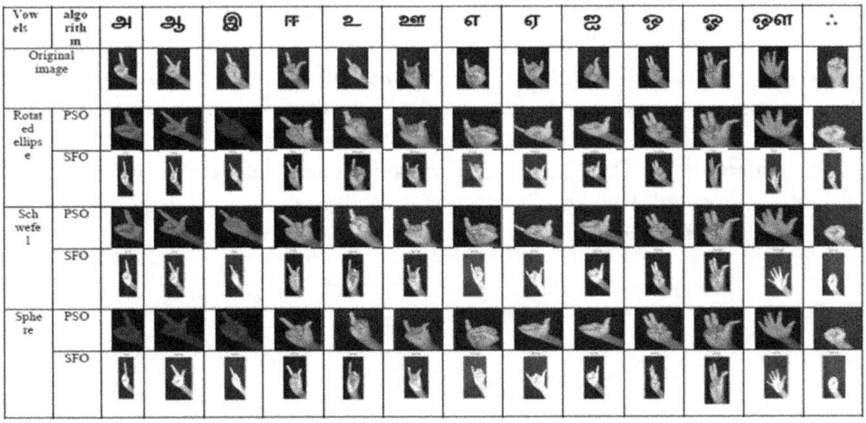

Fig. 3.6 Comparison of OWMF using PSO and OWMF using SFO based on visual assessment

Table 3.1 Comparison of OWMF using PSO and OWMF using SFO based on objective assessment

Tamil vowels	PSO optimized WMF				SFO optimized WMF			
	PSNR	MSE	MAE	CORL	PSNR	MSE	MAE	CORL
அ	63.76	0.10	0.07	0.993	70.33	0.11	0.08	0.9980
ஆ	63.67	0.10	0.99	0.994	70.11	0.10	0.08	0.9981
இ	63.52	0.11	0.07	0.995	69.75	0.11	0.09	0.9980
ஈ	64.27	0.15	0.11	0.997	71.15	0.10	0.07	0.9986
உ	63.86	0.16	0.12	0.996	70.32	0.10	0.08	0.9978
ஊ	64.27	0.15	0.12	0.996	69.71	0.11	0.08	0.9975
எ	63.30	0.17	0.12	0.997	67.09	0.13	0.11	0.9971
ஏ	63.12	0.17	0.13	0.994	67.61	0.13	0.11	0.9981
ஐ	63.73	0.16	0.12	0.997	70.42	0.11	0.08	0.9983
ஒ	64.49	0.15	0.11	0.995	71.03	0.10	0.07	0.9975
ஓ	63.10	0.17	0.14	0.997	69.21	0.12	0.09	0.9978
ஔ	64.25	0.15	0.11	0.996	67.93	0.13	0.10	0.9976
ஃ	62.74	0.18	0.15	0.993	68.14	0.12	0.10	0.9974

3.3.3 Segmentation

Image segmentation is defined as the process of partitioning a digital image into distinct segments for insight assessment and extraction of useful information from it. In this phase, the objective is to propose a global optimization algorithm optimized edge detection method to reduce the number of broken edges and to increase the localization accuracy of edge detection in real time digital images. The goal was successfully achieved by using PSO for optimal search of threshold values for the canny edge detector algorithm. More accuracy in detecting the edge is attained by a new variant of Cellular PSO algorithm which was inspired from the biological phenomena of fibroblast organism for selection of optimal threshold [21]. The objective and subjective results produced shows that the new proposed hybrid optimization approach has performed better than the canny and conventional edge detection algorithms. The experimental results reveal that the better results of similarity index and Pearson Correlation Co-efficient metrics demonstrates the visual quality of images [22].

3.3.3.1 Conventional Edge Detection Methods

Sobel Edge Detector

The Sobel operator is performed edge detection by measuring the intensity of the two dimensional image using discrete differentiation operator. It is applied as mask values in both x and y coordinates and the image details are combined into one single metric [16].

Roberts Edge Detector

The Roberts edge detector is performed spatial gradients corresponds to the pixel of digital images for edge detection. The process of masking is similar to Sobel operator. It estimated absolute magnitude level of the spatial gradient at each pixel values to represent the output image.

$$Sobel \begin{bmatrix} -1 & 0 & 1 \\ -2 & 0 & 2 \\ -1 & 0 & 1 \end{bmatrix} \begin{bmatrix} 1 & 1 & 1 \\ 0 & 0 & 0 \\ -1 & -2 & -1 \end{bmatrix} \quad Roberts \begin{bmatrix} 0 & 0 & 1 \\ 0 & -1 & 0 \\ 0 & 0 & 0 \end{bmatrix} \begin{bmatrix} 1 & 0 & 0 \\ 0 & -1 & 0 \\ 0 & 0 & 0 \end{bmatrix}$$

Prewitt Edge Detector

Prewitt operator calculates the magnitude and orientation of pixel values based on the kernel with respect to the maximum response for edge detection. The set of kernel values used in prewitt is extended up to 8 possible orientation. It is similar to Roberts operator which is used the same convolution kernels [23].

$$Prewitt \begin{bmatrix} -1 & 0 & 1 \\ -1 & 0 & 1 \\ -1 & 0 & 1 \end{bmatrix} \begin{bmatrix} 1 & 1 & 1 \\ 0 & 0 & 0 \\ -1 & -1 & -1 \end{bmatrix}$$

Canny Edge Detector

In this method, the output resultant is to identify slight linear features and the pixels are associated with tracking of edges. The selection of best threshold is done with introducing hysteresis thresholding method. The thickness of edges in the images are varies with threshold and measuring fittest threshold is given as the argument values. It is the most efficient method for removal of noise and it has the capability to identify weak edges [24]. Henceforth, the operator is able to find best edge detector. The discrete approximation of Gaussian value is taken as $\sigma = 1.4$.

Discrete approximation σ = .

2	4	5	4	2
4	9	12	9	4
5	12	15	12	5
4	9	12	9	4
2	4	5	4	2

The Sobel edge detector is implemented to choose the most approximate absolute gradient magnitude level at every pixel with the convolution operation of masking values (G_x and G_y).

$G_x =$

−1	0	1
−2	0	2
−1	0	1

$G_y =$ tt

1	2	1
0	0	0
−1	2	−1

There are two threshold values are incorporated in canny operator. The high threshold value is chosen as H and the low threshold value is taken as L. It tracks the intensity values of pixels where pixel value is minimum than low threshold L is discarded and pixel value is higher than high threshold H is preserved as edge values. Similarly, the edge between L and H is also taken as continuous edges and identifies the connected edges is considered as contours. Figure 3.7 depicts the subjective results of the conventional edge detection methods.

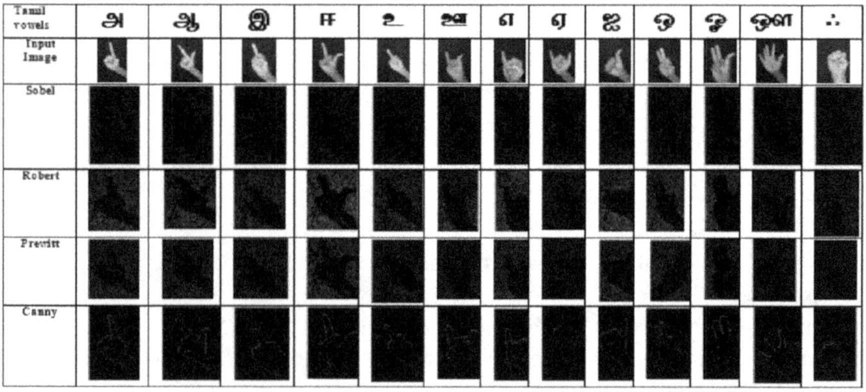

Fig. 3.7 Visual assessment of the traditional edge detection operators

3.3.3.2 A New Variant Motility Factor Based Cellular Particle Swarm Optimization Algorithm for Improved Canny Edges

Cellular based PSO algorithm is applied to choose fittest threshold values which are utilized for detection of edges present in the regional sign language digital images. The range of threshold values are obtained from the measuring the histogram representation of sign language dataset [4]. The algorithmic steps of CPSO based canny detector is described below in algorithm 1.

Algorithm 1:

Step 1: Regional sign language digital images I (u,v) are given as input.

Step 2: Smoothing operation—Gaussian filter is implemented to remove noisy elements present in digital images. The convolution process of kernel values are performed with a core of Gaussian filters using variance of = 1.4 is shown in the following Eq. (3.8) given below:

$$B = \frac{1}{159}[2\,4\,5\,4\,2;\ 4\,9\,12\,9\,4;\ 5\,12\,15\,12\,5;4\,9\,12\,9\,4;2\,4\,5\,4\,2;] \tag{3.8}$$

Step 3: Appropriate selection of gradients—Track the intensity values of pixels which leads to the determination of gradients in the images. Gradient value of each pixel is estimated using Sobel method [25]. The obtained gradients are applied in the horizontal and vertical orientations of the smoothed image using the kernel values denoted in Eqs. (3.9) and (3.10).

$$G_x = [-1\,0\,1;\ -2\,0\,2;\ -1\,0-1;] \tag{3.9}$$

$$G_y = [1\,2\,1;0\,0\,0;\ -1\ -2\ -1;] \tag{3.10}$$

where G_x and G_y are gradient values in the horizontal (x) and vertical (y) coordinates of plane respectively.

The gradient magnitude values are calculated by measuring Euclidean distance value integrates with the Pythagoras law given below (3.11).

$$|G| = \sqrt{G_x^2 + G_y^2} \tag{3.11}$$

Step 4: Non-maximum suppression task—The blurred edges occurred in the gradient images are converted into thick edges to preserve the local maxima and degradation of other noisy pixels present in the digital image. It consists of the following three steps:

- The gradient direction is smoothed to the nearby 45 which was respect to the 8 connected adjacent pixel values of the image.
- The gradient magnitude values of the current pixels are compared with the gradient magnitude of the pixels found in both positive and the negative direction of the smoothed image.

- Suppose if the obtained gradient magnitude of the current pixel is high, then, conserve the intensity value of the current pixel to identifies the thick edges present in the digital image.

Step 5: Double thresholding task—The obtained edge pixel values retrieve noisy pixel values. The canny method uses double thresholding (high T_1 and low T_2) method in order to further degrade the noise content as well as conserve the true edges of an image. The pixel value (T) is greater than high threshold (T_1) are given as strong edge and pixel value (T) is lower than low threshold (T_2) are discarded. Edge pixel values (T) are lie between T_1 and T_2 are taken as weak edges. Choosing the optimal threshold values (T_1 and T_2) are viewed as a non-linear optimization problem. In this step, motility factor based Cellular based Particle Swarm Optimization (m-CPSO) algorithm is applied to select optimal threshold value obtained from given regional sign language images.

Step 6: Edge detection using hysteresis thresholding method—The final output image is considered as the partitioned image and it is used for additional investigation.

The following Procedural step 1 and 2 are performed to identify the range and choose optimal threshold that are implemented in canny edge detector for detection of edges in digital images.

The following procedural step 1 is followed to identify the range of threshold values based on the analysis of histogram method:

Procedural step 1:

Step 1: Regional sign language image dataset are given as input and Region of Interest (ROI) method is applied to determine the edges present in images.

Step 2: The given input image is transformed into gray scale image for further minimize the size of the image.

Step 3: Repeat: The histogram h(z) is calculated to obtain the segmented image.

Step 4: The probability value of a pixel is denoted in the Eq. (3.12):

$$p(z) = \left(\frac{z}{background\ pixel}\right) \times P(background\ pixel) + p\left(\frac{z}{object\ pixel}\right) \times P(object\ pixel)$$
or
$$p(z) = P_b \frac{1}{\sqrt{2\pi\sigma_b}} \mathrm{e}^{-\frac{(z-\mu_b)^2}{z\sigma_b^2}} + P_o \frac{1}{\sqrt{2\pi\sigma_o}} \mathrm{e}^{-\frac{(z-\mu_0)^2}{2\sigma_0^2}} \tag{3.12}$$

where

$p_b(z)$, $p_o(z)$	probability distribution value of background pixel and object pixel
μ_b, μ_o	mean distributions of the background pixel and object pixel
σ_b, σ_o	standard deviations distributions of the background pixel and object pixel
P_b, P_o	a priori probability values of background pixel and object pixel

Step 5: The probability of misclassification rate of a pixel considered as background pixel is represented in Eq. (3.13).

$$E_o(T) = \int_{-\infty}^{T} p_o(z)dz \tag{3.13}$$

Step 6: The mathematical representation for the classification of probability of incorrectly choosing background pixels as object pixels is presented in Eq. (3.14).

$$E_b(T) = \int_{T}^{\infty} p_b(z)dz \tag{3.14}$$

Step 7: The mathematical representation for choosing threshold value is determined by optimization of the above expression is represented in Eq. (3.15).

$$T = \frac{\mu_b + \mu_o}{2} \tag{3.15}$$

Step 8: Until the threshold values are obtained for the images.

The given below procedural step 2: To choose optimal threshold values (Low threshold value (L) and High threshold value (H))

Procedural step 2:

Step 1: Initialization—Swarm of threshold values, says p, with its size m ($0.0 \leq n \leq 1.0$) are initialized with random generation of position (x_p) and velocity (v_p).

Step 2: Fitness evaluation—The fitness of particle is evaluated with standard benchmark function of griewank. The characteristics of griewank is continuous, differentiable, non-separable, scalable and multimodal which are well suited with properties of threshold values. The mathematical representation of griewank is denoted in Eq. (3.16).

$$f(x) = \sum_{i=1}^{n} \frac{x_i^2}{4000} - \Pi \cos\left(\frac{x_i}{\sqrt{i}}\right) + 1 \tag{3.16}$$

Step 3: Choosing an social best value—From the fitness evaluation of threshold value (minima), a neighborhood (g_{best}) value is selected at each cycle.

Step 4: The previous social best value ($g_{besti-1}$) is compared with the current best (g_{best}) value of the particle.

if ($g_{best-1} < g_{best}$)

Set $G_{best} = g_{best}$;

Else Set $G_{best.} = g_{best-1}$;

Step 5: Identify the position of social best—The position of the neighborhood (social) particle in the population are selected and it is assigned to index value (position) g.

Step 6: Updating velocity and position equation—The movement of a particle are updated using the mathematical equation of velocity and position is expressed in Eqs. (3.17) and (3.18):

$$V_{ij}^{(t+1)} = \omega * K * V_{ij}^{(t)} + \left(\varphi_1 * R_1 * \left(P_{best} - x_p^{(t)}\right) + \sigma * \left(\varphi_2 * R_2 * \left(G_{best} - p_g^{(t)}\right)\right)\right) \tag{3.17}$$

$$X_{ij}^{(t+1)} = X_{ij}^{(t)} + V_{ij}^{(t+1)} \tag{3.18}$$

where

$V_{ij}^{(t)}$	velocity representation of jth particle in ith execution cycle
$X_{ij}^{(t)}$	position value of jth particle in ith iteration
φ_1, φ_2	random numbers within range of 0 and 1
P_{best}	Local best (individual) particle
G_{best}	Global best (neighborhood) particle
x_P, p_g	Index value of local and neighborhood particles
t	current execution cycle
t + 1	next execution cycle

$$\omega(\text{inertia weight}) = 0.5 - r()/2$$

where r() function denotes uniform distribution of random number which lies between 0 and 1.

$$K(\textit{constriction co-efficient}) = \frac{2}{\left|2 - \varphi + \varphi\sqrt{\varphi^2 - 4\varphi}\right|}, \quad \varphi = \varphi_1 + \varphi_2$$

$$\sigma(\text{motility factor}) = \psi a^2 e^{-a/asat}$$

where

ψ	$10^4 - 1$
a_{sat}	1.1
e	exponential () function

$$a = \left[\frac{d||c||}{dt}\right] = (p_{c-d_c}||c||) \sum_{i=1}^{M} w\left(f^i - x\right)$$

where

p_c and d_c	positive constants
$\|c\|$	the range of numbers are set from 0 to 0.208
w	from 0.33 to 1
f^i	location of ith particle
x	position of a particle

Step 7: The above steps from 3 to 7 are repeatedly performed either the predetermined number of execution cycles or preset conditions are to be satisfied.

Step 8: Evolution of swarm in search space—The particles are evaluation with the fitness function of 10,000 iterations and CPSO algorithm offer Gbest candidate solution (low and high threshold values).

Step 9: The fittest threshold values (high [T1] and low [T2]) using hysteresis thresholding method is implemented to track the edges in regional sign language image dataset.

The proposed method is significantly enhanced the efficiency of PSO algorithm to obtain promising results in the detection of edges in sign language images. The visual assessment of traditional canny Method, PSO optimized canny and CPSO based canny method are depicted in Figs. 3.8, 3.9 and Table 3.2 portrayed the visual assessment of the methods.

3.3.4 Feature Extraction

Feature extraction is a critical task in solving the classification problem. Several research works have been contributed for feature extraction process, the development of sign recognition system for computer vision application requires a more reliable and efficient approach to achieve high level of accuracy. The efficiency of

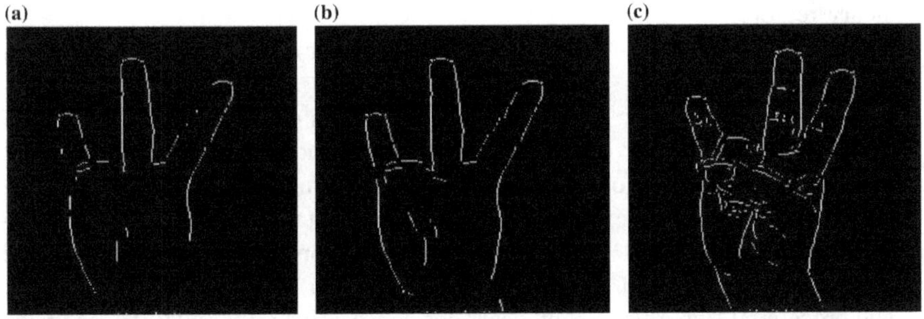

Fig. 3.8 Subjective assessment of edge detection using canny method and optimized canny methods

Bounding box	:	0.50 1.5 256 255
Centroid	:	6.16
Area	:	129.8
Perimeter	:	4
Equidistance	:	2.80
Roundness	:	4.83
Boundary	:	239
Angles	:	41.8
Distances	:	90.19

Fig. 3.9 Structural features

classification process mainly depends on useful and efficient features which are extracted from digital images. Region-based analysis exploits both boundary and interior pixels of an image. The structural features of the sign images include bounding box, area, centroid, perimeter, equidistance, roundness, number of possible boundaries, angles and distance. The following structural features are extracted from the sign digital images.

Bounding Box—A bounding box is the physical shape for a given object. It defines through the different in size and shape than the object's visual appearance.

Centroid—The Centroid is the mean position of all the points in all of the coordinate directions.

Area—The definite amount of pixels found in the region of the object present.

Perimeter—The sum of the pixels present around the edge of each area in the image.

Equiv distance—A length of shortest path between two points, it may refer to a physical length of an object.

Roundness—It calculates the connected shape of an object enclosed in the boundary region.

Number of boundaries—Trace the possible number of boundaries of regions in image.

Angles—Angles are also formed by the intersection of two planes in an object.

Distance—The distinct value that denotes the distance between the actual pixel position and the adjacent non-zero pixel in the given image.

The structural features extracted from the segmented images as portrayed in Fig. 3.9 and the entire the 9 features which are therefore utilized for the proposed classification technique and further used in recognition process.

The procedural step given below describes consecutive steps followed for image classification:

Step 1: Tamil Sign Language digital image I (u, v) is loaded for analysis as an initial step.

Table 3.2 Comparative study of introduced method based on objective evaluation

Tamil constants (Each sign with 10 different signers)	Similarity Index (mean)			Pearson correlation coefficient (mean)		
	Canny	PSO optimized canny	CPSO optimized canny	Canny	PSO optimized canny	CPSO optimized canny
க்	0.5750	0.5752	0.75514	–0.7139	0.2680	0.7186
ங்	0.0735	0.0737	0.8370	–0.4585	0.1082	0.7793
ச்	0.4473	0.4476	0.7678	–0.7026	0.2778	0.7844
ஞ்	0.4929	0.4930	0.7572	–0.5259	0.3666	0.8917
ட்	0.2132	0.5116	0.7415	–0.4583	0.4498	0.7342
ண்	0.3297	0.4486	0.6851	–0.7213	0.1030	0.7800
த்	0.4353	0.3425	0.4457	–0.5948	0.1514	0.6131
ந்	0.2232	0.2819	0.7311	–0.6134	0.1649	0.6248
ப்	0.1094	0.1096	0.6100	–0.4802	0.1278	0.8325
ம்	0.2320	0.3341	0.7098	–0.3101	0.2441	0.6134
ய்	0.0508	0.0776	0.7044	–0.3300	0.3148	0.8615
ர்	0.0280	0.0282	0.3565	–0.4201	0.1866	0.8375
ல்	0.4568	0.4570	0.6112	–0.3133	0.2491	0.5777
வ்	0.0977	0.0416	0.5094	–0.5540	0.2892	0.7656
ழ்	0.1318	0.2320	0.6816	–0.3161	0.2448	0.6104
ள்	0.0868	0.2376	0.6851	–0.5021	0.2609	0.7824
ற்	0.0613	0.1218	0.6127	–0.4191	0.2127	0.8114
ன்	0.1266	0.2616	0.6747	–0.4303	0.2424	0.6927

Step 2: Feature set which encompasses Bounding Box, Area, Centroid, Perimeter, Equiv Distance, Roundness, Number of Possible Boundaries, Angles, and Distance had been selected from images with the aim of analyzing and extracting the hidden useful knowledge from it.

Step 3: The features extracted from images have been applied to learn the training data in back propagation neural network (BPN) classifier to attain better classification accuracy of image (test data). BPN consists of input pattern, processing elements, activation function, weighted connections and output pattern. The learning process of this work deals with binary classification task. The classifier model is constructed with three layers. Each layer consists of processing elements (PEs) which acquire values from its own input connections, carry out the predetermined mathematical operation (sigmoid activation function) and its corresponding single output value is obtained. Each processing element is associated with connection weight. It acts as both a label and a value which store the information or knowledge in a network. The values chosen by the connection weights have major impact on the learning adaptation procedural step of a neural network. It is randomly chosen through the neural network that is able to adapt. When an input pattern is given, the BPN classification system evokes the information to be updated for each processing element. Henceforth, the selection of optimal weight has major influence on the effectiveness and the performance of back propagation neural network. Here, the weight selection process is considered as a non-linear complication problem. It is formulized as an optimization problem in which the value of connection weight falls between 0 and 1. PSO algorithm is implemented to to select optimal weights to improve the learning efficiency of BPN which is portrayed in the following steps [20].

Procedural step for optimal selection of weights applied to classifier:

Step 1: Initialization—A set of population weighted candidate solutions says p, of total m ($0.0 \leq n \leq 1.0$) are generated with arbitrary position (x_p) and velocity (v_p) and parameters such as constriction coefficient, random numbers has to be defined.

Step 2: Fitness estimation—All solutions are evaluated with objective function (Ackley) for stochastically select the best weights. The mathematical representation of Ackley objective function is given in Eq. 3.19 [17].

$$f(x) = -20e^{0.02}\sqrt{D^{-1}\sum_{i=1}^{D} x_i^2} - e^{D-1}\sum_{i=1}^{D}\cos(2\pi x_i) + 20 + e \tag{3.19}$$

Step 3: Identify the cognitive best—PSO iteratively chosen an individual best particle (p_{best}) for the maximum number of iterations.

Step 4: The present element (I_i) is compared with the preceding best element ($I_{besti-1}$) at each step.

if ($I_{besti-1} < I_i$)

Set $I_{best.} = I_{i;}$

else

Set $I_{best} = I_{besti-1}$;

Step 5: Identify the neighborhood best (N_{best}) solution—The fittest social particle is found and it is assigned to index variable g.

Step 6: Updating the velocity and position—The velocity and position of swarm is simplified for every iteration using the Eqs. (3.20) and (3.21) given below:

$$V_{ij}^{(t+1)} = \omega * V_{ij} + \left(c_1 * r_1 * \left(I_{best} - x_p\right) + \left(c_2 * r_2 * \left(i_g - x_p\right)\right)\right) \quad (3.20)$$

$$X_{ij}^{(t+1)} = X_{ij} + V_{ij} \quad (3.21)$$

V_{ij}	velocity of jth particle at ith iteration
X_{ij}	position of jth particle at ith iteration
ω	0.5 – rand ()/2 (rand () function produce a disseminated random number from 0 to 1)
$c_1 = c_2$	2.0
r_1, r_2	range of random values from 0 to 1
I_{best}	Individual best
i_P, x_g	index location of local and neighborhood particles

Step 7: This process has been repeatedly executed for the predefined times (1000, 5000 and 10,000) and predetermined conditions to be met.

Step 8: Progression of swarm in the spatial coordinates—PSO algorithm give Individual best (I_{best}) and Social best (N_{best}) best solution (fittest weights).

Step 9: The obtained fittest weights had applied to processing elements (PEs) which is the core element of the classifier model, it performs the computational task. PE activation function, is also called as threshold function or squashing function. It maps PE's infinite search space to a pre-specified series. To do classification task, a tuple (Tamil sign language image—training data) had given as input pattern, processing element sigmoid (S-shaped) activation function was used to perform computation in the network and the output nodes had predicted the class labels such as Class 1 (correct classification) and Class 0 (incorrect classification). Table 3.3 depicts the classification accuracy of the algorithms.

3.3.5 Classification

In solving a real-time classification problems, the probability of attaining the highest level of classification is not only dependent on the necessarily mean a better model but also the most appropriate selection of classification technique, fine tuning

Table 3.3 Classification accuracy of algorithms

Tamil sign language images	PNN	SVM	BPN	BPN with PSO					
				1000 Iterations		5000 Iterations		10000 Iterations	
				P_{best}	G_{best}	P_{best}	G_{best}	P_{best}	G_{best}
Vowels	76.27	65.38	62.3	90	43.84	86.15	92	78.46	71.46
Constants	180	44.4	48.3	76.1	94.4	95	73.8	95	73.8

of parameters and input dataset. Sign language recognition is an assistive system that acts an interface between deaf and speech impaired people and ordinary people. TSL is the only mode for communication for south Indian people, thereby making learning easier when the system is automated. In this classification phase, the inaccuracy in classifying the Tamil Sign Language digital images is clearly addressed by evident experimentation, using neural networks and optimization techniques. The dataset taken for the study is only static hand with limited number of signers. The result of the classification techniques is evaluated and it is found that the BPN-PSO mechanism fit better than the existing classifiers such as Probabilistic Neural Network (PNN), Support Vector Machines (SVM) and Backpropagation Neural Network (BPN). To improve the quality and classification accuracy a novel approach of PSO algorithm with BPN technique is combined and implemented in TSL images. PNN, SVM and BPN are implemented using MATLAB (R2013a) toolbox. This approach is significant and contributed to construct a framework for Automated Tamil Sign Language Recognition System. PNN computes the probabilistic distances incorporated in patterns of TSL dataset from the input vectors to the training vectors and classifies K objects into the corresponding target class labels. SVM is a binary classifier which partitions TSL dataset into predefined two groups class 1 (correctly identified pattern) and class 0 (not correctly identified pattern). But the gradient is needed to calculate the weight function that minimizes the error discrepancy between identification of patterns.

3.3.6 Recognition

Self-organizing map (SOM) is an unsupervised learning network with the two dimensional lattice. The continuous high dimensional input space is mapped into discrete low dimensional output space with adjustable weights [26]. It is a typical neural network model which has the ability to detect and recognize complex patterns efficiently. Prior to the training, the weights associated with each input node which are connected to the output node of SOM architecture are randomly generated. The probability discriminant function (Euclidean distance) measured between input vector space and weight vector is compared with each node of the network. The node with minimum Euclidean distance difference between itself and the input vector space is considered as winning neuron. The weights are iteratively updated for the winning node and neighborhood nodes converge to the input pattern that characterizes the data can be obtained. Throughout the cycle, the distance between winning neuron and neighborhood nodes are decreased, which in turn, enhance the learning rate, so that the input samples are progressively refined. Once the patterns have been identified, SOM can be used to categorize the input pattern samples into the distinct class labels. Tables 3.4, 3.5 and 3.6 illustrated confusion matrix constructed for validating the efficiency of SFO optimized SOM using standard performance metrics, such as, accuracy, specificity, sensitivity, false positive rate, false negative rate, precision, recall, geometric mean and F-measure and compared with traditional SOM and PSO optimized SOM. The developed pattern recognition model was trained and tested to classify and recognize 31 Tamil Sign Language (TSL) letters including 18 consonants, 12 vowels and 1 Aayutha Ezhuthu. The experimental results presented in Tables 3.7 and 3.8 demonstrated the classification and recognition accuracy of the compared methods. From the empirical analysis, it is identified that the redundant and irrelevant patterns found in the feature vector space such as area, average angle and average distance degraded the accuracy and efficiency of conventional SOM architecture, especially, the high misclassification rate are obtained in the recognition of TSL letters includes ச், ஞ், ண், த், ழ், ள், வ் (consonants) and அ, இ, ஈ, உ, ஊ, ஒ, (vowels). The unique subsets of patterns for the remaining TSL letters have achieved better classification and recognition accuracy of the learning model. The squared Euclidean distance (centroid) between input vector and the weight vector for each neuron are randomly generated, which maximize the probability of winning neuron in SOM architecture. The effect of each learning weight update correlated with winning neuron is mapped to the output vector space. While stochastic selection of

Table 3.4 Confusion matrix of conventional SOM

Actual	Measured	
	Negative	Positive
Negative	30 a	10 b
Positive	550 c	2510 d

Table 3.5 Confusion matrix of PSO optimized SOM

Actual	Measured	
	Negative	Positive
Negative	70 a	40 b
Positive	360 c	2630 d

Table 3.6 Confusion matrix of SFO optimized SOM

Actual	Measured	
	Negative	Positive
Negative	170 a	40 b
Positive	170 c	2720 d

Table 3.7 Performance metrics of classifiers

Metrics	Conventional SOM	PSO optimized SOM	SFO optimized SOM
Accuracy (%)	81.94	87.1	93.23
Sensitivity (%)	82.03	87.96	94.12
False positive rate (%)	25	36.36	19.05
Specificity (%)	75	63.64	80.95
False negative rate (%)	17.97	12.04	5.88
Precision	2761	2695	2788
Recall	0.91	0.96	0.96
g-mean1	47.59	48.69	51.23
g-mean2	0.78	0.75	0.87
F-measure	1.82	1.95	1.96
Error rate (%)	18.06	12.9	6.77

distance function, the negligible focus on prominent feature during the non-cooperative and less adaptation process would affect the efficiency of conventional SOM. In order to improve the efficiency of conventional SOM, the exhaustive metaheuristic search followed by PSO, a population based global search algorithm is implemented with predetermined objective function as probability discriminant function for finely tuned centroid value associated with the input feature vector space. For every iteration, SOM calculated optimized Euclidean distance value for each input node and it is compared with the neighborhood nodes, which is mapped to identify the similar output nodes. From the experimental results, it is confirmed that SFO algorithm significantly calibrated the behavior of SOM by minimizing the distance function and weight positions given in Figs. 3.10 and 3.11. The wide range of non-distinct patterns found in area and average distance feature vector space diminish the performance of SOM in certain situation. Moreover, the quick convergence problem of PSO algorithm retrieved local optimum solution of distance value, would affect the fittest selection of probability density function, that degraded the classification and recognition accuracy of SOM architecture. To overcome this problem, the collaborative and self adaptation behavior exhibited by the novel Synergistic Fibroblast Optimization (SFO) algorithm is applied, to choose optimal probability discriminant function and weight positions for every node, which could significantly improve the performance of SOM architecture. The non-unique pattern presented in average distance feature slightly gives poor results. The examined results confirm that SFO optimized SOM model gives acceptable recognition accuracy for TSL (vowels and consonants) patterns than existing classifiers. Receiver Operating Characteristics (ROC) graph are widely useful for visualizing the hits and pitfalls of classifier models [27]. The trade-off between the True Positive Rate (dependent variable) and False Positive Rate (independent variable) of conventional SOM and novel SOM architectures

Table 3.8 Classification accuracy of conventional SOM

TSL letters	Correctly classified	Image	Misclassified	Wrongly recognized image	Wrongly recognized letters
ச்	90		10		ஞ்
ஞ்	60		40	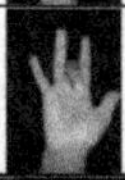	ச்
ண்	60		40	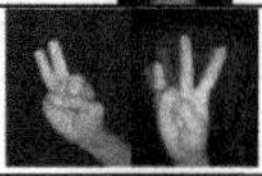	ஒ, த்
த்	70		30		ண்
ழ்	60		40		ள்
ள்	70	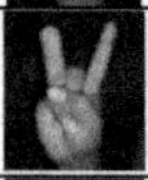	30	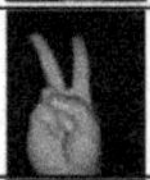	வ்
வ்	70	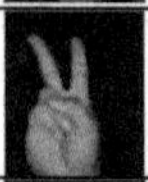	30		ள்
அ	80		20		இ
இ	70		30		அ
ஈ	50		50	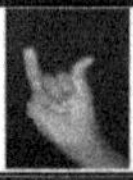	ஊ
உ	70	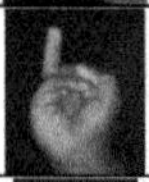	30		அ
ஊ	60		40		ஈ
ஒ	70		30		ள்

depicted in Fig. 3.12 clearly shows that SFO optimized SOM diminishes the false ratio of misclassification rate, which in turn, enhanced the recognition accuracy of SOM architecture. The mathematical representation of performance metrics are given in Eqs. 3.13–3.20. It is constructed with four variables, namely, a denotes

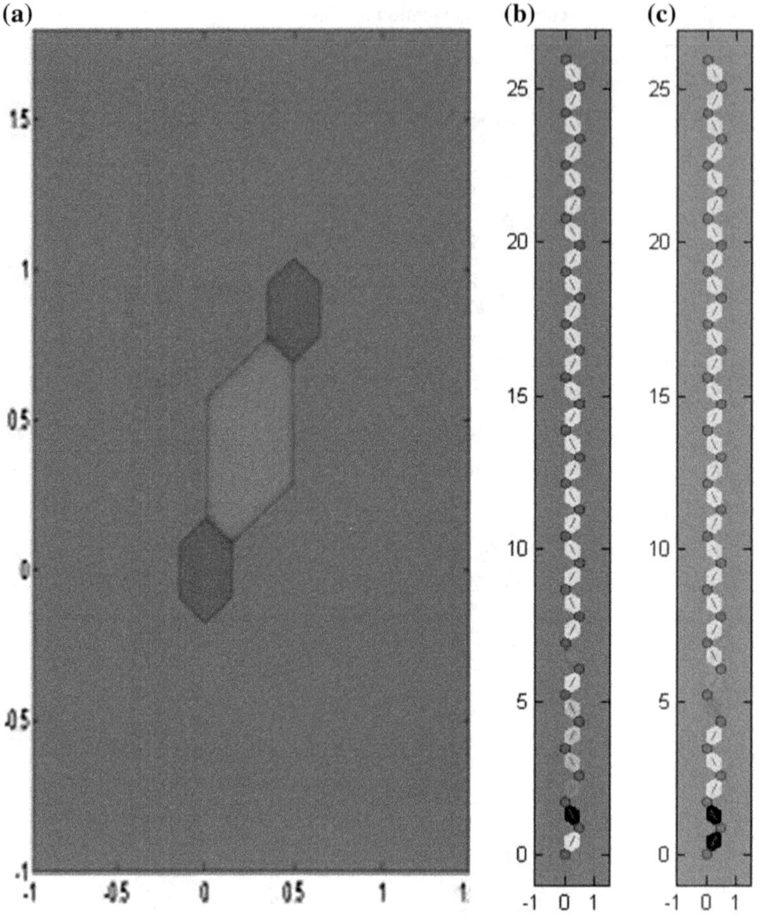

Fig. 3.10 Neighbor distance of conventional SOM, PSO optimized SOM and SFO optimized model

true negative, b designates false positive, c denotes false negative and d represents true positive which are expressed as follows. The mathematical representation of performance metrics are given in Eqs. 3.13–3.20. It is constructed with four variables, namely, a denotes true negative, b designates false positive, c denotes false negative and d represents true positive which are expressed as follows.

Accuracy (AC) is defined as the proportion of the total number of predictions that are correct. It is determined by using the equation:

$$AC = (a + d)/(a + b + c + d) \tag{3.22}$$

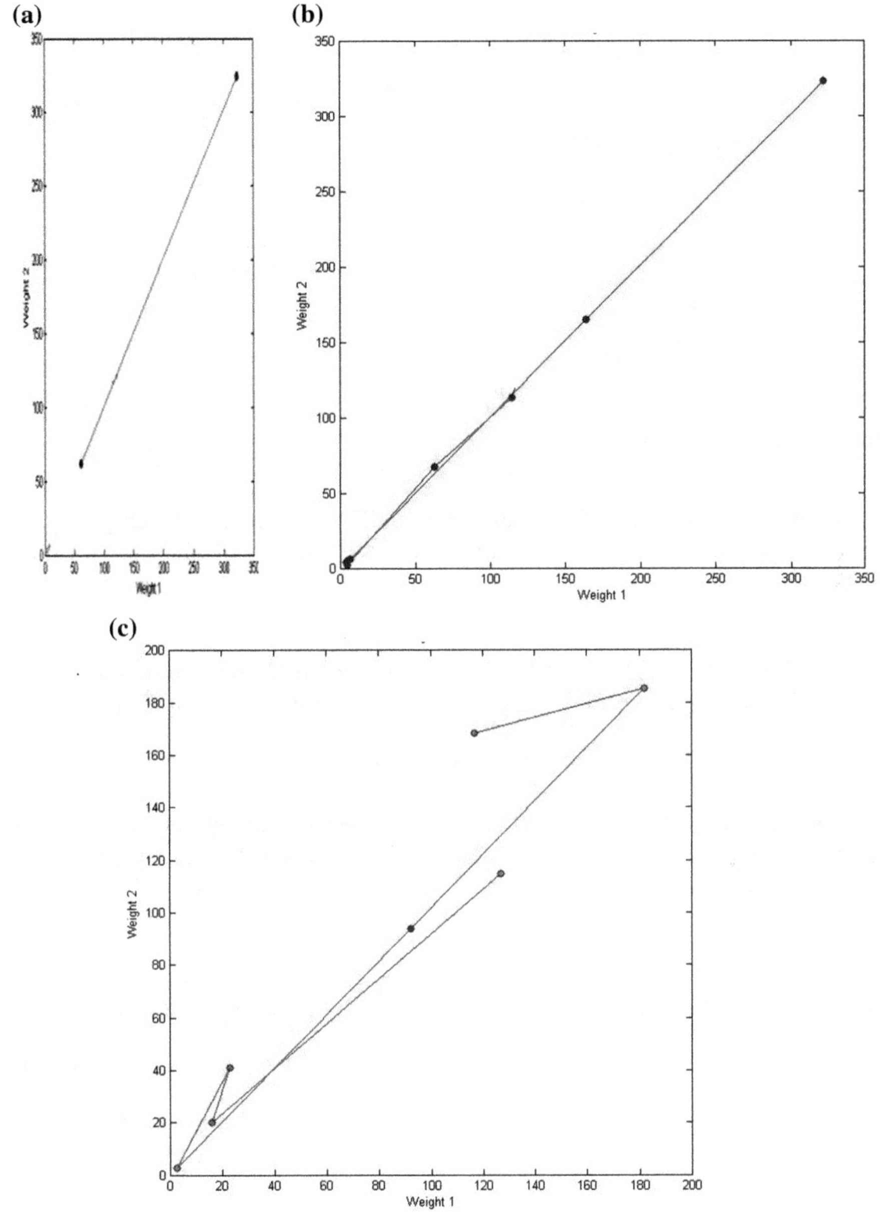

Fig. 3.11 Optimal Euclidean distance weights chosen by SOM models

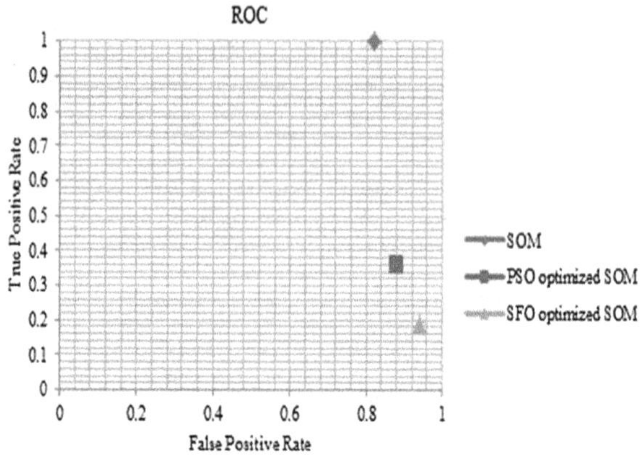

Fig. 3.12 ROC curve of classifiers

Sensitivity or Recall or True Positive rate (TP) is distinct as the proportion of positive cases that are correctly identified, which is calculated using the equation:

$$TP = d/(c+d) \tag{3.23}$$

False Positive rate (FP) is defined as the proportion of negatives cases that are incorrectly classified as positive. It is measured using the following equation:

$$FP = b/(a+b) \tag{3.24}$$

Specificity or True Negative rate (TN) is defined as the proportion of negatives cases that are classified correctly as negative. It is calculated using the equation:

$$TN = a/(a+b) \tag{3.25}$$

False Negative rate (FN) is defined as the proportion of positives cases that are incorrectly classified as negative. It is determined by using the equation:

$$FN = c/c+d \tag{3.26}$$

Geometric mean (g-mean) is defined as mean of product of n numbers and nth root of this product.

$$\begin{aligned} g(mean_1) &= \sqrt{TP} * P \\ g(mean_2) &= \sqrt{TP} * TN \end{aligned} \tag{3.27}$$

Precision (P) is distinct as the proportion of the predicted positive cases that are correct, as calculated using the equation:

$$P = d/(b + d) \tag{3.28}$$

F-measure is defined as an average of weighted harmonic mean of Recall & Precision (R & P). It is also referred as F-score or F1 metric. Since β = 1 is set for most general cases, the value of β is defined to 1.

$$F = (\beta^2 + 1) * P * TP/\beta^2 * P + TP \tag{3.29}$$

Error Rate defined as the proportion of true positive and true negative cases which are predicted incorrectly (misclassification rate).

$$Error\ Rate = (1 - AC) \tag{3.30}$$

3.4 Conclusion

This chapter propose an Automatic Tamil Sign Language Recognition (ATSLR) system for acquisition and dissemination of knowledge which is beneficial to the hearing impaired people in machine vision application. Intially, the digital images are preprocessed with different conventional median filters and higly improvised using optimization techniques, which is then evaluated with its performance metrices. The segmentation is handled with various edge detection techniques for finding the possible edges of the hand sign encompassed in the captured images, by identifying the segment in the images, introducing the PSO variant for improvising the detection of edges. In feature extraction phase, the main objective is to find the structural features in hand image. These features have a set of impact on the performance of image recognition systems. By combining those nine features such as bounding box, area, centroid, perimeter, equidistance, direction, distance, angle and number of possible boundaries of the hand, the enhancement in the performance of the system is accomplished considerably. It is proven that structural features of an image are always considered to be an important attribute in recognition system. The result also illustrated that the fusion of fine tuned features delivers more accuracy which is proved by different classification techniques. Opimization is introduced to achieve high accuracy of results, and the most significant feature of the chapter is that, a novel optimization technique SFO is introduced and evaluted with the existing hybird techniques. The neural network is trained with different signs and classification accuracy is achieved higher using PSO technique. Henceforth, a complete recognition system is been developed using unsupervised algorithm where the accuracy is improvised using PSO and SFO optimization techniques. This proposed framework will be very much useful for effective communication between the hearing impaired people with normal people in learning Tamil language signs which is region based. It has set a platform to develop an automatic Tamil Sign Language for an education and recognition platform for deaf students of India. The system can substantially provided services that can be used to correct, enable, maintain, or improve the functional capabilities of individuals with disabilities in the primary/vocational/higher education of hearing impaired students and people of India.

References

1. Singha, J., Das, K.: Indian sign language recognition using eigen value weighted euclidean distance based classification technique. Int. J. Adv. Comput. Sci. Appl. (IJACSA) **4**(2013), 188–195 (2013)
2. Lee, P.Y.: Geometric optimization for computer vision. Ph.D. dissertation, Australian National University, pp. 1–144 (2005)
3. Jadhav, C.M., Shitalkumar, S.B.: Devnagari sign language recognition using image processing for hearing impaired indian students. Int. J. Eng. Comput. Sci. (IJECS) **4**(2015), 14239–14243 (2015)
4. Krishnaveni, M., Subashini, P., Dhivyaprabha, T.T.: A new optimization approach—SFO for denoising digital images. In: IEEE International Conference on Computational Systems and Information Systems for Sustainable Solutions, pp. 34–39 (2016). http://dx.doi.org/10.1109/CSITSS.2016.7779436
5. Lang, S.: Sign language recognition with kinect. Thesis, Freie Universitat Berlin, pp. 1–62 (2011)
6. Subha Rajam, P., Balakrishnan, G.: Recognition of tamil sign language alphabet using image processing to aid deaf-dumb people. Elsevier Procedia Eng. **30**, 861–868 (2012)
7. Caridakis, G., Diamanti, O., Karpouzis, K., Maragos, P.: Automatic sign language recognition: vision based feature extraction and probabilistic recognition scheme from multiple cues. In: Proceedings of ACM 1st International Conference on Pervasive Technologies Related to Assistive Environments, pp. 1–8 (2008). http://dx.doi.org/10.1145/1389586.1389687
8. Krishnaveni, M., Subashini, P., Dhivyaprabha, T.T.: PSO based canny technique for efficient boundary detection in tamil sign language digital images. Int. J. Comput. Sci. Appl. **7**, 312–318 (2016)
9. Lungociu, C.: Real time sign language recognition using artificial neural networks. Babes-Bolyai Informatica **56**(2011), 75–84 (2011)
10. Taunk, S., Sharma, D.K., Giri, R.N.: Static gesture recognition of devnagari sign language using feed-forward neural network. Int. J. Adv. Res. Comput. Eng. Technol. **3**(2014), 3388–3392 (2014)
11. Ravikiran, J., Mahesh, K., Mahishi, S., Dheeraj, R., Sudheender, S., Nitin Pujari, V.: Finger detection for sign language recognition. In: Proceedings of the International Multi Conference of Engineers and Computer Scientists (IMECS), vol. 1, pp. 1–5 (2009)
12. Khanduja, D., Nain, N., Panwar, S.: A hybrid feature extraction algorithm for devanagari script. ACM Trans. Asian Low-Resource Lang. Inf. Process. **15**, 2:1–2:10 (2015)
13. Pathak, M., Bhagyashree, K., Ravi, P., Rahul, S., Nitin, S.: Marathi sign language recognition using dynamic approach. Int. J. Sci. Adv. Res. **2**(2016), 20–23 (2016)
14. Dasgupta, T., Shukla, S., Kumar, S., Diwakar, S., Basu, A.: A multilingual multimedia indian sign language dictionary tool. In: The 6th Workshop on Asian Language Resources, pp. 57–64 (2008)
15. Johnson, H., Georgsson, F.: Vision-based segmentation of hand regions for purpose of tracking gestures. Thesis in Computer Science, pp. 1–85 (2008)

16. Gao, W., Yang, L., Zhang, X., Liu, H.: An improved sobel edge detection. In: Third IEEE International Conference on Computer Science and Information Technology (ICCSIT), pp. 67–71 (2010)
17. Jamil, M., Yang, X.-S.: A literature survey of benchmark functions for global optimization problems. Int. J. Math. Model. Numer. Optim. **4**, 150–194 (2013). https://doi.org/10.1504/IJMMNO.2013.055204
18. Subashini, P., Dhivyaprabha, T.T., Krishnaveni, M.: Synergistic fibroblast optimization. Proc. Springer Artif. Intell. Evol. Comput. Eng. Syst **517**, 293–302 (2017)
19. Poli, R., Kennedy, J., Blackwell, T.: Particle swarm optimization an overview. Springer Sci. Swarm Intell. **1**, 33–57 (2007)
20. Krishnaveni, M., Subashini, P., Dhivyaprabha, T.T., Sathya Priya, A: Optimized classification based on particle swarm optimization algorithm for tamil sign language digital images. In: IEEE International Conference on Computational Systems and Information Systems for Sustainable Solutions, pp. 53–57 (2016)
21. Krishnaveni, M., Subashini, P., Dhivyaprabha, T.T.: Improved canny edges using cellular based particle swarm optimization technique for tamil sign digital images. Int. J. Electr. Comput. Eng. **6**, 2158–2166 (2016)
22. Kaur, A., Kaur, L., Gupta, S.: Image recognition using coefficient of correlation and structural similarity index in uncontrolled environment. Int. J. Comput. Appl. **59**, 32–39 (2012)
23. Narain Ponraj, D., Evangelin Jenifer, M., Pongodi, P., Samuel Monoharan, J.: A survey on the preprocessing techniques of mammogram for the detection of breast cancer. J. Emerg. Trends Comput. Inf. Sci. **2**, 656–664 (2011)
24. Canny, J.: A computational approach to edge detection. IEEE Trans. Patt. Anal. Mach. Intell. **8**, 679–698 (1986)
25. Sujatha, P., Sudha, K.K.: Performance analysis of different edge detection techniques for image segmentation. Ind. J. Sci. Technol. **8**, 1–6 (2015)
26. Ghorpade, S., Ghorpade, J., Mantri, S.: Pattern recognition using neural networks. Int, J. Comput. Sci. Inf. Technol. (IJCSIT) **2**, 92–98 (2010)
27. Fawcett, T.: An introduction to ROC analysis. Elsevier Pattern Recogn. Letter **27**, 861–875 (2005)

Chapter 4
Improved Detection of Steganographic Algorithms in Spatial LSB Stego Images Using Hybrid GRASP-BGWO Optimisation

S. T. Veena, S. Arivazhagan and W. Sylvia Lilly Jebarani

Abstract With the success of passive steganalysis, active steganalysis proceeds with its first step to reveal the steganographic algorithms being used to create the stego images. This process needs to be modelled as a multi-class classification problem. Stem to stern analysis of the literature points out that the existing universal steganalytic features are a thorn in the flesh because of their dimensionality curse (34,671). Hence this work concentrates on detection of steganographic algorithms by optimal novel features christened Local Residual Pattern (LRP) and Local Distance Pattern (LDiP). LRP captures first order derivatives of the high pass filtered output, while LDiP exploits the multi scaled radii neighbourhood to capture deformities at a distance. Acquiring LRP and LDiP from fifteen different kernels, this work focuses to find optimal features by the proposed hybrid technique of Greedy Randomised Adaptive Search—Binary Grey Wolf Optimisation (GRASP-BGWO). Deriving confidence from the bio-inspired algorithm and the divide and conquer approach of the proposed optimisation, this work succeeds in improving the performance of the employed ensemble logistic regression classifier with minimal features. Experimentations conducted using five representative algorithms of spatial Least Significant Bit (LSB) embedding category for eight different payloads show that the developed minimal feature steganalyser outperforms the state-of-the-art steganalysers.

Keywords Active steganalysis · Hybrid optimisation · Local descriptors
Multi-class classification · Ensemble classifier

S. T. Veena (✉) · S. Arivazhagan · W. Sylvia Lilly Jebarani
Department of Electronics and Communication Engineering,
Centre for Image Processing and Pattern Recognition,
Mepco Schlenk Engineering College, Sivakasi, Tamilnadu, India
e-mail: veena_st@mepcoeng.ac.in

S. Arivazhagan
e-mail: sarivu@mepcoeng.ac.in

W. Sylvia Lilly Jebarani
e-mail: wsylvia@mepcoeng.ac.in

J. Hemanth and V. E. Balas (eds.), *Nature Inspired Optimization Techniques for Image Processing Applications*, Intelligent Systems Reference Library 150,
https://doi.org/10.1007/978-3-319-96002-9_4

4.1 Introduction

In the age of digitisation, it is easier to hide secret messages into any image using steganography which can be a challenge to national or international security. Globalisation has made a large number of steganographic tools freely and easily available even to illegitimate users for example a single website [1] contains more than 110 free steganographic tools. It is the job of steganalysis to surveil these secret communications. Steganalysis starts off with simple detection of stego images from the innocent cover images and proceeds to extract or decipher the secret hidden within them. The former task is known as passive steganalysis and the latter processes are collectively termed as active steganalysis. A large number of literature exists for passive steganalysis of both targeted and universal nature [2–5]. Though the targeted steganalysers are found to be more accurate, the universal steganalysers enjoy favouritism in the context of being able to work on a large range of steganographic algorithms. Particularly, universal steganalysis of spatial LSB steganography in raw image formats has attracted researchers because of their very low embedding change rates and poses a tougher challenge than the JPEG steganalysis. The low volume payload and the content adaptive LSB steganography are two open challenges in spatial LSB steganalysis [6, 7]. Also, there are not many literary works in active steganalysis as in passive steganalysis. And the first task of active steganalysis is that of identification of tools or the algorithm involved in creating the stego images. Identification of the tool is taken up as a branch of forensic study and most of them are signature based steganalysis [8, 9], while identification of algorithms is handled as a pattern recognition process. However, not much of work in literature supports identification of the steganographic algorithm involved.

The first step in this direction of detecting algorithm used in creating stego images was that of classification of the JPEG steganographic techniques [10]. Here, the Discrete Cosine Transform (DCT) features previously developed were used along with a multi-class Support Vector Machine (SVM) with Gaussian kernel trained with images from four JPEG techniques namely F5, MB1, MB2 and Outguess. The multi-class classifier was built on one against one strategy and named in the paper as Max-Wins. They were able to classify images with large messages reliably and when tested with new schemes, they were assigned to closely-related trained schemes. The authors extended their work to double compressed JPEG images with six techniques using calibrated DCT features with the same classifier [11]. They reported that as the JPEG quality factor of compression increases, the reliability of the classifier deteriorates. The technique with low embedding rate was the worst to be identified amongst all. They also inferred that due to two similar embedding algorithms, there may be a merging of results making them indistinguishable. The training sets need to be very dense in the context of techniques and quality factors to give a more reliable result.

Later, Pevny and Fridrich used the average of the DCT features along with Markov features instead of simple concatenation of features to develop a reduced set of features to classify embedding technique in JPEG images [12]. The challenge that the ability of a classifier trained on diverse algorithms may fail to identify unseen images from closely related methods, even as stego, was the inspiration. They built a forerunner for estimating the quality factor and this bi layered double compression detector was followed by the multi-class classifier. They made an interesting note that the multi classifier will not be able to detect steganographic methods with entirely different types of embedding changes. Dong et al. proposed run-length feature based SVM multi classifier for classification of algorithms in both spatial and JPEG images [13]. They also studied hierarchical and non-hierarchical multi-class schemes. In the hierarchical scheme, a separation of the cover and stego images was done followed by separation of the stego classes. The results of the experimentation conducted showed that the hierarchical scheme performed better. The misclassification mostly existed within the intra domain techniques rather than within the inter domain. This was the first scheme that included tested images on the spatial domain.

In [14], the multi-class classification was also carried out with Logistic Regression (LR) classifier and five classes (cover + four spatial algorithms—LSBR, LSBM, LSBR2, LSBRmod5) on three databases. The authors used Subtractive Pixel Adjacency Matrix (SPAM) features and t-test to validate the detection accuracy. They found that LIBSVM was more efficient than LR in passive steganalysis but LR was the best for multi classification. The single bit and multi bit embedding made no difference in the performance with SPAM features. The authors caution that claim on improvement should be on equal footing in all aspects of steganalysis. Zhu et al. suggested an ensemble multi-class classifier for steganalysis of JPEG images with Cartesian Calibrated JPEG domain Rich Model (CC–JRM) features with linear SVM as the base classifier [15]. They used two schemes for ensemble classification and claim less computation cost than other classifiers.

All the reported works or literature for algorithm detection were for JPEG images and the only literature that exists for spatial LSB is that of Lubenko and Ker, which suggest the difficulty of the task in spite of its need. This stays as a motivation to perform algorithm detection in spatial LSB stego images using machine learning. The existing passive steganalytic features [2, 16, 17] are mostly extracted from residuals such that it is rich in stego content and devoid of the cover content. Then, co-occurrence matrices from the quantised and thresholded residual, is formed as a pattern to distinguish stego from cover. However, while moving to the higher order, the co-occurrence matrices become sparsely populated; truncation and quantisation lead to the loss of the minute changes produced by steganographic embedding. Shi et al. suggested Local Binary Pattern as more capable operator than co-occurrence matrices [18]. Following this course, this paper presents a residual based local descriptor for steganalysis.

Similarly, the performance of classification is improved by simple union or concatenation of diverse individual models [16]. However, this leads to a feature which is very huge in dimension. One of the existing state-of-art steganalytic features—Spatial Rich Model (SRM) formed using this technique has a very huge dimension of 34,671. This makes classification task difficult by requiring special classifiers to handle that dimensionality. Also, it was shown by Lyu and Farid that type and number of features being concatenated are crucial to improve the quality of performance and a simple concatenated feature model will not yield optimal efficiency [19]. Therefore it is necessary to obtain both optimally concatenated model from individual models and also to reduce the dimensionality of the so obtained concatenated model for algorithm steganalysis. Hence, optimisation is done in this paper in two phases or as a hybrid. The first phase of optimisation finds out the optimal combination of discriminant individual feature models and the second phase of optimisation proceeds to reduce dimension within the obtained combination of features. The authors in their previous ventures proposed a similar hybrid optimisation algorithm–Greedy Randomised Adaptive Search—Recursive Feature Elimination (GRASP-RFE (GR)) for selection/reduction of features which are based on the principle of divide and conquer to estimate the size of payload in spatial LSB stego images. The proposed GRASP-RFE was found to be very efficient; however the limitation was that the dimension is user defined [20]. Therefore a dynamic hybrid optimisation–Greedy Randomised Adaptive Search—Binary Grey Wolf Optimisation (GRASP-BGWO (GB)) involving a more powerful bio-inspired algorithm is proposed in this paper. This hybrid optimisation is applied for algorithm detection steganalysis of spatial LSB algorithms using the proposed local descriptors. Thus, necessary and tough task of identification of spatial LSB algorithms using minimal optimally concatenated features of novel local descriptors by the proposed hybrid feature selection of GRASP-BGWO is presented in this paper. The paper is organised as follows: The basics of the steganographic algorithms to be detected is presented in Sect. 4.2. Section 4.3 explains the proposed features and Sect. 4.4 presents the proposed hybrid optimisation technique in detail. The experiments conducted and the results are discussed in Sect. 4.5. The paper concludes in Sect. 4.6 with scope for future enhancements.

4.2 Basics of Spatial LSB Algorithms

This section introduces five spatial LSB algorithms—LSB Replacement (LSBR), LSB Matching (LSBM), LSBM Revisited (LSBMR), Two bit LSBR (LSBR2 or 2LSB) and Modulo 5 LSBR (LSBRmod5). In LSBR, a random secret data bit replaces the LSB of the cover image to give the corresponding stego image, while in LSBR2, the last two least significant bits are replaced [14, 21]. Embedding in LSB leads to inherent asymmetry with even values either unchanged or increased by 1 and odd values either unchanged or decreased by 1. To counteract this, LSBM (also known as ± 1 embedding) embeds 1 randomly by either adding to or subtracting from the cover image, if the secret data bit does not match the LSB of the cover image [22]. In LSBMR, the embedding is performed using a pair of pixels as a unit so that fewer pixel change rate is encountered than LSBM [23]. In LSBRmod5 embedding, the least significant digits are adjusted such that the remainder of dividing stego pixel by 5 gives the embedding secret digit [14, 24]. The models are represented in Eq. (4.1).

$$\begin{aligned} LSBR(X) &= 2 \times \lfloor X/2 \rfloor + M \\ LSBM(X) &= 2 \times \lfloor X/2 \rfloor \pm M \\ LSBMR(X) &= LSBR(f(p,q)) \\ LSBR2(X) &= 4 \times \lfloor X/4 \rfloor + M \\ LSBRmod5(X) &= argmin_{Ymod5=M}|X - Y| \end{aligned} \tag{4.1}$$

where X, Y are pixels $\in \{0, 1, \ldots, 255\}$, M is the secret message in bits and $f(p,q)$ is the function defined on pixel pairs (p,q). All the LSB based algorithms embed the secret at random location based on the stego key.

4.3 Proposed Features

The victory of textural co-occurrence features [2] in spatial LSB steganalysis led to the search of other textural features that may help in steganalysis. Local Binary Pattern (LBP) is one such textural feature used in various applications, but its application in steganalysis is not fully exploited [25, 26]. Also, Shi et al. [18] demonstrated that LBP features are better than the co-occurrence features since they are sensitive to noise and are able to capture the deformities of the embedding algorithm in a local neighbourhood. But LBP is a first order statistic and is non-directional in the sense that it encodes the first order derivative difference in all directions. The authors of the paper in their previous venture have proposed a local descriptor called Local Filter Pattern (LFP) for passive steganalysis and found it effective [5]. So, a local descriptor Local Residue Pattern (LRP) that captures LSB distortion using directional and high order information is proposed for LSB steganalysis. To capture subtle distortion patterns that exist within a neighbourhood at varying distance, Local Distance Pattern (LDiP) is proposed. The features are explained in detail in the following subsection.

4.3.1 Local Residue Pattern (LRP)

A local descriptor which acts upon the residues of high pass filters is presented for steganalysis of LSB based steganography. High pass filtering plays an inevitable role in steganalysis since stego signals are additive noises and the image content is suppressed by filtering. Thus, a residue *Re* is formed from the high passed filtered output, which is independent of the image content but contains the noise or the information embedded inside it.

$$Re = I * k \tag{4.2}$$

where I is the input image, k is the high pass filter and $*$ is the convolution operation. The proposed LRP is developed on this residue as magnitude LRP and sign LFP as extended forms of local filter pattern [5] with additional kernels. The first order derivative differences between the residue values are captured by the magnitude LRPs (mLRPs). The sign LRPs (sLRPs) capture the first order derivative differences of the sign (direction) change in residue.

Various linear high pass filter kernels for computing residues used in this research are shown in Fig. 4.1.

The choice of the kernels has been found suitable for steganalysis in various available literature [18, 27, 28]. For the computation of LRP, the first step is to find the residues Re_θ of the image using filter kernels $k1$ to $k15$ in different directions θ using Eq. (4.2).

In case of kernels $k1$ to $k10$—out of the eight different directions, only four of them—horizontal, vertical, major and minor diagonal directions are considered (i.e.) $\theta = \{0°, 45°, 90°, 135°\}$ because of the symmetric nature of residues. In case of residuals from kernels $k11$ to $k14$, two possible directions are considered (i.e.) $\theta = \{0°, 180°\}$. In case of kernel $k15$, processing in a single direction is considered (i.e.) $\theta = \{0°\}$. Then, the magnitude part of LRP (mLRP) is encoded on the residue output $Re_{\theta,c}$ of a local neighbourhood (pixels in a local window) with c as its centre pixel as shown by Eq. (4.3).

$$mLRP_{B,R}\left(Re_{\theta,c}\right) = \sum_{i=1}^{B} f\left(Re_{\theta,i} - Re_{\theta,c}\right)2^{i-1}$$

$$where,\tag{4.3}$$

$$f\left(Re_{\theta,i} - Re_{\theta,c}\right) = \begin{cases} 0 & if \quad Re_{\theta,i} < Re_{\theta,c} \\ 1 & otherwise \end{cases}$$

$\begin{bmatrix} -1 & 1 \end{bmatrix}$ (a) k1

$\begin{bmatrix} -1 & 0 & 1 \end{bmatrix}$ (b) k2

$\begin{bmatrix} 1 & -2 & 1 \end{bmatrix}$ (c) k3

$\begin{bmatrix} 1 & -3 & 3 & -1 \end{bmatrix}$ (d) k4

$\begin{bmatrix} -1 & 4 & -6 & 4 & -1 \end{bmatrix}$ (e) k5

$\begin{bmatrix} -1 & -2 & 0 & 2 & 1 \end{bmatrix}$ (f) k6

$\begin{bmatrix} -1 & 0 & 2 & 0 & -1 \end{bmatrix}$ (g) k7

$\begin{bmatrix} -1 & 2 & 0 & -2 & 1 \end{bmatrix}$ (h) k8

$\begin{bmatrix} -1 & 5 & -10 & 10 & -5 & 1 \end{bmatrix}$ (i) k9

$\begin{bmatrix} 1 & -6 & 15 & -20 & 15 & -6 & 1 \end{bmatrix}$ (j) k10

$\begin{bmatrix} -1 & 2 & -1 \\ 2 & -4 & 2 \\ 0 & 0 & 0 \end{bmatrix}$ (k) k11

$\begin{bmatrix} -1 & 2 & 0 \\ 2 & -4 & 0 \\ -1 & 2 & 0 \end{bmatrix}$ (l) k12

$\begin{bmatrix} -1 & 2 & -1 \\ 0 & -1 & 2 \\ 0 & 0 & -1 \end{bmatrix}$ (m) k13

$\begin{bmatrix} -1 & 2 & -1 \\ 2 & -1 & 0 \\ -1 & 0 & 0 \end{bmatrix}$ (n) k14

$\begin{bmatrix} -1 & 2 & -1 \\ 2 & -4 & 2 \\ -1 & 2 & -1 \end{bmatrix}$ (o) k15

Fig. 4.1 Various high pass filter kernels used

and θ in D and $D = \{0°, 45°, 90°, 135°\}$ or $D = \{0°, 180°\}$ or $D = \{0°\}$ depending on the kernel, B is the number of neighbours in the local window considered and R is the radius of the local neighbourhood from its centre pixel for which binary coding is done using function f. An example illustrating the LRP binary coding is given as Fig. 4.2.

The histogram of the mLRP, $Hist(mLRP_{B,R})$ is the image feature that is constructed by concatenating the encoded output from all applicable directions and binning the occurrences of the concatenated output as given by Eq. (4.4).

$$Hist\left(mLRP_{B,R},j\right) = Hist\left(\left\{mLRP_{B,R}\left(Re_{\theta,c}\right)|\theta \in D\right\},j\right) \tag{4.4}$$

In this study, the value of B and R is taken as 8 and 1 respectively. As a result, the feature vector is 256 in dimension. To further reduce the dimension, rotation invariant form of LBP is also used, since the starting order of the binary sequence is immaterial for steganalysis. The histogram of the rotation invariant form $mLRP^{ri}_{B,R}$ given by Eq. (4.5) has a dimension of 36.

$$mLRP^{ri}_{B,R}\left(Re_{\theta,c}\right) = \min_{0 \le i \le 2^{B-1}} \mathrm{ROR}\left(mLRP_{B,R}\left(Re_{\theta,c}\right), i\right) \tag{4.5}$$

where $ROR(x,i)$ denotes 'i' right bitwise rotations on number 'x'. Thus, a total of 30 (15 rotation variant and 15 rotation invariant) mLRPs are proposed as feature sets for mLRP feature model.

Similarly, the sign or direction based LRP (sLRP) also known as Local Filter Pattern (LFP) [5] is defined as shown in Eq. (4.6).

$$sLRP_{B,R}\left(Re_{\theta,c}\right) = \sum_{i=1}^{B} f'\left(Re_{\theta,i}, Re_{\theta,c}\right)2^{i-1} \tag{4.6}$$

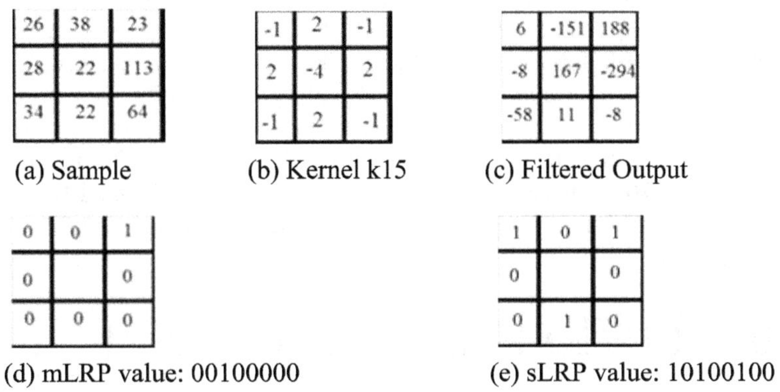

Fig. 4.2 Illustration of LRP binary encoding

where $D = \{0°, 45°, 90°, 135°\}$ or $D = \{0°, 180°\}$ or $D = \{0°\}$ depending on the kernel. The histogram for sLRP is encoded in the same way as mLRP using Eq. (4.4). The rotation invariant form of sLRP, $sLRP^{ri}_{B,R}$ is given by Eq. (4.5) replacing mLRP with sLRP. Thus, thirty (15 + 15) feature models of sLRP capture the higher order gradient information from the residuals. Thus, a total of 60 feature sets exist for LRP feature model.

4.3.2 *Local Distance Pattern (LDiP)*

To further capture the dependencies that exist between pixels within a distance, the following arrangement of neighbouring pixels are considered as shown in Fig. 4.3b. The value indicates the sequence of the neighbours in forming the binary pattern. This rectangular pattern of considering neighbours rather than the conventional square type helps in capturing dependencies that exist over sequential neighbours at a distance. Also, alternate left and right numbering of neighbours, help in giving weightage to the neighbour dependencies directly proportional to their distance from centre pixel. Thus, pixels near to the centre pixel will form Most Significant Bits in the binary pattern, thereby contributing more to capturing distortions by embedding changes. The vertical, horizontal and two diagonal directions of the operator are indicated by 0LDiP, 90LDiP, 45LDiP and 135LDiP. The sign and magnitude form of LDiP are constructed in the same way as LRP as in Eqs. (4.3 and 4.6). An example illustrating the LDiP binary encoding is given in Fig. 4.3.

The histograms of LDiP are constructed using Eq. (4.7).

$$Hist(\theta LDiP_B, j) = Hist(LDiP_B(Re_{\theta,c}), j) \tag{4.7}$$

The rotation invariant form of LDiP is also constructed. Here again, the value of B is taken to be 8. Total of 16 feature sets (8 rotation variant + 8 rotation invariant) are formed as LDiP feature sets.

The LRP and LDiP represent the histogram features of the LRP and LDiP (both sign and magnitude) respectively, while LRPri and LDiPri represent the rotation invariant LRP and LDiP histogram features. The sign or magnitude representation is done by '*s*' or '*m*' preceding them. In case of LRP, the kernel from which the feature has been arrived is represented at the posterior. While in LDiP, the direction is represented preceding the sign or magnitude representation. The naming convention and the 76 proposed feature sets formed using LRP and LDiP feature models are summarised in Table 4.1 with their dimensions along with other LBPs found in literature.

12	5	6	22	11	1	28	32	2

(a) Sample

8	6	4	2	0	1	3	5	7

(b) Sequence of Neighbours in LDiP

1	0	0	1		0	1	1	0

(c) 0mLDiP value: 01101001

Fig. 4.3 Illustration of LDiP binary encoding

Table 4.1 Summary of the proposed 76 feature models with their dimensionality along with other existing LBP models

Feature models	Dimensionality
$\{s,m\}LRP_{8,1}-k\{1-15\}$, $\{0,45,90,135\}\{s,m\}LDiP_{8,1}$	256
$\{s,m\}LRP^{ri}_{8,1}-k\{1-15\}$, $\{0,45,90,135\}\{s,m\}LDiP^{ri}_{8,1}$	36
LBP [29]	256
LBP^{u2} [30]	59
LBP^{ri} [30]	36
LBP^{riu2} [30]	10

4.4 Proposed Feature Selection Technique

Universal steganalysis is generally done by combining features from different models to form a mega model. This is because a single model generally leads to under populated bins, which hampers the task of universally detecting a wide spectrum of embedding algorithms. However, forming a mega model introduces curse of dimensionality. Optimisation techniques help to reduce dimensionality and thereby save CPU time [31, 32]. Global optimisation techniques like evolutionary algorithms are powerful and robust [33], but consume high CPU time and are poor in terms of convergence. On the other hand, local search algorithms converge faster, but get caught in local minima/maxima. A hybrid or bi-level technique proves to be strong in terms of converging time, thus reducing computation time, at the same time increasing solution quality [34, 35]. A Bi-level optimisation approach (Greedy Random Adaptive Search Procedure–Recursive Feature Elimination (GRASP-RFE)) was proposed by the authors for quantitative steganalysis and was found to be effective. However, the RFE method suffers from two main limitations. The first one is that the dimensionality of the selected features is user defined and second it is time consuming. So, a hybrid algorithm using GRASP and a bio inspired evolutionary algorithm—Binary Grey Wolf Optimisation (BGWO) is proposed. The GRASP algorithm is used for obtaining the optimal concatenated model and is explained in detail in [20]. The second level of the proposed optimisation, the Binary grey wolf optimisation is explained in the following subsection.

4.4.1 Binary Grey Wolf Optimisation (BGWO)

Nature inspired Meta heuristic algorithms are best suited for feature selection which leads to dimensionality reduction. Grey Wolf optimisation technique is a recent swarm-based technique which imitates the leadership ranking and hunting strategy of the Grey Wolf pack [36]. The detailed Binary Grey Wolf Optimisation (BGWO) is given by Algorithm 1.

Algorithm 1 BGWO
INPUT: N - Number of Grey Wolf in the pack
MaxIter- Number of Iterations,
OUTPUT: BestPos - Optimal Grey Wolf binary positions

1: function BGWO (N,MaxIter)
2: Initialise a population of *N* Grey Wolves whose positions is $Pos_{i,j}$ where i = {1, 2,..., N} and j = {1, 2, ..., dim}
3: Find the alpha, beta and gamma wolves based on the fitness function given in Eq. (4.11)
4: Initialise a = 2, and calculate A and C as per Eq. (4.8)
5: for iter = 1 to MaxIter do
6: for each Wolf 'i' in the Pack do
7: Update $Pos_{i,:}$ according to Eq. (4.10)
8: end for
9: Update alpha, beta and gamma wolves based on previous step
10: end for
11: BestPos ← $Pos_{alpha,:}$
12: return BestPos
13: end function

COMMENT: rand () produces random number in the range (0,1].

Here the pack is led by social ordering of wolves–alpha, beta, delta and omega. Alpha wolves are the dominant ones and they lead the pack. Beta and delta wolves assist alpha in hunting decisions and omega are the followers. In other words, the best wolf is the alpha followed by beta (second), delta (third) and lastly by omega (others). The encircling of prey during hunting is determined by adjusting the position of the *k*th wolf, X_k with respect to the prey *p* in *i* + 1th iteration given by Eq. (4.8).

$$\begin{aligned} X_{p,k}(i+1) &= X_p - A \times |C \times (X_p - X_k(i))| \\ \text{where,} \quad A &= 2 \times a \times rand() - a \\ C &= 2 \times rand() \end{aligned} \tag{4.8}$$

and the parameters—*a* is linearly decreased from 2 to 0 for each iteration, *A* and *C* help to converge the algorithm globally and *rand()* is a random number (0,1] generation function. Since alpha, beta and delta are the best wolves that give the best position from the prey, the optimum location of prey is determined by alpha, beta and delta wolves' positions and the positions of all wolves are updated according to Eq. (4.9).

$$X_k(i+1) = \frac{X_{alpha,k}(i) + X_{beta,k}(i) + X_{delta,k}(i)}{3} \tag{4.9}$$

This bio inspired technique has been remodelled for feature selection by Emary et al. using two squashing functions [37]. The role of the squashing function is to retain the population position values as binary. One of them is the sigmoid squash function which helps to maintain the binary input needed, has more potential for the feature selection process in steganalysis and given by Eq. (4.10) is used.

$$BinX_k(i+1) = \begin{cases} 1 & \text{if} \quad sigmoid(X_k(i+1)) \geq rand() \\ 0 & \text{otherwise} \end{cases}$$
$$\text{where, } sigmoid(j) = \frac{1}{1+e^{-10(j-0.5)}} \tag{4.10}$$

The fitness function for BGWO is the selection of the best feature (i.e.) the one with maximum classification accuracy and minimum number of features. So, the fitness function f for classification, is set as in Eq. (4.11)

$$f = \alpha \times Accuracy + \beta \times \frac{|T-L|}{T} \tag{4.11}$$

where, *Accuracy* is the classification accuracy using the features, T is the total number of features and L is the length of the selected features. α and β are the two parameters that determine the quality of the classification and length respectively, where $\alpha = 0.99$ and $\beta = 1 - \alpha$ as in base paper [37].

4.5 Experimental Results and Discussion

4.5.1 Experimental Setup

The goal is to establish a universal low complex steganalytic feature for identifying the commonly used (Traditional LSB) spatial steganography. Bossbase v1.01 [38] images of size 512 × 512 embedded with five LSB algorithms and eight different payloads are used. The samples of the stego images for one such random cover image for various algorithms and the statistical metrics—Mean Square Error (MSE) and Entropy show the embedding distortion caused are given in Table 4.2.

It can be inferred from Table 4.2 that even with a high volume secret payload of 1.0 bpp (262,144 bits), the stego images are not visually distinguishable from the cover image and amongst themselves. The variations in the measures are also so small which depicts the challenge in identifying the algorithms of same nature using their stego images. The train-test ratio for the experimentation is fixed as 50%, i.e., for each payload, random 500 images of each cover and stego images of each algorithm (500 + 500 × 5 = 3000) are trained using ensemble One Against One (OAO) Logistic Regression (LR) classifier and the remaining unseen 3000 images are tested. The statistics for all the experiments are the median of the statistics collected by repeating the experiment ten times with different random train/test datasets.

Table 4.2 Comparison of the stego images from various algorithms for a payload of 1.0 bpp from its cover image

	Cover image	LSBR stego image	LSBM stego image	LSBMR stego image	LSBR2 stego image	LSBRmod5 stego image
Visual comparison			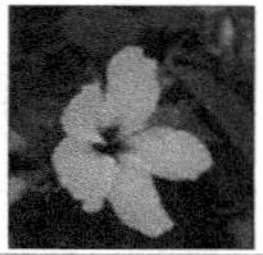			
MSE	0	0.5003	0.4994	0.3760	3.2641	0.8512
Entropy	6.6207	6.6248	6.6276	6.6263	6.6426	6.5844

4.5.2 Algorithm Detection Using Individual LRP and LDiP Feature Models

The 76 individual feature sets discussed in Sect. 4.2 are trained and tested individually on stego images with a low volume payload of 0.1 bits per pixel (bpp). The Receiver Operating Curve (ROC) plots of the individual group by micro averaging, portray the experimental results in Fig. 4.4 along with Area Under Curve (AUC) measures.

Figure 4.4 shows the excellency of LRP over the LDiP features and rotation variant forms are slightly better than rotation invariant forms. Particularly, considering sign LRP and magnitude LRP, the mLRPs are more contributing than the sLRPs. It is because the considered training set consists of images from both single and multi-bit embedding algorithms. It can be noted that *mLRP* with kernel *k*14 is the best among the proposed models with an accuracy of 58.23%. It is important to note the difference between the accuracy and the AUC shown in ROC plots. This is because in multi-class classification, the number of negative classes is greater than the number of positive classes. The feature models that give the maximum accuracy for other payloads are given in Table 4.3.

As payload increases, the magnitude LDiP feature captured in vertical direction (*90mLDiP*) is better performing because of its spatial multi resolution property. Figure 4.5 shows the individual class ROC plot of both the features.

It can be seen that as payload increases, the order of detection of algorithms becomes different. In low payload (Fig. 4.5a), LSBR2 is easily detected, followed by LSBRmod5, Cover, LSBR, LSBM and lastly by LSBMR, while in high payload (Fig. 4.5b), the order is LSBR2, LSBR, Cover, LSBRmod5, LSBM and LSBMR. Thus, universal active steganalysis with similar algorithms is a true challenge with detection accuracy slightly greater than random choice. This stretches the experiment to move towards the improvement in performance which is sought by optimal concatenation of features.

4.5.3 Algorithm Detection Using Optimally Concatenated LRP + LDiP Features

The concatenation of LRP and LDiP feature models by GRASP is done to improve the detection accuracy of the individual feature models. The optimal solution obtained from experimentation on a low volume payload of 0.1 bpp is the concatenation of features–*LRPri-k3, k4, k15, mLRPri-k1, k3, k7, k8, k9, k11, k12, k13, sLRPri-k2, 90mLRPri* with a dimensionality of 576. An increase of 12.5% detection accuracy is achieved for 0.1 bpp payload and the ROC of the proposed concatenated model (LRP + LDiP) for various payloads is given in Fig. 4.6.

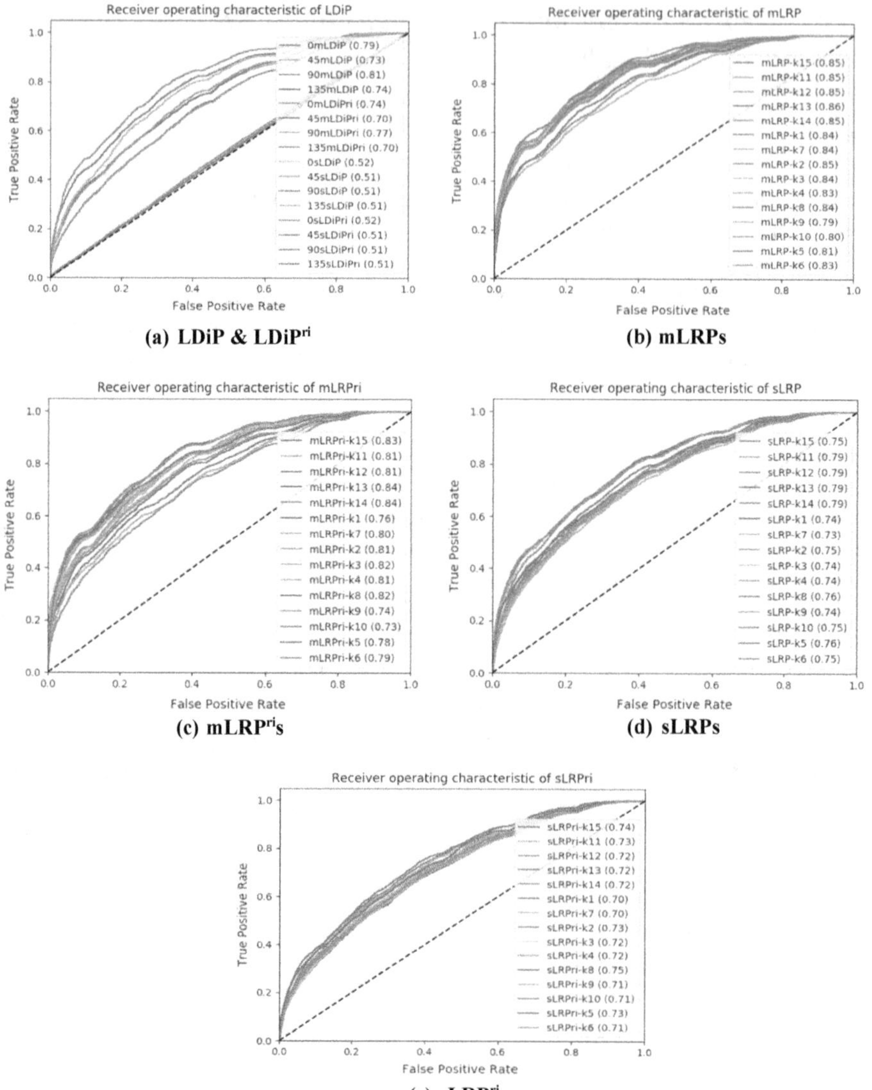

(a) LDiP & LDiP^{ri} (b) mLRPs

(c) mLRP^{ri}s (d) sLRPs

(e) sLRP^{ri}s

Fig. 4.4 ROC plots for algorithm steganalysis of traditional LSB using LRP and LDiP features

Performance analysis of the proposed LRP + LDiP feature in varying groups: To illustrate the difficulty of the algorithm detection task and its dependence on choice of algorithms chosen for training, three groups of LSB algorithms have been designed–First group G1 consists of images from all the above said algorithms, the second group G2 consists of the cover and LSBR, LSBM, LSBR2 and LSBmod5 stego images (most difficult algorithm removed) and the last group G3 consists of the cover, stego images from LSBR, LSBM and LSBMR algorithms

Table 4.3 Traditional LSB Algorithm detection for various payloads using proposed individual features

Payload (bpp)	Overall accuracy in (%)	Max. feature
0.1	58.23	mLRP-k14
0.2	66.8	mLRP-k1
0.25	70.33	90mLDiP
0.3	73.37	90mLDiP
0.4	79.03	90mLDiP
0.5	82.63	90mLDiP
0.75	88.4	90mLDiP
1.0	90.2	90mLDiP

(only single bit embedding). The ROC plots for these groups embedded with a low volume payload of 0.1 bpp are given in Fig. 4.7.

It can be seen from Fig. 4.7 that the most difficult is the group consisting of single bit schemes (G3 accuracy 61.35%, Cover–429, LSBR&LSBMR-281, LSBM-236), followed by the scheme where all algorithms are considered (G1 accuracy 70.73%, Cover-410, LSBR-286, LSBM-226, LSBMR-273, LSBR2-483, LSBRmod5-444). The easiest is the scheme where LSBMR is exempted (G2 accuracy 77.68%, Cover-420, LSBR-294, LSBM-302, LSBR2-482, LSBRmod5-444).

The confusion matrix of the LRP + LDiP feature in algorithm classification for a low volume payload of 0.1 bpp using different groups is given as Table 4.4.

Thus, the choice of the training stego algorithms mainly affects the algorithm detection process and the intermediate group G1 is considered for further experimentation.

Performance analysis of Classifier: To compare the effectiveness of classifier against other classifiers in LSB steganographic algorithm detection, experimentations are carried out with various classifiers on the obtained optimal concatenated feature model of dimensionality 576. Two groups of classifiers are considered. The first group is the simple classifier models. The classifiers considered are Logistic Regression (LR), Naive Bayes, K-Nearest Neighbour (KNN), Linear Support Vector Machine (LinearSVM) and Decision Tree. The second group consists of ensemble classifiers like Random Forest, Extremely Randomised Tree (Extra Tree), Adaboost, Gradient boosting and Bagging with the default base learner and One Against All (OAA) with LR as base learner.

The results tabulated in Table 4.5 show that among simple classifiers, LR provides twice more accurate results than other simple classifiers. However, the ensemble form of LR (OAO) produces 5% more accuracy in low volume of 0.1 bpp than simple LR. Again, among various ensemble classifiers like Trees, Boosting and Bagging, OAO (LR) is better and gives nearly twice more accuracy than the tree based algorithms and about 7% more accuracy than Gradient Boosting, the best among Boosting and Bagging classifiers. Though OAA (LR) is a simpler model compared to OAO, it performs at par only for 1.0 bpp payload. Thus, OAO (LR) is better suited for LSB algorithm classification.

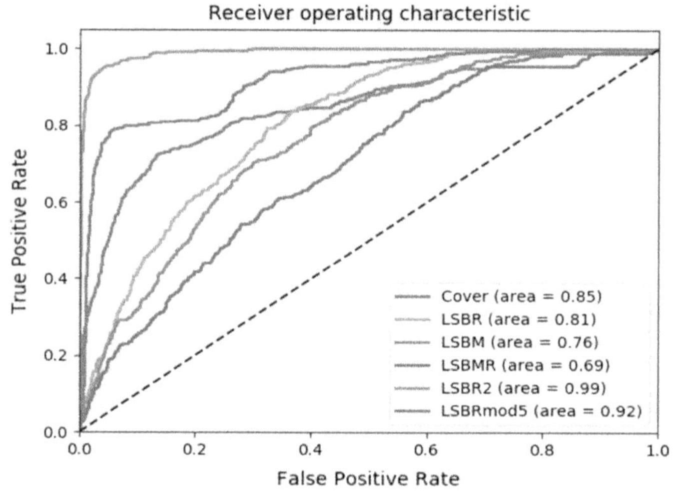

(a) mLRP-k14 for 0.1 bpp

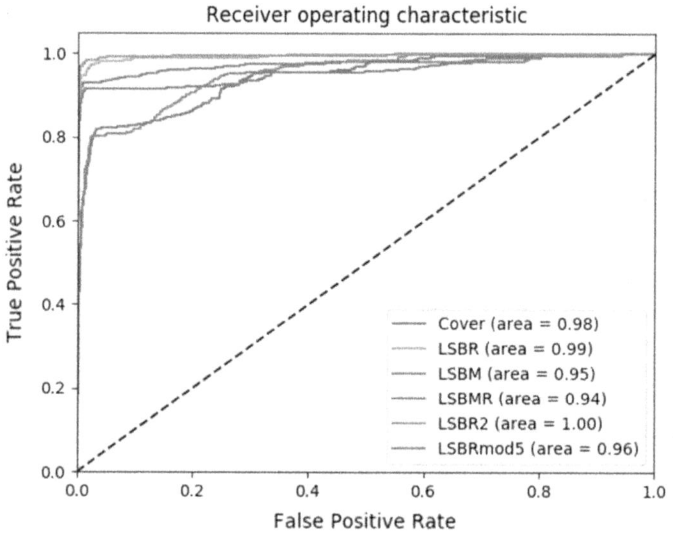

(b) 90mLDiP for 1.0 bpp

Fig. 4.5 ROC plots of the feature for each class of algorithm detection

4.5.4 Algorithm Detection Using Optimised LRP + LDiP Feature Using GB

Though performance of the model has been increased from 58.23 to 70.73% for 0.1 bpp payload, this has indirectly led to the increase of dimensionality from 256

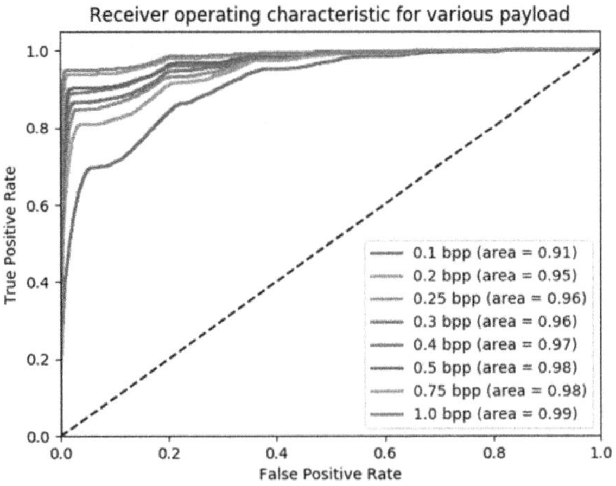

Fig. 4.6 ROC plots of optimised LRP + LDiP features for various payloads of algorithm detection

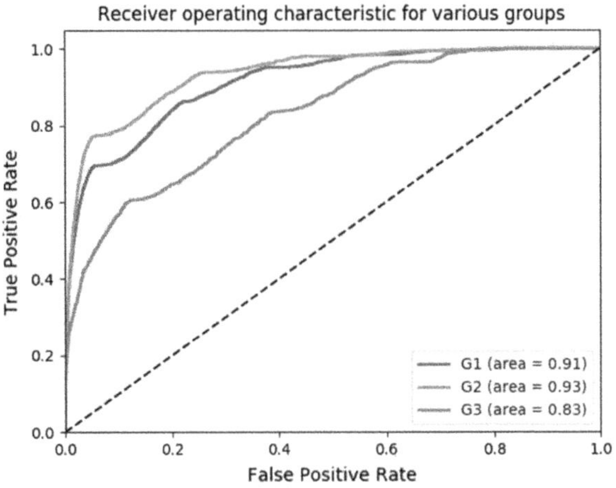

Fig. 4.7 ROC plots of LRP + LDiP for various groups of algorithm detection

to 576. So, the dimensionality of the features obtained are reduced by use of two feature selection algorithms–GRASP-RFE (GR) and GRASP-BGWO (GB) using OAO (LR) classifier. The RFE feature selection method in scikit-learn 0.18.1 package [39] is applied as a dimensionality reducer. The default dimensionality reduction is half of the given features (288). The desired dimensions can be set by the user and are set from 100 to 500 in steps of 100 and the dimension where the best accuracy is obtained is reported. The improved results are tabulated in Table 4.6.

The results show that the GB selection process gives better results than the basic GRASP model. It is able to decrease the dimension by nearly 120 features yet increasing the accuracy to nearly 2% for all payloads. Comparing with GR, GB produces a minimum of 1% increase in accuracy for all payloads and additionally enjoys dynamic feature selection. And most of the results saturate after 20th

Table 4.4 Confusion matrix for algorithm detection of various groups using optimally concatenated LRP + LDiP features in 0.1 bpp payload

Group	Class wise accuracy (%)						Overall accuracy (%)
	Cover	LSBR	LSBM	LSBMR	LSBR2	LSBRmod5	
G1	82	57.2	45.2	54.6	96.6	88.8	70.73
G2	84	58.8	60.4	–	96.4	88.8	77.68
G3	85.8	56.2	47.2	56.2	–	–	61.35

Table 4.5 Algorithm detection using LRP + LDiP with various classifiers and payloads

Classifier	Detection accuracy in percentage for given payload in (bpp)							
	0.1	0.2	0.25	0.3	0.4	0.5	0.75	1.0
Naive Bayes	24.23	34.97	37.6	40.03	42.43	43.83	47.4	50.53
KNN	24.13	39.77	41.63	44.07	47.9	51.1	56.97	61.67
Linear SVM	32.03	65.53	75.7	65.53	70.43	79.57	86.13	92.5
Decision tree	37.0	49.73	52.9	55.43	62.4	65.77	71.5	74.13
Logistic Regression	65.1	80.13	83.83	87.03	89.97	91.47	94.4	**96.07**
Random forest	36.9	52.17	58.27	60.63	64.0	67.73	74.5	77.2
Extra trees	37.33	52.63	54.5	59.0	63.17	66.0	72.5	76.9
AdaBoost	38.77	49.5	54.0	53.97	58.27	60.8	60.4	67.73
Bagging	40.7	56.47	59.7	62.93	67.0	70.9	77.1	81.1
Gradient boosting	48.17	62.97	66.6	71.27	75.23	78.33	85.0	88.1
OAA (LR)	62.07	76.63	82.67	85.2	87.23	90.1	93.5	95.23
OAO (LR)	**70.73**	**82.93**	**86.3**	**88.37**	**90.57**	**91.87**	**94.5**	95.67

iteration of BGWO, thus making employment of bio-inspired algorithm better than any other Meta heuristic method in terms of both time and complexity. The obtained results reinstate the toughness of identification of algorithms in stego images and effectiveness of employment of a bio-inspired algorithm in selecting features for universal algorithm steganalysis.

4.5.5 Comparison with Existing Works

There is a scarcity of literature on identification of algorithms in steganalysis. Further difficulty, is finding literature that works with same stego images of same domain and algorithm (most of the literature on algorithm identification is on JPEG images). So, two types of comparisons are done. First, is the comparison with only existing literature [14] for multi-class classification of spatial LSB algorithms as

Table 4.6 Optimisation of features for LSB Algorithm steganalysis for various payloads using GR and GB

Payload in bpp	Optimisation of features by					
	GRASP		GR		GB	
	Dim	Overall accuracy (%)	Dim	Overall accuracy (%)	Dim	Overall accuracy (%)
0.10	576	70.73	500	70.53	452	**71.6**
0.20	576	82.93	400	83.23	474	**83.9**
0.25	576	86.3	400	85.23	460	**87.23**
0.30	576	88.37	300	88.43	443	**89.3**
0.40	576	90.57	400	90.87	439	**92.0**
0.50	576	91.87	500	91.0	443	**93.2**
0.75	576	94.5	400	94.2	464	**95.2**
1.00	576	95.67	300	95.33	321	**96.43**

such, which employs Subtractive Pixel Adjacency Matrix (SPAM) feature set with their LR classifier. Here, the multi-class classification is done only for a group of LSB variant algorithms—LSBR, LSBM, LSBR2 and LSBRmod5 for a payload of 0.5 bpp to yield an accuracy of 82.3%. Clearly, even a single proposed individual feature *90mLDiP* achieves an accuracy (89.68%) greater than the literature with lesser dimension of 256 compared to 686 of SPAM.

Since the latter comparison method is not complete, a second comparison is done by employing the universal state-of-the-art passive steganalysers for algorithms detection. To achieve this, SPAM [2], Spatial Rich Model (SRM) [16] and Projected SRM (PSRM) [17] features are extracted from our database are used for classification by the same OAO classifier. SPAM features were proposed for Markov chain based steganalysis of spatial domain algorithms, particularly for LSBM. Here the spatial pixel differences between adjacent neighbours of first and second order Markov chains were found and the probability transition matrix of the differences formed the 686 features of SPAM.

The SRM features were formed with the strategy of assembling various diverse noise sub models from various linear and non-linear filters. These noise sub models were formed from the joint PDF of neighbours in quantised noise residuals. This led to a huge dimensional feature (34,671) which could steganalyse both non adaptive scheme and content adaptive steganographic schemes. Later Holub and Fridrich [17] proposed another strategy of statistical representation of diverse noise models other than the joint PDF of neighbours. They suggested the projection of the residuals into a set of random vectors and called it the PSRM feature. These representations were advantageous than the co-occurrence matrix because they were able to capture dependencies over a large number of neighbourhood pixels, flexibly adjust the trade-off between accuracy and dimensionality and select random neighbourhood sizes to provide better, diverged and discriminant features. Though PSRM is a more agile model than SRM, the feature extraction time

Table 4.7 Comparison table for algorithm steganalysis for various payloads using LRP + LDiP GB method

Payload in bpp	Detection accuracy in %				
	Proposed method (LRP + LDiP-GB)		SPAM (Dim 686)	PSRM (Dim 12,870)	SRM (Dim 34,671)
	Dim	Overall accuracy (%)	Overall accuracy (%)	Overall accuracy (%)	Overall accuracy (%)
0.10	452	**71.6**	56.2	63.37	69.0
0.20	474	**83.9**	72.43	80.53	82.47
0.25	460	**87.23**	77.53	85.17	85.33
0.30	443	**89.3**	81.33	87.33	88.5
0.40	439	**92.0**	85.43	90.53	91.4
0.50	443	**93.2**	88.6	93.2	94.03
0.75	464	**95.2**	92.3	96.2	96.1
1.00	321	**96.43**	94.27	97.63	97.27

complexity of the PSRM model (approximately 672 s for a single image, SRM–5, SPAM–1 and LRP–0.3 s) makes steganalysis using PSRM highly difficult.

Table 4.7 shows the results of this comparison. The proposed method is better than all the existing methods for low volume payloads. In a low volume payload of 0.1 bpp, 71.6% accuracy is reached with a feature dimension of just 452. The proposed features excel SPAM, the designated steganalyser for traditional LSB steganalysis in all payloads. It also surpasses the PSRM and SRM features with at least 2% more accuracy with a diminutive feature nearly 34,000 less features in payloads less than 0.5 bpp. However, for high volume payloads, SRM and PSRM are able to achieve less than 1% increase at the cost of very huge dimensionality. Thus, the proposed features along with the efficient proposed optimisation technique proves to a boon to the steganalysis of algorithm in spatial LSB stego images.

4.5.6 Algorithm Detection in Content Adaptive Algorithms

From the previous sections, it can be seen that algorithm detection is a tough task with the low volume payload on closely related algorithms. In case of content adaptive steganalysis, it tends to be lot tougher with more similarity among very closely related content adaptive algorithms and very low embedding rates. As far as the authors' knowledge, there exists no literature for detecting algorithm among content adaptive stego images. Stego images from three content adaptive algorithms–Highly Undetectable steGanOgraphy (HUGO), Wavelet Obtained Weights (WOW) and Spatial UNIversalWAvelet Relative Distortion (S-Uniward or SW) are considered. All these algorithms are LSBM and content adaptive algorithms and in addition, WOW and SW obtain the embedding distortion from the same domain (wavelets). The stego images are created using random 1000 images of Bossbase v1.01 with a payload of 0.4 bpp. It is to be noted that though the embedding payload is of 0.4 bpp, the embedding change rates for an image are that of 0.0933, 0.0918 and 0.0703 (HUGO, WOW, SW) (very, very low) which makes steganalysis of content adaptive algorithms with even 0.4 bpp tougher. It is also difficult to identify algorithms with same change rate than with different ones (As seen from Fig. 4.7, G3 was the most difficult). The train-test ratio is maintained at 50:50 and the median of tenfold cross validation result with OAO (LR) classifier is reported.

The proposed features detect content adaptive algorithms with accuracies below 50%, which illustrates the difficulty of the task. The best accuracy is obtained by *sLRP-k15* which offers 35.55% accuracy which is at par with SPAM with twice smaller number of features. However, the SRM and PSRM are better than the individual feature. So, the LRP + LDiP model from previous experimentation is then tested for content adaptive algorithms and the results are tabulated in Table 4.8 along with the existing state-of-the-art steganalysers–SPAM, PSRM and SRM.

Clearly from Table 4.8, it can be inferred that the proposed hybrid optimisation and features perform excellently even in first level of optimisation (GRASP) than the agile PSRM feature model with just 576 features against 12,870. Also, further employment of bio-inspired BGWO helps to increase the performance by 3% with nearly 200 less features. The confusion matrix of the classification is given in Table 4.9. It can be observed that Cover images are better identified followed by HUGO and SW. The most difficult WOW images are the least identified and are mostly misclassified as Cover. A tougher problem of identification of algorithms in content adaptive scenario is thus addressed by a universal feature common to all type of spatial LSB steganographic algorithm and whose performance is improved by the proposed novel hybrid optimisation technique—GRASP-BGWO.

Table 4.8 Detection of content adaptive algorithms for 0.4 bpp payload

Features (dimension)	Accuracy in %
SPAM (686)	35.7
SRM (34,671)	46.5
PSRM (12,870)	44.1
Individual feature mLRP-k15 (256)	35.55
LRP + LDiP-GRASP (576)	45.55
LRP + LDiP-GB (402)	**47.9**

Table 4.9 Confusion matrix for content adaptive algorithm detection using optimal LRP + LDiP–GB features

Embedding algorithm	Classified as			
	Cover	HUGO	WOW	SW
Cover	307	71	54	68
HUGO	76	279	55	90
WOW	157	86	159	98
SW	98	108	80	213

4.6 Conclusion

A low dimensional local steganalytic feature, which is sensitive to the embedding algorithm and that change considerably with payloads, is presented for LSB variant algorithm detection. It was observed that the algorithm detection is highly dependent on the training algorithms, payloads and features. The proposed model excels all existing state-of-the-art steganalysers even in low volume payload. The universal nature of the feature is further established in detecting steganographic algorithms of content adaptive nature. Additionally, the proposed hybrid method of optimisation helps to improve performance by 12–13% with a minimum of 400 features for maximum 6 class classification in spatial LSB images. Thereby, a new low dimensional feature selection using hybrid GRASP-BGWO optimisation is proposed using novel local descriptors for effective universal algorithm steganalysis of spatial LSB images. The future scope is to scale the existing feature models along with the proposed models for much larger number of steganographic algorithms.

Acknowledgements The authors would like to thank Dr. Vojtˇech Holub for providing the necessary code for comparison. They would also like to express their gratitude to the anonymous editors and reviewers for their helpful suggestions and constructive comments. Also, the authors would like to express their sincere thanks to the Management and Principal of MSEC for providing the necessary facilities and support to carry out this research work.

References

1. Johnson, N.F.: Steganography software. Available at http://www.jjtc.com/Steganography/tools.html (2012). Accessed 27 Aug 2015
2. Pevny, T., Bas, P., Fridrich, J.: Steganalysis by subtractive pixel adjacency matrix. IEEE Trans. Inf. Forensics Secur. **5**(2), 215–224 (2010)
3. Cogranne, R., Retraint, F.: An asymptotically uniformly most powerful test for LSB matching detection. IEEE Trans. Inf. Forensics Secur. **8**(3), 464–476 (2013)
4. Fridrich, J., Kodovský, J.: Steganalysis of LSB replacement using parity-aware features. In: Proceedings of Fourteenth International Conference on Information Hiding, pp. 31–45. Springer, Heidelberg (2013)
5. Veena, S.T., Arivazhagan, S.: Local descriptor based steganalysis of spatial LSB variant stego images. Presented at the TEQIP II International Conference on Computation, Communication and Innovation, pp. 54–57. ACCET, Karaikudi (2016)
6. Farhat, F., Ghaemmaghami, S.: Towards blind detection of low-rate spatial embedding in image steganalysis. IET Image Process. **9**(1), 31–42 (2015)
7. Yu, J., Li, F., Cheng, H., Zhang, X.: Spatial steganalysis using contrast of residuals. IEEE Signal Process. Lett. **23**(7), 989–992 (2016)
8. Swaminathan, A., Wu, M., Liu, K.J.R.: Digital image forensics via intrinsic fingerprints. IEEE Trans. Inf. Forensics Sec. **3**(1), 101–117 (2008)
9. Bell, G., Lee, Y.-K.: A method for automatic identification of signatures of steganography software. IEEE Trans. Inf. Forensics Sec. **5**(2), 354–358 (2010)
10. Pevný, T., Fridrich, J.: Towards multi-class blind steganalyzer for JPEG images. In: Digital Watermarking, pp. 39–53. Springer, Heidelberg (2005)

11. Pevny, T., Fridrich, J.: Determining the stego algorithm for JPEG images. IEE Proc. Inf. Sec. **153**(3), 77 (2006)
12. Pevny, T., Fridrich, J.: Merging markov and dct features for multi-class jpeg steganalysis. In: Proceedings of SPIE: Security, Steganography, and Watermarking of Multimedia Contents IX, vol. 6505, pp. 650503–650513 (2007)
13. Dong, J., Wang, W., Tan, T.: Multi-class blind steganalysis based on image run-length analysis. In: Ho, A.T.S., Shi, Y.Q., Kim, H.J., Barni, M. (eds.) Proceedings on 8th International Workshop on Digital Watermarking IWDW 2009, Guildford, UK, August 24–26, pp. 199–210. Springer, Berlin (2009)
14. Lubenko, I., Ker, A.D.: Steganalysis using logistic regression. In: IS&T/SPIE Electronic Imaging, p. 78800 K (2011)
15. Zhu, J., Guan, Q., Zhao, X.: Multi-class JPEG image steganalysis by ensemble linear svm classifier. In: Shi, Y.-Q., Kim, H.J., Pérez-González, F., Yang, C.-N. (eds.) 13th International Workshop on Digital-Forensics and Watermarking (IWDW 2014), Taipei, Taiwan, 1–4 Oct 2014. Revised selected papers, pp. 470–484. Springer International Publishing, Cham (2015)
16. Fridrich, J., Kodovsky, J.: Rich models for steganalysis of digital images. IEEE Trans. Inf. Forensics Sec. **7**(3), 868–882 (2012)
17. Holub, V., Fridrich, J.: Random projections of residuals for digital image steganalysis. IEEE Trans. Inf. Forensics Sec. **8**(12), 1996–2006 (2013)
18. Shi, Y.Q., Sutthiwan, P., Chen, L.: Textural features for steganalysis. In: Information Hiding, pp. 63–77 (2013)
19. Lyu, S., Farid, H.: Steganalysis using higher-order image statistics. IEEE Trans. Inf. Forensics Secur. **1**(1), 111–119 (2006)
20. Veena, S.T., Arivazhagan, S.: Quantitative steganalysis of spatial LSB based stego images using reduced instances and features. Pattern Recognition Letters. Available online on Aug 2017. https://doi.org/10.1016/j.patrec.2017.08.016
21. Ker, A.D.: Steganalysis of embedding in two least-significant bits. IEEE Trans. Inf. Forensics Secur. **2**(1), 46–54 (2007)
22. Sharp, T.: An implementation of key-based digital signal steganography. In: International Workshop on Information Hiding, pp. 13–26. Springer, Berlin (2001)
23. Mielikainen, J.: LSB matching revisited. IEEE Signal Process. Lett. **13**(5), 285–287 (2006)
24. Xu, W.-L., Chang, C.-C., Chen, T.-S., Wang, L.-M.: An improved least-significant-bit substitution method using the modulo three strategy. Displays **42**, 36–42 (2016)
25. Lafferty, P., Ahmed, F.: Texture-based steganalysis: results for color images. In: Proceedings on SPIE 5561, Mathematics of Data/Image Coding, Compression, and Encryption VII, with Applications, pp. 145–151 (2004)
26. Gui, X., Li, X., Yang, B.: Steganalysis of LSB matching based on local binary patterns, pp. 475–480 (2014)
27. Ker, A.D., Böhme, R.: Revisiting weighted stego-image steganalysis. Electron. Imaging **2008**, 681905 (2008)
28. Fridrich, J., Kodovsky̌, J., Holub, V., Goljan, M.: Steganalysis of content-adaptive steganography in spatial domain. In: Information Hiding, pp. 102–117. Springer, Berlin (2011)
29. Ojala, T., Pietikäinen, M., Harwood, D.: A comparative study of texture measures with classification based on featured distributions. Pattern Recogn. **29**(1), 51–59 (1996)
30. Ojala, T., Pietikäinen, M., Mäenpää, T.: Multiresolution gray-scale and rotation invariant texture classification with local binary patterns. IEEE Trans. Pattern Anal. Mach. Intell. **24**(7), 971–987 (2002)
31. Lu, J., Liu, F., Luo, X.: Selection of image features for steganalysis based on the fisher criterion. Digit. Invest. **11**(1), 57–66 (2014)
32. Akhavan, S., Akhaee, M.A., Sarreshtedari, S.: Images steganalysis using GARCH model for feature selection. Sig. Process. Image Commun. **39**(Part A), 75–83 (2015)
33. Mohammadi, F.G., Abadeh, M.S.: Image steganalysis using a bee colony based feature selection algorithm. Eng. Appl. Artif. Intell. **31**, 35–43 (2014)

34. Ben Brahim, A., Limam, M.: A hybrid feature selection method based on instance learning and cooperative subset search. Pattern Recogn. Lett. **69**, 28–34 (2016)
35. Moradi, P., Gholampour, M.: A hybrid particle swarm optimization for feature subset selection by integrating a novel local search strategy. Appl. Soft Comput. **43**, 117–130 (2016)
36. Mirjalili, S., Mirjalili, S.M., Lewis, A.: Grey wolf optimizer. Adv. Eng. Softw. **69**, 46–61 (2014)
37. Emary, E., Zawbaa, H.M., Hassanien, A.E.: Binary grey wolf optimization approaches for feature selection. Neurocomputing **172**, 371–381 (2016)
38. Filler, T., Pevný, T., Craver, S., Ker, A. (eds.) Information Hiding, vol. 6958. Springer, Berlin (2011)
39. Pedregosa, F., Varoquaux, G., Gramfort, A., Michel, V., Thirion, B., Grisel, O., Blondel, M., Prettenhofer, P., Weiss, R., Dubourg, V., Vanderplas, J., Passos, A., Cournapeau, D., Brucher, M., Perrot, M., Duchesnay, E.: Scikit-learn: Machine learning in Python. J. Mach. Learn. Res. **12**, 2825–2830

Chapter 5
Nature Inspired Optimization Techniques for Image Processing—A Short Review

S. R. Jino Ramson, K. Lova Raju, S. Vishnu and Theodoros Anagnostopoulos

Abstract Nature–inspired optimization techniques play an essential role in the field of image processing. It reduces the noise and blurring of images and also improves the image enhancement, image restoration, image segmentation, image edge detections, image generation, image fusion, image pattern recognition, image thresholding and so on. Several optimization techniques have been proposed so far for various applications of image processing. This chapter presents the short review of nature inspired optimization algorithms such as Genetic algorithm, Genetic programming, evolutionary strategies, Grey wolf optimization, Bat optimization, Ant colony optimization, Artificial Bee Colony optimization, Particle swarm optimization, Firefly optimization, Cuckoo Search Algorithm, Elephant Herding optimization, Bumble bees mating, Lion optimization, Water wave optimization, Chemical reaction optimization, Plant optimization, The raven roosting algorithm with the insight of applying optimization algorithms in advanced image processing fields.

Keywords Optimization techniques · Image processing · Short review
Evolutionary algorithms · Swarm intelligence algorithms

5.1 Introduction

Nature inspired optimization techniques play a key function in the field of engineering, business, industrialized designs, image processing and so on. The main objectives of nature inspired optimization technique are to increase the productivity,

---

S. R. Jino Ramson (✉) · K. Lova Raju · S. Vishnu
Department of Electronics and Communication Engineering,
Vignan's Foundation for Science, Technology and Research,
Guntur, Andhra Pradesh, India
e-mail: jinoramson@gmail.com

T. Anagnostopoulos
Department of Infocommunication Technologies, ITMO University,
Saint Petersburg, Russia

J. Hemanth and V. E. Balas (eds.), *Nature Inspired Optimization Techniques for Image Processing Applications*, Intelligent Systems Reference Library 150,
https://doi.org/10.1007/978-3-319-96002-9_5

gain, efficiency, accomplishment and so on, and to underrate the energy use, cost, size and so forth. Digital Images are viewed as a group of picture element, and each picture element containing few values to represent visual property, illumination, tone etc. Generally, image processing defines refine/manage/transfer an image. Also, it uses various algorithms to improve the nature of the image, to obtain confidential data. Nature-inspired optimization techniques play [1] an essential role in image processing. It reduces the noise and blurring of images and also improve the image enhancement/image restoration/image segmentation/image edge detections/image generation/image fusion/image pattern recognition/image thresholding.

5.1.1 *Nature Inspired Optimization Algorithms*

A lot of special approaches were received to perform various works on the image. In recent times various new techniques and algorithms are popularized which are motivated from the nature. The keys which are best surrounded by massive group of solutions are forwarded after formation or after iteration step and inactivity is not needed. The recent algorithms are very effective compared to early Nature Inspired Algorithms. These algorithms have been reached extensive popularity in recent years to handle many tough real world optimization problems. All these comes under the category of meta-heuristics algorithms.

Several nature inspired optimization algorithms have been developed and studied so far. They are, Genetic Algorithm (GA), Simulated annealing (SA), Artificial immune systems (AIS), Boids, Tabu Search, Memetic Algorithm (MA), Ant Colony Optimization Algorithm (ACO), Cultural Algorithms (CA), Particle Swarm Optimization (PSO), Self-propelled Particles, Differential Evolution (DE), Bacterial Foraging Optimization, Harmony Search (HS), Marriage in Honey Bees Optimization (MBO), Artificial Fish School Algorithm, Bacteria Chemotaxis (BC) Algorithm, Social Cognitive Optimization (SCO), Artificial Bee Colony Algorithm, Bees Algorithm, Glowworm Swarm Optimization (GSO), Honey-Bees Mating Optimization (HBMO) Algorithm, Invasive Weed Optimization (IWO), Shuffled Frog Leaping Algorithm (SFLA), Central Force Optimization, Intelligent Water Drops algorithm, River Formation Dynamics, Biogeography-based Optimization (BBO), Roach Infestation Optimization (RIO), Bacterial Evolutionary Algorithm (BEA), Cuckoo Search (CS), Firefly Algorithm (FA), Gravitational Search Algorithm (GSA), Group Search Optimizer, League Championship Algorithm (LCA), Bat Algorithm, Bumble Bees Mating Optimization (BBMO) Algorithm, Eagle Strategy, Fireworks algorithm for optimization, Hunting Search, Altruism Algorithm, Spiral Dynamic Algorithm (SDA), Strawberry Algorithm, Artificial Algae Algorithm (AAA), Bacterial Colony Optimization, Differential Search Algorithm (DS), Flower pollination algorithm (FPA), Krill Herd, Water Cycle Algorithm, Black Holes Algorithm, Cuttlefish Algorithm, Gases Brownian Motion Optimization, Mine blast algorithm, Plant

Propagation Algorithm, Social Spider Optimization (SSO), Spider Monkey Optimization (SMO) algorithm, Animal Migration Optimization (AMO) Algorithm, Artificial Ecosystem Algorithm (AEA), Bird Mating Optimizer, Forest Optimization Algorithm, Golden Ball, Grey Wolf Optimizer, Seed Based Plant Propagation Algorithm, Lion Optimization Algorithm (LOA), Nature-Inspired Meta-heuristic Algorithm, Optics Inspired Optimization (OIO), The Raven Roosting Optimisation Algorithm, Vortex Search Algorithm, Water Wave Optimization, collective animal behavior CAB algorithm, Bumble bees mating optimization (BBMO), Parliamentary optimization algorithm (POA), Artificial Chemical Process Algorithm, Artificial Chemical Reaction Optimization Algorithm, Bull optimization algorithm, Elephant herding optimization (EHO). All the nature inspired optimizations falls under two main classification namely Evolutionary Algorithms and Swarm Intelligence Algorithms. This chapter presents the short review of some Nature-Inspired Optimization Techniques which are efficiently applied for image processing.

5.2 Evolutionary Algorithms

The flow cycle of evolutionary algorithm is shown in Fig. 5.1. The evolutionary algorithms are inspired from biological evolution like reproduction, mutation, recombination, and selection. The optimization technique plays a vital role in estimating accurate solutions or best solutions from a group of solutions. If a group of individual is concerned, each individual will have his own best solution and the global best will be the best among the local best. Evolutionary algorithm achieved victory on many difficult problem solving with the help of fitness function and the stream which is using Evolutionary algorithm as a tool for problem solving is known as Evolutionary Computation. The evolutionary computation is fundamentally based upon the fitness function and improving the fitness function will results in optimal solutions.

5.2.1 Classification of Evolutionary Algorithms

The broad classification of Evolutionary algorithm for image processing is shown in Fig. 5.2.

(a) **Genetic Algorithm (GA)**:

In 1989, Genetic Algorithm (GA) was introduced by D. Goldberg, J. Holl and K. De Jong. Genetic algorithm is a sub-class of evolutionary algorithms which is inspired from the natural selection. This algorithm resembles the operations such as mutation, crossover and selection.

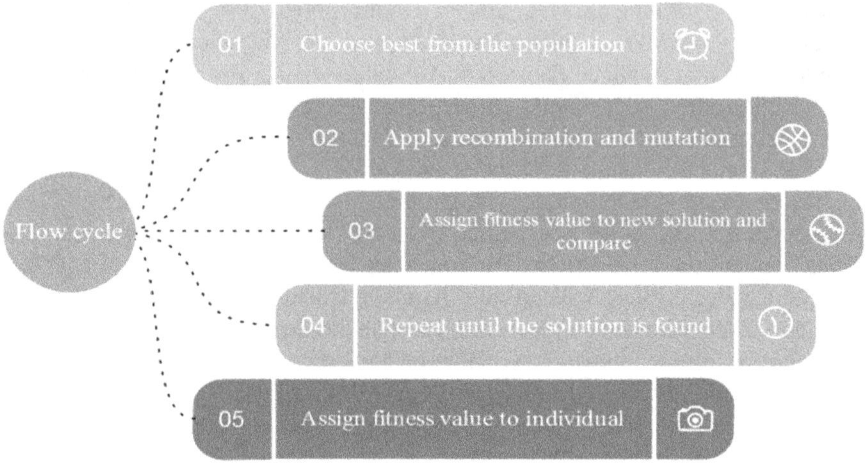

Fig. 5.1 Flow cycle of evolutionary algorithm

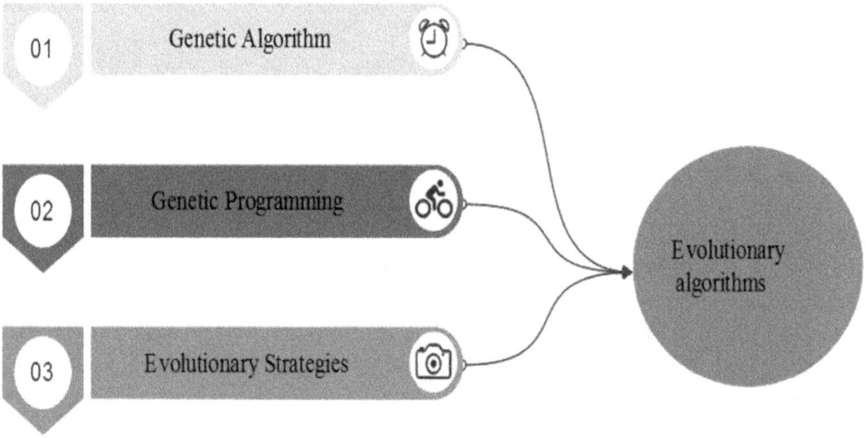

Fig. 5.2 Evolutionary algorithms

GA randomly generates some set of possible solutions to a problem. Each solution is subjected to the fitness function to evaluate each solution. New possible solutions will be generated from the best solutions of the previous step. The process will be continued until the acceptable solution is found. References [2–11] presents the applications of GA in various fields of Image processing and the detailed review is tabulated in Table 5.1.

(b) **Genetic Programming (GP)**:

John Koza has introduced Genetic Programming (GP) in 1992. Genetic programming (GP) is an extension of Genetic Algorithm which is used for testing and

Table 5.1 Review of genetic algorithm

References	Technique used	Applications	Parameters evaluated	System used/software used
Maihami et al. [2]	Genetic-based prototyping	Image annotation	Scale-invariant feature transform and robust hue descriptor	MATLAB programming language in Intel Core i5 CPU 2.4 G and 4 G RAM
Pujari et al. [3]	DNA sequence	Image encryption	–	Matlab
Abbas et al. [4]	Rational ball cubic B-spline representation	Image interpolation	Peak signal to noise ratio (PSNR)	Matlab
Tarigan et al. [5]	Back propagation neural network	Automatic ticketing system for vehicles	Epoch and time (in seconds)	Matlab
Miri et al. [6]	–	Image steganography	Mean square error (MSE), peak signal to noise ratio (PSNR), PSPNR	Matlab
Sukhija et al. [7]	Principal component analysis (PCA)	Face recognition	Number of classes, number of test cases	Matlab
Hung et al. [8]	Parallel fuzzy C-means clustering	MRI segmentation	–	NVIDIA Jetson TK1, Kepler GPU architecture with 192 CUDA cores and 2 GB DDR2 RAM, integrated with an ARM Cortex- A15 CPU with four cores with an Ubuntu Linux Operating System. CUDA version is 6.5
Nagarajan et al. [9]	Diverse density (DD)	Medical image feature extraction	No of cycles, fitness value, probability	Matlab
Zafari et al. [10]	Modified selective computational ghost imaging (SCGI)	Noise filtering	Quality index of the image	Matlab
Sethi et al. [11]	Cryptography	Image hiding	Mean square error (MSE), peak signal to noise ratio (PSNR), capacity	Matlab

Table 5.2 Review of genetic programming

References	Technique used	Application	Parameters evaluated	System/software used
Liang et al. [12]	Support vector machine (SVM)	Figure-ground segmentation	Fitness function, accuracy	Matlab R2014b
Liang et al. [13]	Clustering	Figure-ground segmentation	Mean, variance, skewness, kurtosis, energy, entropy	Matlab R2014b
Iqbal et al. [14]	Transfer learning GP-criptor	Image classification	Kylberg, Brodatz, and Outex data sets	EC Java-based software
Mahmood et al. [15]	Blind image de-convolution	Image acquisition	Root mean square error (RMSE), peak signal to noise ratio (PSNR)	GP simulations are carried out using GPLAB toolbox in MATLAB 7.0

selecting best choice among the set of results. It uses biological evolution to find solutions for complex problems. References [12–15] presents the application of GP in various fields of image processing such as image classification, figure ground segmentation, image segmentation, image acquisition and it is tabulated in Table 5.2.

(c) **Evolutionary Strategies (ES's)**:

In 1960, Evolutionary Strategies were introduced by Schwefel, Rechenberg and Bienert. It follows the process of mutation and recombination for the purpose of obtaining better solutions. Evolutionary Strategies can be classified into three types.

(1 + 1)-ES:

This strategy operates on a parent and its mutant. The mutant becomes a parent if and only if its health is good as its parent. If not the mutant is omitted.

(1 + λ)-ES:

λ mutants are generated with compete with the parent.

(1, λ)-ES:

The best mutant is made by a parent of next generation by neglecting the parent. References [16–20] represents the application of ES in different fields of image processing such as image segmentation, medical images, pattern denoising. Table 5.3 presents the study of various applications, image processing technique used, parameters evaluated and the system used for the implementation.

Table 5.3 Study of different evolutionary strategies

References	Technique used	Application	Parameters evaluated	System used/software
Naidu et al. [16]	Shannon and fuzzy entropy	Image segmentation	Scaling factor (F) and crossover rate (CR)	Matlab 2009b with Intel core i5 processor capacitated 2 GB RAM
Sarkar et al. [17]	Support vector machine (SVM)	Medical images	Distribution index, mutation probability	Matlab R2012a on a workstation with Intel® Core™ i3 3.2 GHz processor
Bhandari et al. [18]	Tsallis entropy based multilevel thresholding	Image segmentation	Peak signal-to-noise ratio (PSNR), mean squared error (MSE), structural-similarity index (SSIM) and feature similarity index similarity metrics (FSIM)	Matlab
Bu et al. [19]	Matching suitable feature construction	Construct synthetic aperture radar (SAR) images	Crossover probability, mutation probability, evolution time	Matlab, 2.4 GB Intel Core2 CUP
Li et al. [20]	Kernel ridge regression	Pattern de-noising	Average value (AVE), standard deviation (STD) and fitness value	Matlab

5.3 Swarm Intelligence Algorithms

In 1989, Swarm Intelligence Algorithm was introduced by Gerardo Beni and Jing Wang. It consists of agents or individuals, interest locally with one another and with environment. They follow rules for individuals, there is no centralized control structure behave individually. The local interaction of agents will make a global behavior implies global intelligence. The broad classification of Swarm intelligence algorithms for image processing is shown in Fig. 5.3.

5.3.1 Gray Wolf Optimization (GWO)

This algorithm is following the leadership and hunting styles of grey wolves and is proposed by Mirjalili, Seyed Mohammad Mirjalili and Andrew Lewis in 2014. Since Gray wolves are considered in top of food chain, they considered as apex predators. The members in the groups are following a very strict social dominant hierarchy (Fig. 5.4).

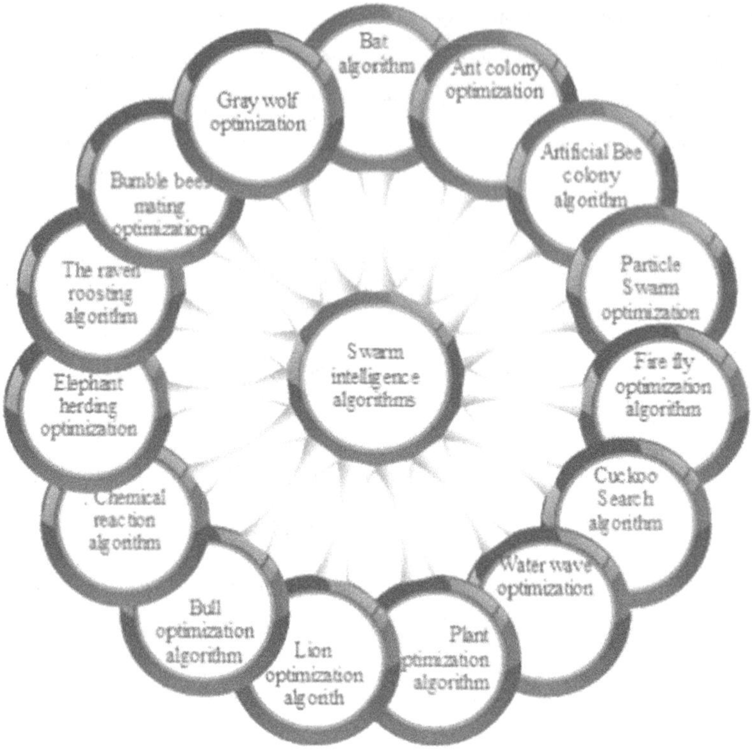

Fig. 5.3 Swarm intelligence algorithms

Fig. 5.4 Types of grey wolf optimization

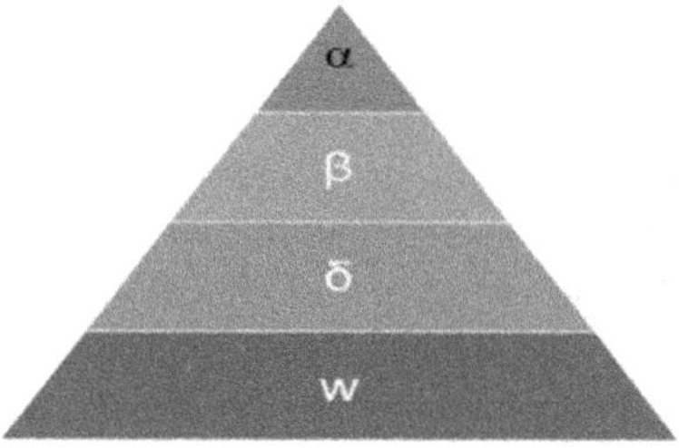

The first level is alpha (α) and they are the leaders, who are responsible for making decisions regarding hunting, sleeping place etc. The second level is known as beta (β). These members are acting like subordinates for the decision making level alpha (α). There beta (β) candidates will be a replacement for the alpha (α) when they are old or passed away. The third level is called delta (δ) which is dominating omega (w). Omega (w) is the lowest level in the hierarchy. Delta wolves will act as subordinates for both alpha (α) and beta (β). In GWO fittest solution will be (α), second will be (β) and third be delta. Remaining all will be

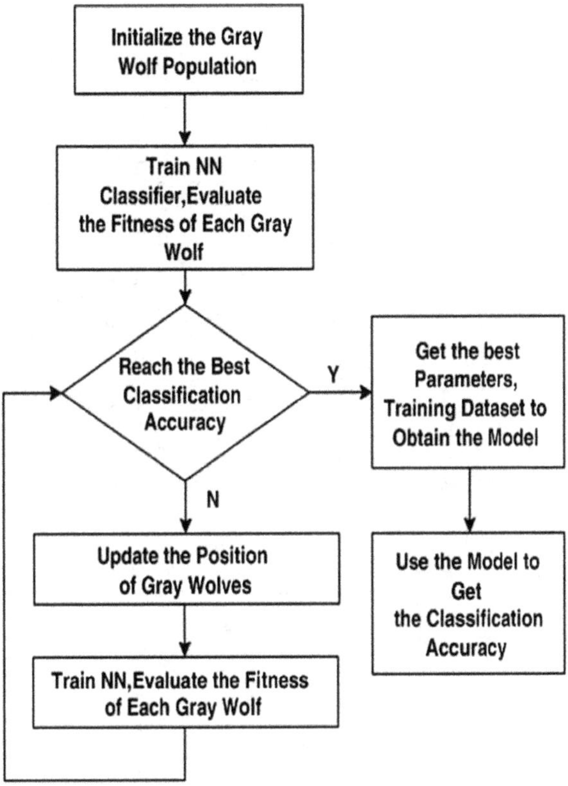

Fig. 5.5 Flowchart of gray wolf optimization

considered as Omega (w). So hunting will by guided by $(\alpha), (\beta)$ and (δ) and (w) follow these three.

The main advantage of GWO is easy to design and GWO has some disadvantages such as slow convergence, low searching ability. To overcome the disadvantages Improved Gray Wolf Optimization (IGWO) is opted.

The flowchart of GWO is shown in Fig. 5.5. The major applications of GWO in image processing are shown in Table 5.4. References [21–28] represents the applications of GWO in various fields of Image processing such as image segmentation, data clustering, medical images, medical image fusion, image edge extraction and it is tabulated in Table 5.4.

5.3.2 *Bat-Algorithm (BA)*

The Bat Algorithm (BA) is an optimization algorithm inspired from the behaviour of micro-bats. It is based on the echolocation behaviour of micro-bats, along with changing pulse rates of emission and loudness. BA comes under Swarm

Table 5.4 A survey of gray wolf optimization

References	Technique used	Application	Parameters evaluated	System used/software used
Ramakrishnan et al. [21]	Support vector machine (SVM)—sequential minimal optimization (SMO)	Image segmentation	Sensitivity, specificity, accuracy, positive predictive value (PPV), negative predictive value (NPV)	Matlab 2012
Khairuzzaman et al. [22]	Multilevel thresholding	Image segmentation	Number of grey wolves, number of iterations, mean structural SIMilarity index (MSSIM)	Matlab R2010a with Intel core-i7 CPU @ 3.40 GHz
Jadhav et al. [23]	Hybridization	Data clustering	F-measure, rand coefficient, Jaccard coefficient, MSE	PC with Intel Core i-3 processor 4 GB RAM and Windows 8 operating system. The experimentation is carried out using the MATLAB
Wang et al. [24]	Kernel extreme learning machine	Bankruptcy prediction	Validation accuracy (Va-acc), training accuracy (Tr-acc), test accuracy (Te-acc)	MATLAB platform. The empirical experiment was conducted on Intel®Core™ i7 4790CPU @ 3.60 GHz with 8 GB of RAM and the system is Windows7. For SVM
Daniel et al. [25]	Optimum laplacian wavelet mask	Medical image	Mean scale value, best scale value	Matlab 2010a with Pentium dual core processor with speed 2.30 GHz
Daniel et al. [26]	Optimum spectrum mask	Medical image fusion	Entropy, standard deviation, mutual information	Matlab 2010a with Pentium dual core processor with speed 2.30 GHz
Zhang et al. [27]	Template matching	Image edge extraction	Average (Ave), CPU average time and correct rate	Matlab R2012a environment and executed on a 4-core Intel Core i5-4200U CPU with 8 GB RAM running at 4 × 1.60 GHz under Windows8.1 operating system
Li et al. [28]	Fuzzy multilevel image thresholding	Image segmentation	Standard deviation (STD), peak signal to noise ratio (PSNR), root mean squared error (RMSE)	MATLAB, Lenovo Laptop with an Intel Core i3 processor and 4 GB memory

Intelligence method. This algorithm was developed by Xin-She Yang in 2010. Few bats have developed a highly sophisticated sensibility of hearing. They emit sounds that consider of objects in their path and send echoes return to bats. According to the bats can determine the size of objects, how they are travelling fast and far away.

The considerations for designing the bat optimization algorithm are as follows:

1. The echolocation strategy of the natural bat is considered with the distance calculation between two objects. It has considered that the bats are knowledgeable to distinguish between the prey and the objects.
2. In this algorithm, it has considered that the bats have flying with the velocity p_i from a position p_{i-1} with a smallest frequencies f_{min} and changing wavelength lambda and loudness L_0 in search of prey. The bats have capable of adapting the frequencies f_i and also the rate of frequencies $f \in (0, 1)$ and the adapting depends on the favorable or failed searching of prey or for the kind (small or large) of prey.
3. The transmitted sound of the natural bat changes according to the social needs, so, the bat optimization algorithm considers that the maximum loudness is L_{max} and minimum is L_{min}. The bat optimization process does not exist where two or more number of bats hunt for the same object. Regarding the frequencies of the bat algorithm, it is considered that the frequency corresponding to the wavelength varies within a fixed range. Now the bat position x_i maneuvers from a position x_{i-1} with the velocity v_i and frequency f_i. The optimization relates to the movement of its position with relative velocities and frequencies. The positional updating formula of the bats are as follows:

$$F_i = f_{max} + (f_{max} - f_{min}) * \beta \tag{5.1}$$

$$V_{i+1} = V_{i-1} + (X_i - X^*) * F_i \tag{5.2}$$

The bat optimization technique is also very powerful tool for many complex problems of the various fields of engineering and science.

The main advantage of Bat algorithm is simplicity and flexibility. It is easy to design. The flowchart of BA is shown in Fig. 5.6. References [29–31] presents the major applications of BA in image processing are listed in Table 5.5.

5.3.3 Ant Colony Optimization (ACO)

The ant colony optimization (ACO) is an algorithm inspired from the behaviour of ants in searching their food. The ants always prefer shortest path between their nest and source of food. These ants are having a indirect communication by releasing pheromone. Once they found food source, then the ant will deposit pheromone when it is travelling towards the nest. So that the fellow ants can easily reach the

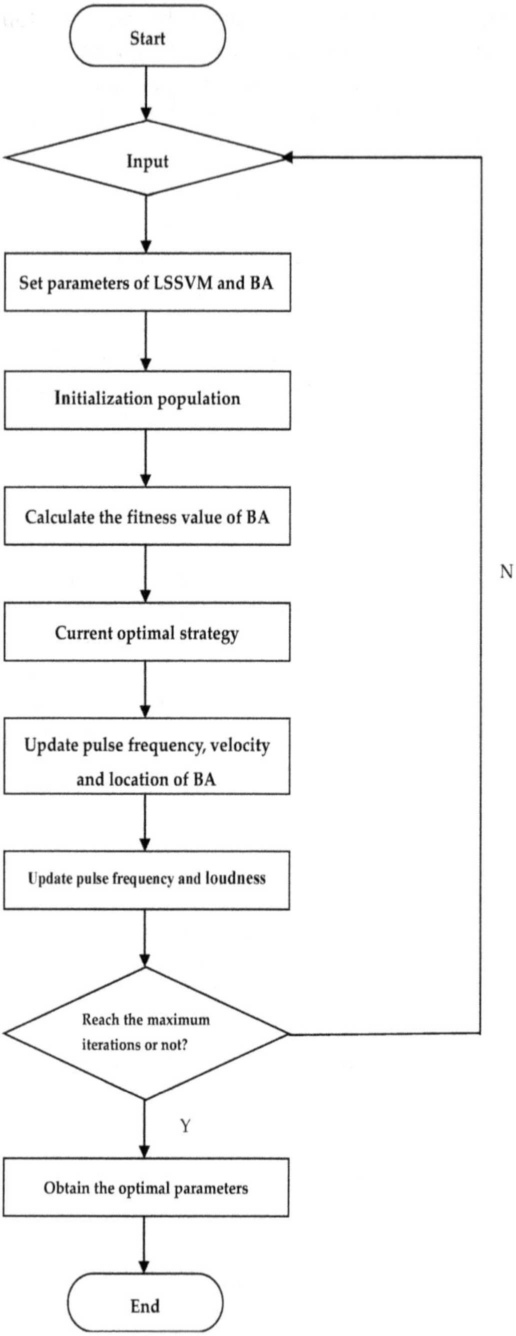

Fig. 5.6 Flowchart of bat algorithm

Table 5.5 A survey on various works over bat algorithm

References	Technique used	Application	Parameters evaluated	System used/software used
Karri et al. [29]	Vector quantization (VQ)	Image compression	Bitrate/bits per pixel, peak-signal to noise ratio (PSNR), mean square error (MSE)	Window XP PC with an Intel® Core ™ i5-2540 machine with 2.60 GHz CPU, and 2.94 GB of RAM. moreover, all the programs are written and compiled on MATLAB version 7.9.0 (R2009b)
Senthilnath et al. [30]	Clustering approach	Multispectral satellite image classification	Maximum generation and population size	Matlab 7.12.0.635, on a system having an i-7 processor and 6-GB RAM
Yang et al. [31]	Echolocation behaviour	Image visualization	Convergence rate, sensitivity	Matlab on a standard 3 GHz desktop computer

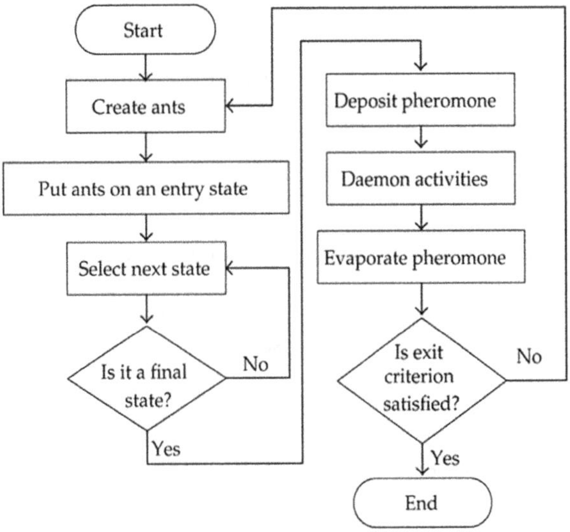

Fig. 5.7 Shows the flowchart of ant colony optimization algorithm

food. When one ant find a short path from the colony towards the source of food, then other ants, also will follow the new path.

ACO comes under Swarm Intelligence method. ACO is proposed by Marco Dongo, A. Colorni and V. Maniezzo in 1991. The ACO are commonly using to a optimal solution in graphical way of problem solving. The major advantage of ACO over genetic algorithm is its ability to handle dynamically changing graph. And also

Table 5.6 A survey of ant colony optimization

References	Technique used	Application	Parameters evaluated	System/software used
JayaBrindha et al. [32]	Cascaded support vector machine (SVM)	The images of sunflower seeds classification	Boundary descriptors, cosine descriptors, fourier descriptor	Matlab
Miria et al. [33]	Discrete cosine transform (DCT)	Medical image de-noising	Peak signal-to-noise ratio (PSNR), structural similarity index measure (SSIM)	Matlab
Kuo et al. [34]	Source optimization (SO)	Image enhancement	Edge placement error (*EPE*)	Matlab with a PC platform at Intel Core i7 (3.4 GHz) with 8 GB of memory
Yin et al. [35]	Support vector machine (SVM)	Very high-resolution (VHR) images	Correctly extracted road pixels (TP), incorrectly extracted (FP) road pixels, missed road pixels (FN)	Matlab
Zhang et al. [36]	Graphics processing units (GPUs)	Hyper spectral images	Root mean square error (RMSE)	Matlab with Intel Xeon X5660 CPU, 12 GB RAM and an NVidia Quadro 5000 GPU

capable of providing an almost good solution. It has robustness and ability to search for a better solution in solving performance.

The flowchart of Ant Colony Optimization Algorithm is shown in Fig. 5.7 and Table 5.6 shows the application of ACO in various areas of image processing.

5.3.4 Artificial Bee Colony Optimization (ABC)

Artificial Bee Colony Optimization algorithm is inspired from the behaviour of honey bees. This was proposed by Karaboga and Basturk in the year 2007. Figure 5.8 shows the flowchart of the ABC algorithm.

In this mode, 3 groups of bees are there. They are

1. Employed bees,
2. Onlooker bees and
3. Scout bees.

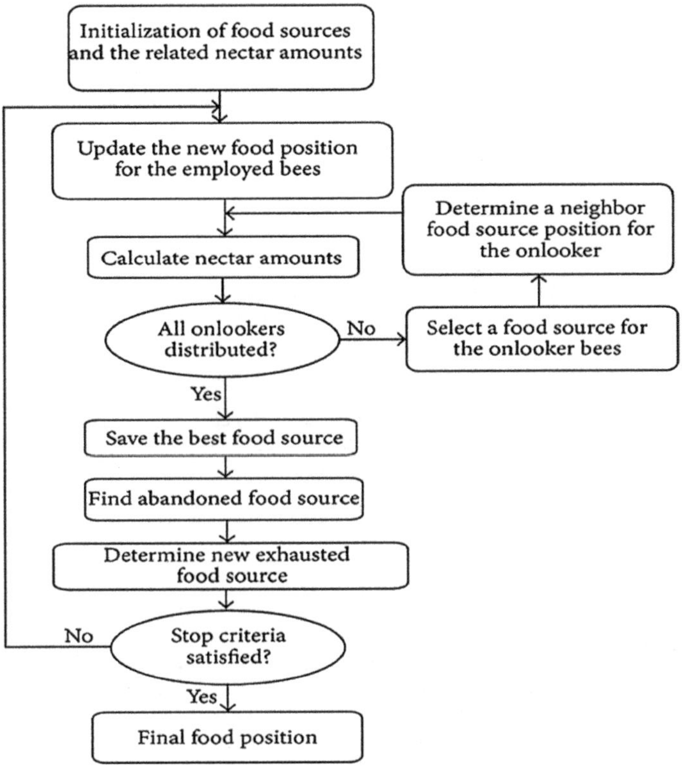

Fig. 5.8 Arificial bee colony optimization algorithm flowchart

Employed bees search food and share this information in the group. The onlooker bees will choose the best food source from the employed bees information. i.e., each employed bees will come with the information of unique food source. The scout bees are a subset of employed bees, whose food source is rejected. References [37–42] represents the applications of ABC in various fields of Image processing like image segmentation, image watermarking, region based image steganalysis and it is tabulated in Table 5.7.

5.3.5 *Particle Swarm Optimization (PSO)*

In 1995, Particle Swarm Optimization (PSO) was proposed by Kennedy and Eberhart. PSO technique is acquired by analyzing the behavior of fish and birds when they are moving as a group. From the given parameters the PSO will do iterations until getting a improved solution for the candidate. Particle also known as candidate can improve its position by considering inertia, personal influence and

Table 5.7 A study of artificial bee colony optimization algorithm

References	Technique used	Application	Parameters evaluated	System used/software used
Gao et al. [37]	Multi-level thresholding	Image segmentation	Accuracy and convergence speed	Matlab
Chen et al. [38]	ABC	Image contrast enhancement	Peak signal-to-noise ratio (PSNR), structural similarity index measure (SSIM), information fidelity criterion (IFC), visual information fidelity (VIF), and visual signal to noise ratio (VSNR)	PC with a Intel Core i5 CPU 3.2 GHz and an 8 GB RAM
Abdelhakim et al. [39]	Watermarking optimization	Image watermarking	Peak signal-to-noise ratio (PSNR), mean square error (MSE)	Matlab
Sajedi et al. [40]	Support vector machine (SVM)	Region based image steganalysis	Pixel value (PV), feature dimension (D), population size (P)	Matlab
Mostafa et al. [41]	Clustering	CT liver segmentation	Similarity index (SI)	Matlab
Goel et al. [42]	Maximum likelihood classifier (MLC)	Image classification	User accuracy, producer accuracy	MATLAB 7 and are executed on a DELL Studio15 computer with the configuration of Intel Core I3 CPU M370 at 2.40 GHz and 4 GB RAM

Fig. 5.9 shows the flowchart of the PSO algorithm

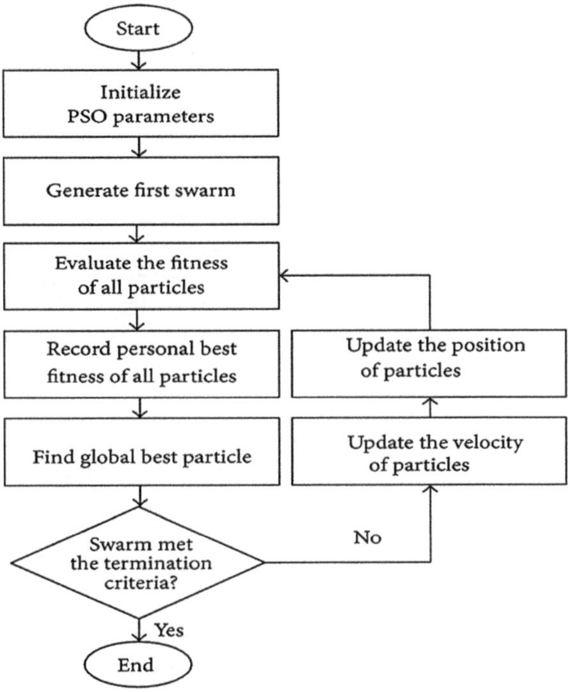

social influence. Main application of PSO is in functional optimization and optimum control in control systems. The main advantage of PSO is easy to implement and it has less parameters, it is used to handle non linear optimization problems. It is flexible to practical applications. Figure 5.9 shows the flowchart of PSO algorithm. References [43–49] presents the areas where PSO is used in image processing and it is tabulated in Table 5.8.

5.3.6 *Firefly Optimization (FFO)*

Firefly Optimization (FFO) algorithm was proposed by Xin-She Yang in 2009. FFO was influenced by the fireflies those having flickering behavior. This algorithm searches the optimal matching patch from left to right and also from top to bottom and finally, it searches the patch. However, if there are a large number of candidate patches, it will leads to heavy workload and inaccuracy. Therefore, the Firefly optimization algorithm has been introduced to search the best matching patch.

The Firefly optimization algorithm is a universal optimization method which is based on the fly foraging behavior and the result of this algorithm is completely depends upon the foraging process. The first is smell search process: using the smell to perceive the various gases in air and determine the food position which is close to

Table 5.8 A survey on different works over particle swarm optimization

References	Technique used	Application	Parameters evaluated	System/software used
Wu et al. [43]	Modal transformation	Remote sensing	Number of correct matches and root mean square error (RMSE)	MATLAB, on an Intel Core i5 machine
Mozaffari et al. [44]	Thresholding	Multilevel image thresholding segmentation	Standard deviation (STD), peak signal to noise ratio (PSNR)	Personal computer with 16 GB RAM, CPU with 7 Cores and 3.4 GHz speed using MATLAB 2016 Software which was installed on a 64-Bit Version of Windows 10
Sabeti et al. [45]	PSO versions	Medical image	Peak signal-to-noise ratio (PSNR), uniformity measure (UM), structural similarity index measure (SSIM)	Matlab
Zhang et al. [46]	2D fuzzy fisher	Image segmentation	Number of iterations, cost function value (cfv)	PC with Inter Core CPU @ 2.40 GHz and 2G memory
Salucci et al. [47]	Inverse scattering	Microwave imaging	Signal to noise ratio (SNR)	Standard Laptop with a Single-core 2.1-GHz CPU
Liu et al. [48]	Adaptive translational motion compensation	Inverse synthetic aperture radar (ISAR) images	Mean squared error (MSE)	Matlab
Xue et al. [49]	Integrating the harmonic analysis (HA), particle swarm optimization (PSO), and support vector machine (SVM)	Airborne visible infrared imaging spectrometer	Minimum noise fraction (MNF), and independent component analysis (ICA)	Matlab R2012b in a desktop PC equipped with an Intel Core i7 CPU (at 3.4 GHz and 64-bit) and 16 GB of RAM

it and the second is visual orientation process: in the visible range, determining the food position accurate and flying to it.

The primary purpose for a firefly's flash is to act as a signal system to kill other fireflies.

FFO based on three idealized rules:

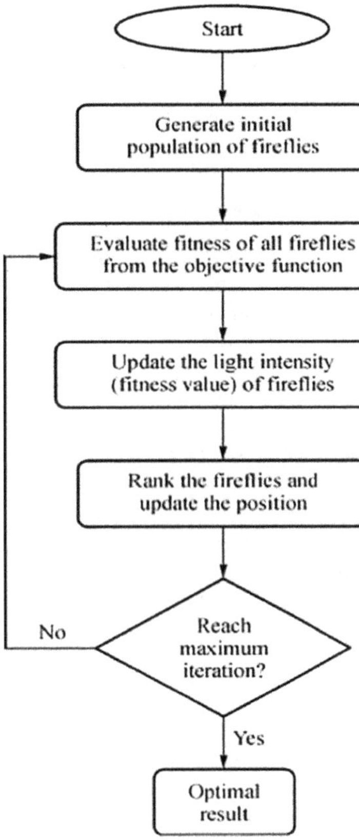

Fig. 5.10 Flowchart of firefly algorithm

- All fireflies are unisexual, so that any individual firefly will be killed to all other fireflies.
- Attractiveness is proportional to their brightness, and for any two fireflies, the less bright one will be killed by the brighter one, but the intensity decrease as their mutual distance increases.
- If there are no fireflies brighter than a given firefly, it will move anyway. The brightness should be associated with the objective function.

The main advantage of FFO is easy to operate and implement, it has less parameters. It is easy to combine with other algorithms to improve the performance of the algorithm.

The flowchart of FFO is shown in Fig. 5.10. References [50–54] presents the major applications of FFO in image processing are shown in Table 5.9.

Table 5.9 Review of fire-fly optimization

References	Technique used	Application	Parameters evaluated	System/software used
Kora et al. [50]	Sequency ordered complex Hadamard transform	ECG based atrial fibrillation detection	Sensitivity (Sen) and specificity (Spe)	Matlab 7.12.0
Pare et al. [51]	Modified fuzzy entropy (MFE)	Image thresholding	Peak signal to noise ratio (PSNR), structural similarity index measures (SSIM), mean square error (MSE), feature similarity index measures (FSIM)	MATLAB R2014b on a PC with 3.4 GHz Intel core-i7 CPU, 4 GB RAM running on Windows 7 system
Zhang et al. [52]	Modified local binary pattern descriptor	facial emotion recognition	Randomization and dynamic parameter	Matlab
Rajinikanth et al. [53]	RGB histogram	Image segmentation	Peak signal to noise ratio (PSNR), structural similarity index measure (SSIM) and CPU time	Matlab R2010a on an Intel Dual Core 1.6 GHz CPU, 1.5 GB RAM running window XP
Nayak et al. [54]	Multilayer perceptron (MLP)	1D/2D predictive image coding	Number of epochs, root mean square error (RMSE)	MATLAB 9.0 on a system with an Intel Core 2 Duo CPU T5800, 2 GHz processor, 2 GB RAM and Microsoft Windows-2007 OS

5.3.7 *Cuckoo Search Algorithm (CS)*

Cuckoo search (CS) is an optimization algorithm. CS was introduced by Suash Deb, Xin-She Yang in 2009. It was influenced by the constrain brood parasitism of few cuckoo species by laying their eggs in the nests of other host birds. Few shelter birds can lease direct discord with the intruding cuckoos. For example, if a host bird finds the eggs are not their own, it will either throw these alien eggs away or build a novel nest in another place. Part of female parasitic cuckoos is generally very particular in the mimicry in colors and design of the eggs of some select accommodates species. Cuckoo search algorithm can be applied for many optimization problems.

Cuckoo search algorithms can be represented as follows:

The goal behind the algorithm is to use the novel and conceivably optimal results to re-establish a not-so-better solution in the nests. In the easiest form, each nest has

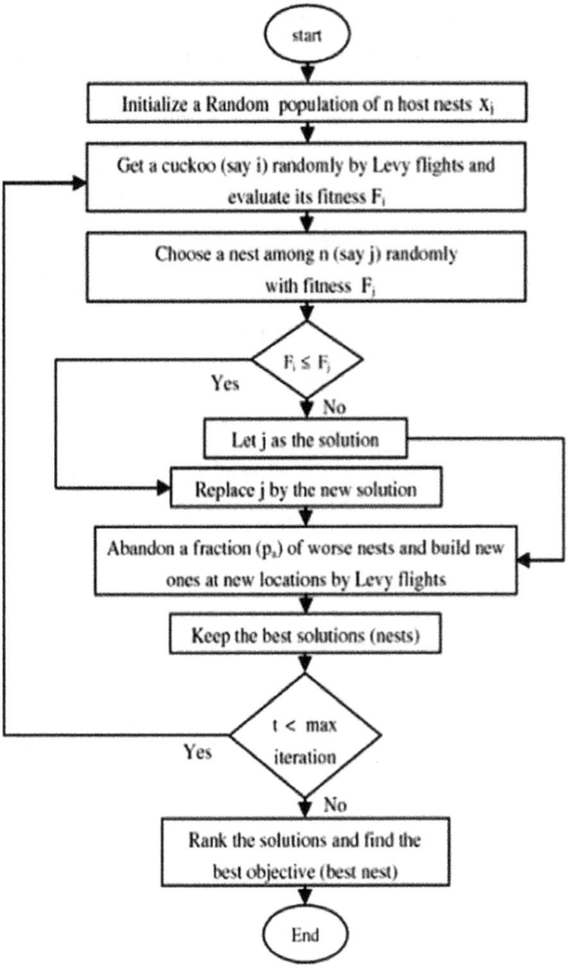

Fig. 5.11 Flowchart of CS algorithm

one egg. The algorithm can be continued to further complex cases in which any nest has multiple eggs characterizing a group of solutions for that specified purpose. The flowchart of CS is shown in Fig. 5.11.

Cuckoo search is based on three idealized rules:

- Any cuckoo lays single egg at a time, and dumps its egg in a selected nest.
- The optimal nests with more quality of eggs will carry over to the next production to improve their population.
- The number of applicable host's nests is established, and the egg place by a cuckoo search is invented by the host bird with a probability $p_a \in (0, 1)$, $p_b \in (0, 1)$. Finding operates on few set of worst nests, and invented solutions dumped from further estimates.

Table 5.10 A review of cuckoo search algorithm

References	Technique used	Application	Parameters evaluated	System/software used
Suresh et al. [55]	Histogram equalization (HE)	Enhancement of satellite images	Color enhancement factor (CEF), structure similarity index measure (SSIM), mean square error (MSE), peak signal to noise ratio (PSNR)	MATLAB R2015a running on an Intel Core i7 PC With 3.40 GHz CPU and 8 GB RAM
Pare et al. [56]	Minimum cross entropy	Image thresholding	Mean and standard deviation (STD), structure similarity index measure (SSIM)	Matlab
Chiranjeevi et al. [57]	Image compression	Vector quantization (VQ)	Skewness and mutation probability, signal to noise ratio (PSNR)	Windows XP operating system with an Intel® Core™ i5-2540 and 2.60 GHz CPU with 2.94 GB RAM. Moreover, MATLAB version 7.9. (R2009b)
Mohammed Ismail et al. [58]	Cuckoo inspired fast search (CIFS)	Fractal image compression	Mean square error (MSE), peak signal to noise ratio (PSNR)	Core i5-368 5200U; 4 GB RAM 1 TB Hard-Disk, 2 GB Graphics with 369 Windows 10. The implementation of proposed method is carried out on NVIDIA Ge Force GTX 480 GPU using CUDA language

References [55–58] presents the major applications of CS in image processing are listed in Table 5.10.

5.3.8 Elephant Herding Optimization (EHO)

Elephant Herding Optimization (EHO) was introduced by Suash Deb, Gai-Ge Wang and Coelho in 2015. EHO is inspired by the herding behavior of elephant group. Elephants are one kind of the biggest mammals on land. The African

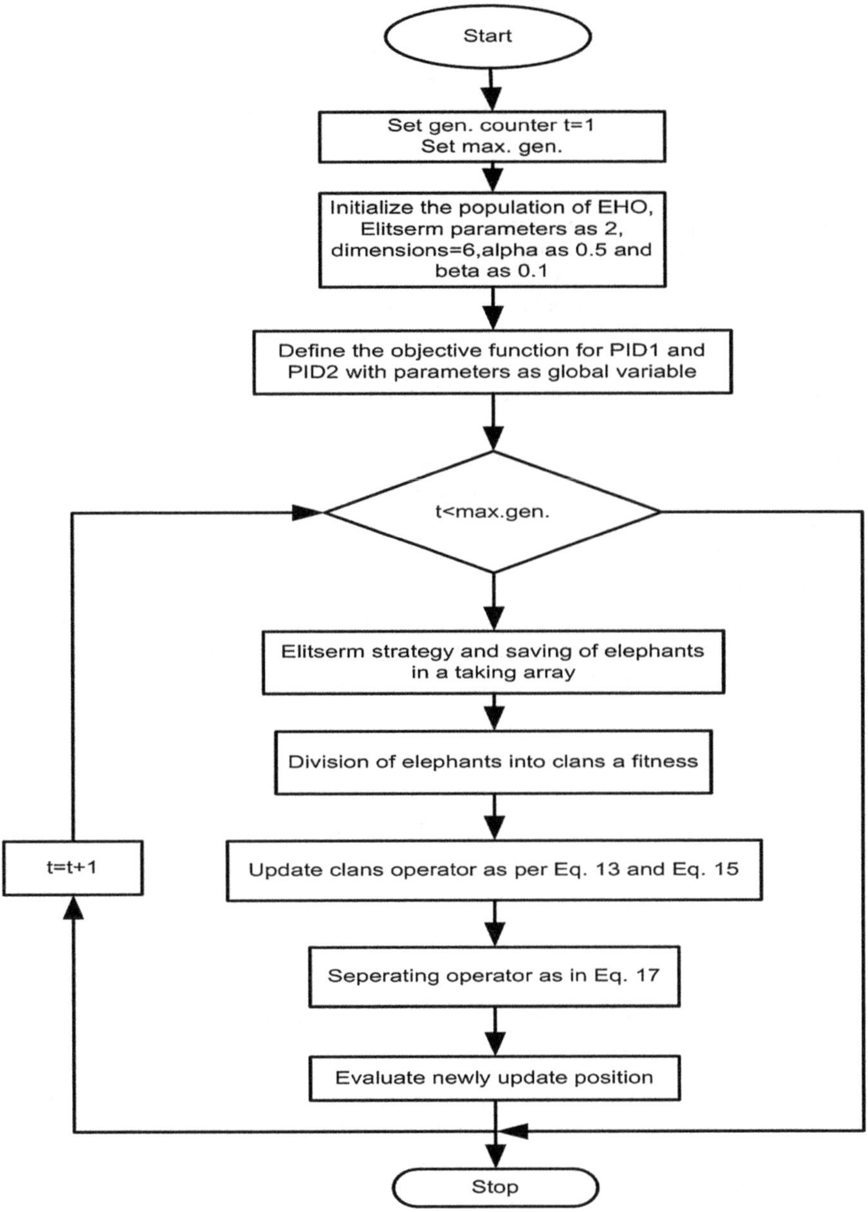

Fig. 5.12 Flowchart of EHO algorithm

elephant and the Asian elephant are two, which are generally identified species. A long trunk is the most typical feature that is multi-purpose, such as breathing, lifting water and grasping objects. In environment, elephants are social animals, and

Table 5.11 Review of EHO algorithm

References	Technique used	Application	Parameters evaluated	System/software used
Tuba et al. [59]	–	Image thresholding	Mean and standard deviation	Matlab 2016a, Intel Core i7-3770K CPU at 4 GHz, 8 GB RAM, Windows 10, Professional OS
Tuba et al. [60]	Support vector machine (SVM)	Automatic diagnosis of different diseases	training and testing	Matlab 2016a, Intel Core i7-3770K CPU at 4 GHz, 8 GB RAM, Windows 10, Professional OS

they have complex social structures of females and calves. An elephant group is composed of several clans under the leadership of a mother, frequently the oldest cow. The flowchart of EHO is shown in Fig. 5.12.

A clan is consisting of one female with her calves or certain related females. Females prefer to live in family groups, while male elephants likely to live in separation, and they will leave their family group when growing up. Though male elephants live away from their family group, they can stay in contact with elephants in their clan through low-frequency vibrations. In this way to prepare the assemble behavior of elephants resolve entire set of world optimization difficulties, it has treated to reduce into the following idealized regulations:

- The elephant population is collected of few clans, and each clan consist permanent number of elephants.
- Permanent number of male elephants will leave their family group and lives alone, far away from the main elephant set at each production.

References [59, 60] presents the major applications of EHO in image processing are tabulated in Table 5.11.

5.3.9 Bumble Bees Mating Optimization (BBMO)

Bumble Bees Mating Optimization (BBMO) algorithm comes under category of meta-heuristic optimization, it is also a population-based search algorithm. It was recommended by F. Comellas and J. Martinez Navarro in 2009. The behavior of honey bee colonies are mimics their food seeking. Bumble Bees Mating Optimization (BBMO) algorithm is a fairly novel swarm intelligence algorithm and which is resembles the mating behavior that a swarm of bumble bees performs. This algorithm is inspired by a novel nature that resembles the mating behavior of the bumble bees, the Bumble Bees Mating Optimization (BBMO) algorithm, is used for solving global corrupted optimization problems. References [61, 62] presents the major applications of BBMO in image processing are shown in Table 5.12.

Table 5.12 A survey on various works over BBMO algorithm

References	Technique used	Application	Parameters evaluated	System used/ software used
Abdelhakim et al. [61]	Robust watermarking	Quality of the watermarked image	Fitness function, number of iterations, peak signal-to-noise ratio (PSNR)	Matlab
Jiang et al. [62]	Histogram thresholding	Image segmentation	Peak signal-to-noise ratio (PSNR), CPU time	PC with 2.40 GHz CPU, 2 GB RAM with window 7 system and MATLAB 7.2 software

5.3.10 Lion Optimization Algorithm (LOA)

Lion Optimization Algorithm (LOA) was introduced by Maziar Yazdani and Fariborz Jolai in 2016. This algorithm is inspired by a novel nature, lion's behavior. Lions are the most socially willing of among wild cat species which show great levels of cooperation and antagonism. Lions are of specific curiosity because of their strong sexual dimorphism in combination of social behavior and appearance. The lion is a wild felid with two types of social organization: residents and nomads. Residents lives in sets, known as pride.

A pride of lions generally adds five females, their cubs of both sexes, and in that one or more adult males. Young males are ignoring from their birth pride when they become sexually grown-up. The flowchart of LOA is shown in Fig. 5.13. References [63, 64] presents the major applications of LOA in image processing are listed in Table 5.13.

5.3.11 Water Wave Optimization (WWO)

Water wave optimization was introduced by Y. J. Zheng in 2015. Water wave optimization (WWO) comes under a novel meta-heuristic technique, it is used for global optimization problems. It shows that how graceful phenomena of water waves, like propagation, refraction, and breaking, can be used to derive effective mechanisms for searching in a high-dimensional solution domain. In general, the algorithmic scheme of WWO is does not complicated and get clear design with a least-size population and only a several control constants. WWO is tested on a diverse set of standard problems, and applied WWO to a real-world high-speed train scheduling problem in China. The computational results show that WWO is much aggressive with state-of-art evolutionary algorithms. References [65, 66]

Fig. 5.13 Flow chart of LOA algorithm

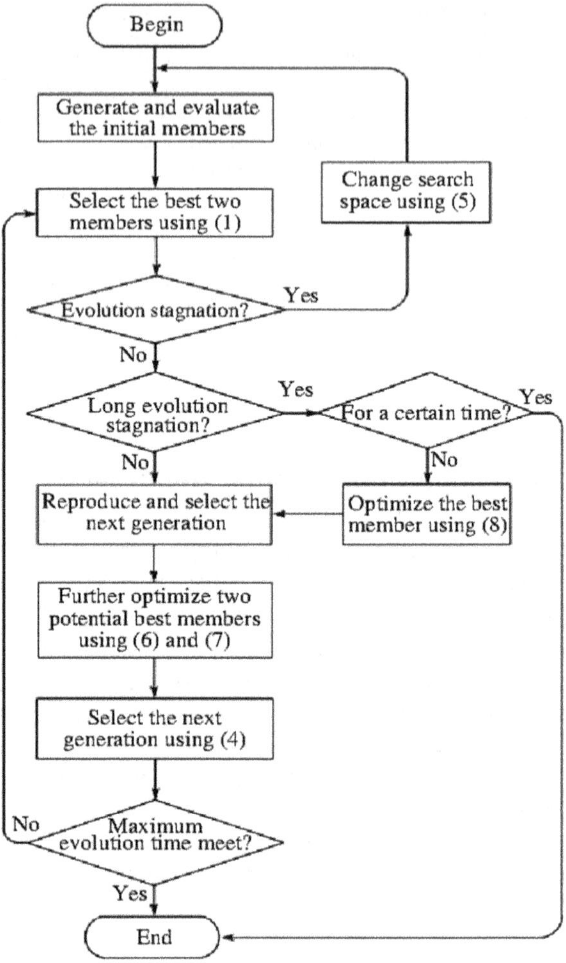

Table 5.13 Review of LOA algorithm

References	Technique used	Application	Parameters evaluated	System/ software used
Kanimozhi Suguna et al. [63]	Machine learning (ML)	Medical imaging	Mean, entropy, energy and CPU time	Matlab
Yazdani et al. [64]	Clustering	Image and video processing (IVP)	Median, STD	Matlab

Table 5.14 Study of WWO algorithm

References	Technique used	Application	Parameters evaluated	System/software used
Xu et al. [65]	Gray correlation analysis	Pattern recognition	Correct matching rate and average running time	Matlab 2014b on a personal computer with a 3.20 GHz CPU, 4.00 G RAM under Windows 7 system
Wu et al. [66]	Elite opposition-based learning (EOBL)	Global optimization	Maximum and minimum fitness and wavelength	PC of Intel® with 3.5 GHz Xeon CPU and 8 GB of memory, Windows 7, Matlab 2012a

presents the major applications of WWO in image processing are shown in Table 5.14.

5.3.12 Chemical Reaction Optimization Algorithm (CRO)

CRO is a currently introduced general-purpose meta-heuristic. In 2011, the CRO was recommended by Lam, Bilal Alatas, it was originally designed for solving conjunctional optimization problems. Chemical reaction optimization (CRO) comes under a population-based meta-heuristic algorithm, which is based on the principles of chemical reaction. A chemical reaction is a method of sending the reactants or molecules through a sequence of reactions into products. This process of transformation is designed in the CRO algorithm to resolve optimization problems. Chemistry is a domain in science and it was managed studies with respect to the chemical properties like matter and its structure. Chemical reactions discontinuity chemical bonds into molecules and form novel bonds using molecules participating in reaction. References [67, 68] Shows the application of CRO in various areas of image processing and it is listed in Table 5.15.

Table 5.15 Review of CRO algorithm

References	Technique used	Application	Parameters evaluated	System/software used
Asanambigai et al. [67]	Fuzzy C means	Medical image processing	Sensitivity, specificity, Jaccard index and dice coefficients	Matlab
Duan et al. [68]	Contour matching	Remote sensing applications	rotation angle (θ), and scaling factor (s)	PC with Intel Core i5, 2.6-GHz CPU, 4-GB memory, and 32-b Windows 7, using Matlab 8.0.0.783 (R2012b)

5.3.13 Plant Optimization Algorithm (POA)

Plant optimization algorithm was introduced by Jun Li, Zhihua Cui and Zhongzhi Shi in 2012. Plant optimization algorithm (POA) comes under a novel meta-heuristic algorithm, influenced by tree's growing process. POA is nature inspired, it follows the path plants, in specific the strawberry plant, propagate. A basic POA has been expressed and tested on one objective as well as many objective continuous optimization problems. The test problems though standard are least dimension. The results displayed that POA has advantages and get more investigation on greater dimensional problem cases as well as problems proceeding in practice, these are frequently very challenging. POA is good-looking because, between other things, it is simple to illustrate and design small size population. POA has been implemented to solve many known hard forced optimization problems arising in the field of engineering design with continuous disciplines. POA established either adjacent good known solutions or optimal ones to all of them. Reference [69] presents the application of POA in image segmentation. The technique used in this paper is molecular biology and the parameters evaluated are Shape, Colour and Texture, identification rate.

5.3.14 The Raven Roosting Algorithm (RRO)

Raven Roosting Optimization was popularized by Anthony Brabazon, Wei Cui, Michael O'Neill in 2014. This algorithm influenced by the social roosting behaviour of raven or a bird. This social roosting is exhibited by especially by the birds. in order to maintain the communication between the members about the food sources and nearby threats we will use social roots as hubs. It is deals with the mimic behaviour of ravens or Foraging behaviour of bird species, the natural raven, and it take influence from this to design a new optimisation algorithm which is called as the raven roosting optimisation algorithm (RRO). Birds of Heterogeneity, insects enroll in roosting. In raven Roosting, Roosts are information centers or can say servers and scrounge feature of common ravens inspired to solve problems. This technique is good enough to handle number of overloaded tasks transfer on Virtual Machines (VMs) by determining the availability of VMs capacity. Raven Roosting Optimization (RRO) random allocation of VMs to Cloudlet results huge change in make span with respect to VM to which allocated. The flow chart of RRO is shown in Fig. 5.14 and [70] shows the application of RRO in heterogeneity of birds. The parameters evaluated are average response time, average waiting time.

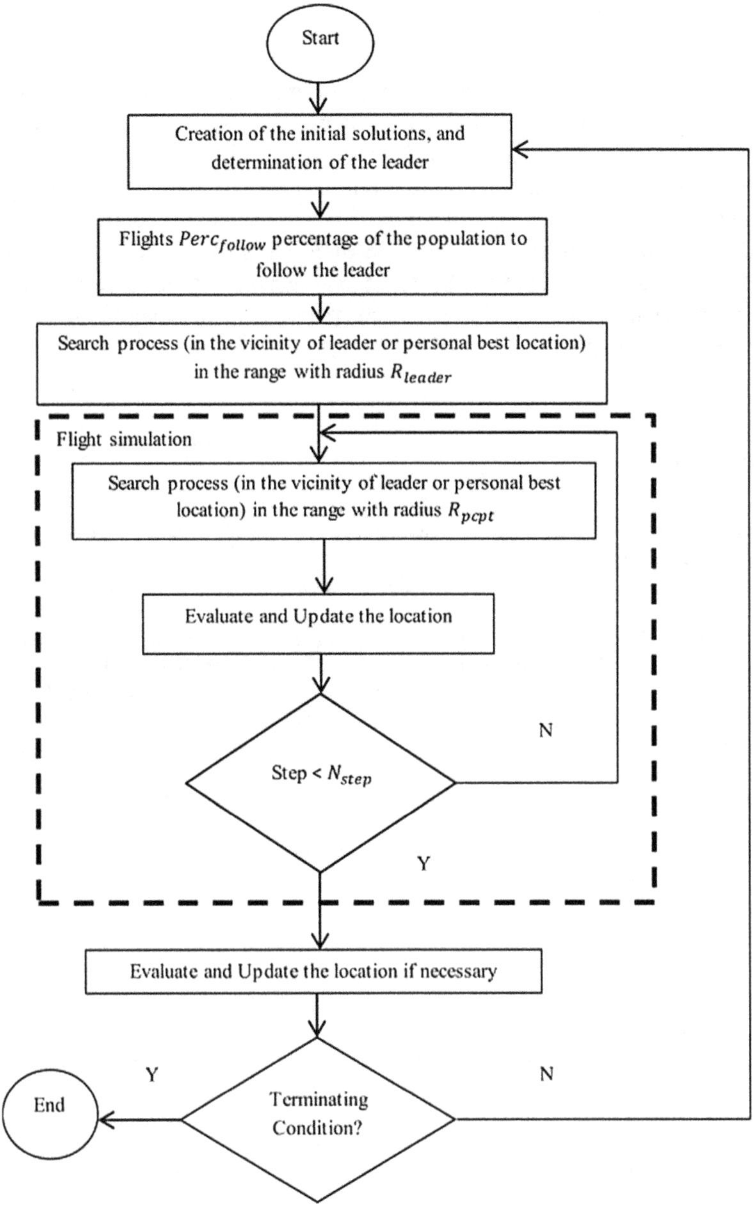

Fig. 5.14 Flowchart of RRO algorithm

5.4 Conclusion

A short review of Genetic algorithm, Genetic programming, evolutionary strategies, Grey wolf optimization, Bat optimization, Ant colony optimization, Artificial Bee Colony optimization, Particle swarm optimization, Firefly optimization, Cuckoo Search Algorithm, Elephant Herding optimization, Bumble bees mating, Lion optimization, Water wave optimization, Chemical reaction optimization, Plant optimization, The raven roosting nature inspired algorithm have been described in this chapter. Also, the various image processing applications of each algorithm, different image processing technique used, parameters evaluated and the system used have been compared and studied.

References

1. Gonzalez, R.C., Woods, R.E.: Digital Image Processing. Addison-Wesley Publishing Company, Inc, New York (2007)
2. Maihami, V., Yaghmaee, F.: A Genetic-Based Prototyping for Automatic Image Annotation, pp. 1–13. Elsevier, New York (2017)
3. Pujari, S.K., Bhatta Charjee, C., Bhoi, S.: A Hybridized Model for Image Encryption Through Genetic Algorithm and DNA Sequence, pp. 165–171. Elsevier, New York (2017)
4. Abbas, S., Hussain, M.Z., Irshad, M.: Image Interpolation by Rational Ball Cubic B-spline Representation and Genetic Algorithm, pp. 3–7. Elsevier, New York (2017)
5. Tarigan, J., Nadia, Diedan, R., Suryana, Y.: Plate Recognition Using Backpropagation Neural Network and Genetic Algorithm, pp. 365–372. Elsevier, New York (2017)
6. Miri, A., Faez, K.: Adaptive Image Steganography Based on Transform Domain via Genetic Algorithm. Optics 1–21 (2017)
7. Sukhija, P., Behal, S., Singh, P.: Face Recognition System Using Genetic Algorithm, pp. 410–417. Elsevier, New York (2016)
8. Hung, C.-L., Wu, Y.-H.: Parallel Genetic-Based Algorithm on Multiple Embedded Graphic Processing Units for Brain Magnetic Resonance Imaging Segmentation, pp. 1–11, Elsevier, New York (2016)
9. Nagarajan, G., Minu, R.I., Muthukumar, B., Vedanarayan, V., Sundarsingh, S.D.: Hybrid Genetic Algorithm for Medical Image Feature Extraction and Selection, pp. 455–462. Elsevier, New York (2016)
10. Zafari, M., Ahmadi-Kandjani, S., Kheradmand, R.: Noise Reduction in Selective Computational Ghost Imaging Using Genetic Algorithm, pp. 182–187. Elsevier, New York (2016)
11. Sethi, P., Kapoor, V.: A Proposed Novel Architecture for Information Hiding in Image Steganography by Using Genetic Algorithm and Cryptography, pp. 61–66. Elsevier, New York (2016)
12. Liang, Y., Zhang, M., Browne, W.N.: Image Feature Selection Using Genetic Programming for Figure-Ground Segmentation. Eng. Appl. Artif. Intell. **62**, 96–108 (2017) (Elsevier)
13. Liang, Y., Zhang, M., Browne, W. N.: Genetic Programming for Evolving Figure Ground Segmentors from Multiple Features, pp. 1–33. Elsevier, New York (2016)
14. Iqbal, M., Xue, B., Al-Sahaf, H., Zhang, M.: Cross-domain reuse of extracted knowledge in genetic programming for image classification. IEEE Trans. Evol. Comput. **21**(4), 569–587 (2017)

15. Mahmooda, M.T., Majid, A., Han, J., Choi, Y.K.: Genetic programming based blind image deconvolution for surveillance systems. Eng. Appl. Artif. Intell. **26**, 1115–1123 (2013) (Elsevier)
16. Naidu, M.S.R., Rajesh Kumar, P., Chiranjeevi, K.: Shannon and Fuzzy Entropy Based Evolutionary Image Thresholding for Image Segmentation, pp. 1–13. Elsevier, New York (2017)
17. Sarkar, S., Das, S., Chaudhuri, S.S.: Multi-level thresholding with a decomposition-based multi-objective evolutionary algorithm for segmenting natural and medical images. Appl. Soft Comput. **50**, 142–157 (2016) (Elsevier)
18. Bhandari, A.K., Kumar, A., Singh, G.K.: Tsallis entropy based multilevel thresholding for colored satellite image segmentation using evolutionary algorithms. Expert Syst. Appl. **42**, 1–24 (2015) (Elsevier)
19. Bu, Y., Tang, G., Liu, H., Pan, L.: Matching suitable feature construction for SAR images based on evolutionary synthesis strategy. Chin. J. Aeronaut. **26**(6), 1488–1497 (2013)
20. Li, J., Su, L., Cheng, C.: Finding pre-images via evolution strategies. Appl. Soft Comput. **11**, 4183–4194 (2011) (Elsevier)
21. Ramakrishnan, T., Sankaragomathi, B.: A professional estimate on the computed tomography brain tumor images using SVM-SMO for classification and MRG-GWO for segmentation. Pattern Recogn. Lett. **163**, 1–12 (2017)
22. Khairuzzaman, A.K.M., Chaudhury, S.: Multilevel thresholding using grey wolf optimizer for image segmentation. Expert Syst. Appl. **86**, 1–34 (2017)
23. Jadhav, A.N., Gomathi, N.: WGC: Hybridization of Exponential Grey Wolf Optimizer with Whale Optimization for Data Clustering, pp. 1–16. Elsevier, New York (2017)
24. Wang, M., Chen, H., Li, H., Cai, Z., Zhao, X., Tong, C., Li, J., Xu, X.: Grey wolf optimization evolving kernel extreme learning machine: application to bankruptcy prediction. Eng. Appl. Artif. Intell. **63**, 54–68 (2017) (Elsevier)
25. Daniel, E., Anitha, J., Gnanaraj, J.: Optimum Laplacian wavelet mask based medical image using hybrid cuckoo search—grey wolf optimization algorithm. Knowl. Syst. **131**, 58–59 (2017) (Elsevier)
26. Daniel, E., Anitha, J., Kamaleshwaran, K.K., Rani, I.: Optimum spectrum mask based medical image fusion using gray wolf optimization. Biomed. Signal Process. Control **34**, 36–43 (2017) (Elsevier)
27. Zhang, S., Zhou, Y.: Template matching using grey wolf optimizer with lateral inhibition. Optik **130**, 1229–1243 (2016) (Elsevier)
28. Li, L., Sun, L., Kang, W., Guo, J., Han, C., Li, S.: Fuzzy multilevel image thresholding based on modified discrete grey wolf optimizer and local information aggregation. IEEE Access. **4**, 6438–6450 (2016)
29. Karri, C., Jena, U.: Fast vector quantization using a Bat algorithm for image compression. Eng. Sci. Technol. Int. J. **19**, 769–781 (2017) (Elsevier)
30. Senthilnath, J., Kulkarni, S., Benediktsson, J.A., Yang, X.S.: A novel approach for multispectral satellite image classification based on the bat algorithm. IEEE Geosci. Remote Sens. Lett. **13**(4), 599–603 (2016)
31. Yang, X.S.: A New Metaheuristic Bat-Inspired Algorithm, pp. 1–10 (2010)
32. JayaBrindha, G., Gopi Subbu, E.S.: Ant colony technique for optimizing the order of cascaded SVM classifier for sunflower seed classification. IEEE Trans. Emerg. Top. Comput. Intell. **2**(1), 78–88 (2018)
33. Miria, A., Sharifian, S., Rashidi, S., Ghods, M.: Medical image denoising based on 2D discrete cosine transform via ant colony optimization. Optik (Optics) **156**, 938–948 (2018)
34. Kuo, H.-F., Frederick, C.Y.H.: Ant colony optimization-based freeform sources for enhancing nanolithographic imaging performance. IEEE Trans. Nanotechnol. **15**(4), 599–606 (2016)
35. Yin, D., Du, S., Wang, S., Guo, Z.: A direction-guided ant colony optimization method for extraction of urban road information from very-high-resolution images. IEEE J. Sel. Top. Appl. Earth Obs. Remote Sens. **8**(10), 4785–4794 (2015)

36. Zhang, B., Gao, J., Gao, L., Sun, S.: Improvements in the ant colony optimization algorithm for endmember extraction from hyperspectral images. IEEE J. Sel. Top. Appl. Earth Obs. Remote Sens. **6**(2), 522–530 (2013)
37. Gao, H., Fu, Z., Pun, C.-M., Hu, H., Lan, R.: A multi-level thresholding image segmentation based on an improved artificial bee colony algorithm. Comput. Electr. Eng. 1–8 (2017) (Elsevier)
38. Chen, J., Yu, W., Tian, J., Chenb, L., Zhou, Z.: Image contrast enhancement using an artificial bee colony algorithm. Swarm Evol. Comput. 1–8 (2017) (Elsevier)
39. Abdelhakim, A.M., Saleh, H.I. Nassar, A.M.: A quality guaranteed robust image watermarking optimization with artificial bee colony. Expert Syst. Appl. 1–10 (2016)
40. Sajedi, H., Ghareh Mohammadi, F.: Region based image steganalysis using artificial bee colony. J. Vis. Commun. Image Represent. 1–25 (2016)
41. Mostafa, A., Fouad, A., Elfattah, M.A., Hassanien, A.E., Hefny, H., Zhu, S.Y., Schaefer, G.: CT liver segmentation using artificial bee colony. Proc. Comput. Sci. **60**, 1622–1630 (2015) (Elsevier)
42. Goel, S., Gaur, M., Jain, E.: Nature inspired algorithm in remote sensing image classification. Proc. Comput. Sci. **57**, 377–384 (2015) (Elsevier)
43. Wu, Y., Miao, Q., Ma, W., Gong, M., Wang, S.: PSOSAC: particle swarm optimization sample consensus algorithm for remote sensing image registration. IEEE Geosci. Remote Sens. Lett. **15**(2), 242–246 (2018)
44. Mozaffari, M.H., Lee, W.S.: Convergent heterogeneous particle swarm optimization for multilevel image thresholding segmentation. IET Image Processing. IET J. 605–619 (2017)
45. Sabeti, M., Boostani, R., Davoodi, B.: Improved particle swarm optimization to estimate bone age. IET Image Processing. IET J. 179–187 (2017)
46. Zhang, C., Xie, Y., Liu, D., Wang, L.: Fast threshold image segmentation based on 2D fuzzy fisher and random local optimized QPSO. IEEE Trans. Image Process. **26**(3), 1355–1362 (2017)
47. Salucci, M., Poli, L., Anselmi, N., Massa, A.: Multifrequency particle swarm optimization for enhanced multiresolution GPR microwave imaging. IEEE Trans. Geosci. Remote Sens. **55**(3), 1305–1317 (2017)
48. Liu, L., Zhou, F., Tao, M., Sun, P., Zhang, Z.: Adaptive translational motion compensation method for ISA imaging under low SNR based on particle swarm optimization. IEEE J. Sel. Top. Appl. Earth Obs. Remote Sens. **8**(11), 5146–5157 (2015)
49. Xue, Z., Du, P., Su, H.: Harmonic analysis for hyperspectral image classification integrated with PSO optimized SVM. IEEE J. Sel. Top. Appl. Earth Obs. Remote Sens. **7**(6), 2131–2146 (2014)
50. Kora, P., Annavarapu, A., Yadlapalli, P., Sri Rama Krishna, K., Somalaraju, V.: ECG based atrial fibrillation detection using sequency ordered complex Hadamard transform and hybrid firefly algorithm. Eng. Sci. Technol. Int. J. **20**, 1084–1091 (2017) (Elsevier)
51. Pare, S., Bhandari, A.K., Singh, G.K.: A new technique for multilevel color image thresholding based on modified fuzzy entropy and Levy flight firefly algorithm. Comput. Electr. Eng. 1–20 (2017) (Elsevier)
52. Zhang, L., Mistry, K., Neob, S.C., Liun, C.P.: Intelligent facial emotion recognition using moth-firefly optimization. Knowl. Syst. **111**, 248–267 (2016) (Elsevier)
53. Rajinikanth, V., Couceiro, M.S.: RGB histogram based color image segmentation using firefly algorithm. Proc. Comput. Sci. **46**, 1449–1457 (2015) (Elsevier)
54. Nayak, J., Naik, B., Behera, H.S.: A novel nature inspired firefly algorithm with higher order neural network: performance analysis. Eng. Sci. Technol. Int. J. **19**, 197–211 (2015) (Elsevier)
55. Suresh, S., Lal, S., Reddy, C.S., Kiran, M.S.: A novel adaptive cuckoo search algorithm for contrast enhancement of satellite images. IEEE J. Sel. Top. Appl. Earth Obs. Remote Sens. **10** (8), 3665–3676 (2017)

56. Pare, S., Bhandari, A.K., Kumar, A., Singh, G.K.: An optimal color image multilevel thresholding technique using grey-level co-occurrence matrix. Embed. Syst. Appl. 1–46 (2017)
57. Chiranjeevi, K., Jena, U.R.: Image compression based on vector quantization using cuckoo search optimization technique. Ain Shams Eng. J. 1–15 (2016)
58. Mohammed Ismail, B., Eswara Reddy, B., Bhaskara Reddy, T.: Cuckoo inspired fast search algorithm for fractal image encoding. J. King Saud Univ. Comput. Inf. Sci. 1–8 (2016)
59. Tuba, E., Ribic, I., Capor-Hrosik, R., Tuba, M.: Support vector machine optimized by elephant herding algorithm for erythemato-squamous diseases detection. Proc. Comput. Sci. **122**, 916–923 (2017) (Elsevier)
60. Tuba, E., Alihodzic, A., Tuba, M.: Multilevel image thresholding using elephant herding optimization algorithm. 240–243 (2017)
61. Abdelhakim, A.M., Saleh, H.I., Nassar, A.M.: Quality metric-based fitness function for robust watermarking optimisation with Bees algorithm. IET image processing. IET J. 247–252 (2015)
62. Jiang, Y., Huang, C.-L., Deng, S., Yang, J., Wang, Y., He, H.: Multi-threshold image segmentation using histogram thresholding-bayesian honey bee mating algorithm. IEEE Congr. Evol. Comput. (CEC) 2729–2736 (2015)
63. Kanimozhi Suguna, S., Ranganathan, R.: A new evolutionary-based optimization algorithm for mammogram image processing. Int. J. Pure Appl. Math. **117**(Special Issue), 241–247 (2017)
64. Yazdani, M., Jolai, F.: Lion optimization algorithm (LOA): a nature-inspired metaheuristic algorithm. J. Comput. Design Eng. **3**, 24–36 (2017) (Elsevier)
65. Xu, W., Ye, Z., Hou, Y.: A fast image match method based on water wave optimization and gray relational analysis. IEEE International Conference on Intelligent Data Acquisition and Advanced Computing Systems. 771–776 (2017)
66. Wu, X., Zhou, Y., Lu, Y.: Elite opposition-based water wave optimization algorithm for global optimization. Hindawi Mathematical Problems in Engineering. Research article, 1–26 (2017)
67. Asanambigai, V., Sasikala, J.: Adaptive chemical reaction based spatial fuzzy clustering for level set segmentation of medical images. Ain Shams Eng. J. 1–12 (2016)
68. Duan, H.: Elitist chemical reaction optimization for contour-based target recognition in aerial images. IEEE Trans. Geosci. Remote Sens. **53**(5), 2845–2859 (2015)
69. Jamil, N., Hussin, N.A.C., Nordin, S., Awang, K.: Automatic plant identification: is shape the key feature? Proc. Comput. Sci. **76**, 436–442 (2015) (Elsevier)
70. Rani, E., Kaur, H.: Efficient load balancing task scheduling in cloud computing using raven roosting optimization algorithm. Int. J. Adv. Res. Comput. Sci. **8**(5), 2419–2424 (2017)

Chapter 6
Application of Ant Colony Optimization for Enhancement of Visual Cryptography Images

G. Germine Mary and M. Mary Shanthi Rani

Abstract Visual Cryptography is a method that shows the idea of maintaining secrecy by concealing secrets in images. An image may be separated into k shares that can be stacked together to recover the first image approximately. This secret sharing scheme enables distribution of a secret amongst *n* persons, such that only predefined approved persons will be able to recreate the secret. In Visual Cryptography, the secret can be remade visually by superimposing shares. One of the fundamental disadvantage of conventional Visual Cryptography is the pixel expansion, where every pixel is substituted by *m* sub-pixels in each share that results in the loss of resolution. Thus enhancing the visual nature of Visual Cryptography is a generally researched area. The proposed technique improves the visual quality and resolution of Visual Cryptography utilizing the Ant Colony Optimization Algorithm and it takes into account a wide range of images, color and also gray. The proposed technique builds the quality and sharpness of the image. It is assessed subjectively regarding human visual perception and quantitatively utilizing standard measurements.

Keywords Ant colony optimization · Visual cryptography · Image enhancement Image security · Secret sharing · Human visual perception · Pheromone trail

---

G. G. Mary (✉)
Fatima College, Madurai, Tamil Nadu, India
e-mail: germinemary@yahoo.co.in

M. M. S. Rani
Gandhigram Rural Institute—Deemed University, Gandhigram, Dindigul, Tamil Nadu, India
e-mail: drmaryshanthi@gmail.com

J. Hemanth and V. E. Balas (eds.), *Nature Inspired Optimization Techniques for Image Processing Applications*, Intelligent Systems Reference Library 150,
https://doi.org/10.1007/978-3-319-96002-9_6

6.1 Introduction

Cryptography is a science with complex mathematics and logic to design strong encryption methods to safeguard the security of secret information such as confidentiality, integrity, data security, entity authentication etc. Visual Cryptography (VC) is a secret sharing technique proposed by Naor and Shamir in 1994 [1], where information is encrypted in such a way that their decryption can be performed by the human visual system.

Visual cryptography utilizes the idea of hiding secrets in images. An image can be divided into *n* shares that can be stacked together to recover the original image approximately. A secret sharing scheme enables distribution of a secret amongst *n* persons, such that only predefined authorized persons will be able to recreate the secret [2, 3]. In VC, the secret can be reconstructed visually by superimposing shares. No computation is necessary in order to decrypt the information [4].

In VC, the decrypted image encounters the main setback of degraded image quality owing to pixel addition. This drawback of VC can be resolved by enhancing the shares using Ant Colony Optimization (ACO). This improves the quality and sharpness of the image. The functioning of the proposed method is compared with other available customary techniques like histogram equalization. The proposed method is assessed qualitatively in terms of human visual perception and quantitatively using standard metrics like Discrete Entropy, Contrast Improvement Index, Histogram, Peak Signal-to-Noise Ratio, Number of Edges detected, Universal Image Quality Index, Absolute Mean Brightness Error, and Image Enhancement Factor, to support the dominance of the ACO algorithm.

6.2 Review of Literature

Image enhancement is a procedure that emphasizes or sharpens image features to make the image more intelligible for analysis. The enhancement increases the dynamic range of the chosen features so that they can be detected easily. In vision perception and in many applications Image enhancement plays a significant role. Many novel enhancement techniques are available for enhancing the digital images. The contrast enhancement methods are based on either spatial domain or Transform domain techniques [5]. Contrast of the images has been enhanced effectively using Nature-Inspired Algorithms such as Genetic Algorithm (GA), Ant Colony Optimization (ACO), and Particle Swarm Optimization (PSO).

Pourya and Shayesteh [6] have suggested a hybrid method which is a combination of GA, ACO, and Simulated Annealing. ACO is used to create a function to map the input intensities to the output. A local search method is used to modify this function by simulated annealing. The evolutionary process of ants' characteristics is managed by GA. The fitness function operates repeatedly and tends to offer a balance between contrast and naturalness of images.

A novel approach for the enhancement of high dynamic range color images using Artificial Ant Colony System (ACS) and fuzzy logic techniques was proposed by Om Prakash et al. [7]. The lower and the upper thresholds are defined to provide an estimate of the over-exposed, under-exposed and mixed-exposed regions in the image. The RGB color space is converted into HSV color space in order to preserve the chromatic information. An objective function consists of Shannon entropy function as the information factor and visual influence factor is optimized using Artificial ACS to ascertain the parameters needed for the enhancement of a particular image.

The clustered objects present in the image are a challenge for image analysis. In order to solve this problem, Katteda et al. [8] have proposed an algorithm called Ant Colony Optimization and Fuzzy logic based technique. By using fuzzy logic, rules are formed and by using ACO each pixel intensity value is collected separately. Pixels are grouped together based on the fuzzy logic rules in order to retrieve the structure, and it is proved to be very complex to form fuzzy rules.

A new approach proposed by Gupta and Gupta [9] suggests automatic image enhancement using real-coded ACO is implemented by specifying a suitable fitness function to increase the number and intensity of the edgel pixels. At the same time, it also aims to improve the entropic measure of the image. To maximize the total number of pixels in the edges thus being able to visualize more details in the images is the prime objective of this approach.

The principal aim of the image enhancement technique is to modify the attributes in an image to make it more suitable for the given task and specific purpose. In the method proposed by Kanchan and Gurjot [10], Weirner filter with ACO is used to enhance the image. In medical images, the presence of multiple objects overlapping in an image and the closeness of adjacent pixel values make the diagnostic process a difficult task. In order to overcome these limitations, a new algorithm is proposed by Kumar et al. [11], which utilize ACO ability to find optimistic adjustment factor for better fuzzy based enhancement.

Biao [12] has proposed a novel image enhancement algorithm based on genetic-ant colony mixed methods, by integrating the advantages of GA and ACO algorithm in finding the optimal solution. This algorithm has improved efficiency and time complexity and it is capable of automatically locating the optimal value of the non-linear transformation function.

In all the above literature, the quality of the degraded image is improved to a great extent. The existing enhancement techniques can be used to enhance the decrypted SI. However, this will introduce an additional overhead to the receiver. Hence this study focuses on improving the features of decrypted VC image by enhancing VC shares before passing it to the receiver using ACO, which are made up of only two intensity values for each of RGB color channel. The appropriate fitness function is applied to trace the pheromones deposited by the ants in the entire iteration to calculate the intensity values of the pixels in VC shares. The proposed technique is designed and implemented in MATLAB 7.10.0.

6.3 Image Enhancement of VC Shares Using ACO

6.3.1 Image Enhancement in VC Shares

The concept in enhancement techniques is to bring out features that are obscured and to emphasize certain important aspects in an image. Histogram equalization (HE) is one of the best known image enhancement techniques and is popular for its simplicity and effectiveness [13].

In VC share enhancement, the secrecy of the shares must be maintained by limiting its RGB intensities to merely two values. Each of the VC shares has pixels with intensity values of either maximum (255) or minimum (0) for all the RGB color channels. Therefore, the existing enhancement techniques are not suitable for enhancing VC shares. We know only 8 color shades are possible in VC shares as shown in Fig. 6.1a, which results in low contrast and decreases the quality of the image. Figure 6.1b shows the actual shares created based on Naor and Shamir's algorithm [1], where pixels of the decrypted image attain any one of the eight colors as shown in Fig. 6.1a.

After performing enhancement, assume, we get the minimum and maximum pixel value for the red channel in the two shares as follows.

Red share1 takes the value 60 or 180
Red Share2 takes the value 15 or 240.

When we perform the bitor function to overlap the two shares, there are 4 probable combinations.

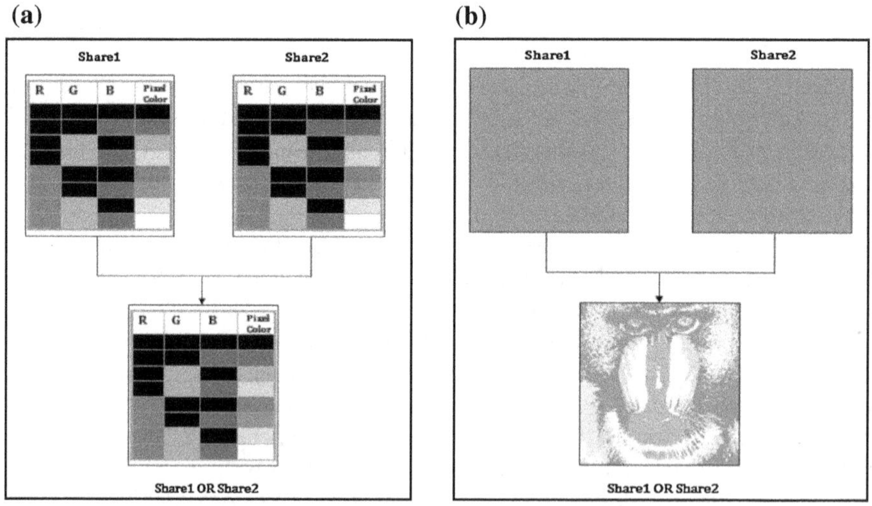

Fig. 6.1 **a** Color map of VC share, **b** VC shares and the decrypted SI

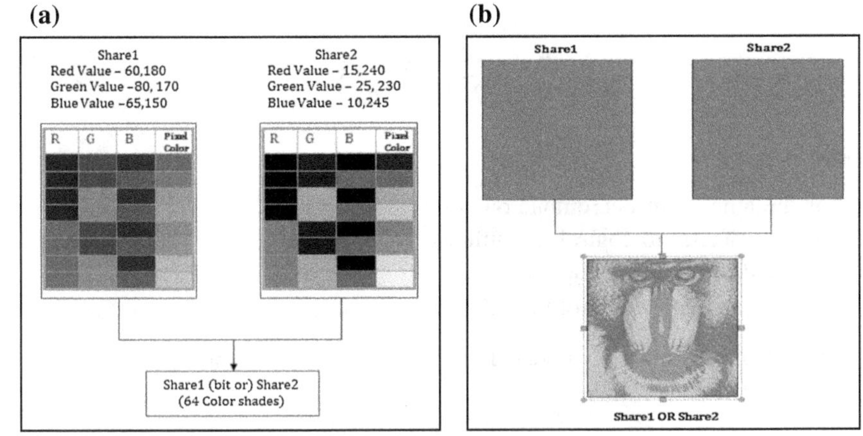

Fig. 6.2 **a** ACO enhanced color map. **b** ACO enhanced VC shares and decrypted SI

$$\begin{array}{lllll}\text{Namely} & 60(\text{bitor})\ 15 \rightarrow 63 & \text{and} & 60\ (\text{bitor})\ 240 \rightarrow 252 \\ & 180(\text{bitor})\ 15 \rightarrow 191 & \text{and} & 180\ (\text{bitor})\ 240 \rightarrow 244\end{array}$$

This will result in 4 different shades for the red channel. Likewise, we get four different shades for green and blue channels respectively. Now this RGB channels will combine randomly to give 64 color shades as shown in Fig. 6.2a. The significance of this enhancement is that the two shares created are made up of just two intensities for each RGB channel as in Original VC shares as shown in Fig. 6.2b, thus preserving the secrecy of the shares.

6.3.2 Basics of ACO

ACO developed by Gambardella Dorigo in 1997 is a Probabilistic Metaheuristic optimization technique, searching for an optimal path in the graph based on the behavior of ants seeking a path between their colony and source of food. Ants which are blind navigate from nest to food source. The shortest path is discovered via pheromone trails which each ant deposits as it moves at random. More pheromone on path increases the probability of path being followed by more ants. The virtual trail is accumulated on path segments. Ant selects next node in the path at random based on the amount of "trail" present on possible paths from starting node. On reaching the next node, selects next path and continues until it reaches end node. The finished tour is considered as a solution and is analyzed for optimality [14].

An ant will move from node i to node j with probability

$$p_{i,j} = \frac{\left(\tau_{i,j}^{\alpha}\right)\left(\eta_{i,j}^{\beta}\right)}{\sum\left(\tau_{i,j}^{\alpha}\right)\left(\eta_{i,j}^{\beta}\right)} \quad (1)$$

where

$\tau_{i,j}$ is the amount of pheromone on edge *i*, *j*
α is a parameter to control the influence of $\tau_{i,j}$
$\eta_{i,j}$ is the desirability of edge *i*, *j*
β is a parameter to control the influence of $\eta_{i,j}$.

Amount of pheromone is updated according to the equation

$$\tau_{i,j} = (1 - \rho)\tau_{i,j} + \Delta\tau_{i,j} \quad (2)$$

where

$\tau_{i,j}$ is the amount of pheromone on a given edge i, j
ρ is the rate of pheromone evaporation.

The density of pheromone deposited on the edge (i, j) by *m* ants at that instant is

$$\Delta\tau_{ij} = \sum_{k=1}^{m} \Delta\tau_{ij}^{k} \quad (3)$$

$$\Delta\tau_{ij} = Q/L_k$$

if *k*th ant uses edge (*i*, *j*) in its tour, else 0. Q is a constant and L_k is the length of *k*'s tour. $\Delta\tau_{ij}^{k}$ is Pheromone density for *k*'s tour [15].

This procedure is followed to enhance the VC shares using ACO.

6.3.3 VC Share Enhancement Using ACO

Various parameters used in this algorithm are

τ_{init}, the initial value of the pheromone is the inverse of the mean square error of the image.

$\alpha = 1$, the weighing factor of the pheromone Information.
$\beta = 0.1$, the weighing factor of the heuristic information.
$\rho = 0.1$, the evaporation rate.

$\eta_{i,j}$, heuristic information of the pixel is initialized with a random value.

Number of ants considered = 25.
Number of Iteration = 75.

The given SI is converted into VC shares using the procedure described in [1–4] and are used as input for the ACO algorithm

3.1. VC Share Enhancement using ACO - Algorithmic Description

Input : Original Image and Six VC shares created (2 shares for each color RGB)

Output : Enhanced VC Share1 and VC Share2

```
BEGIN
  1. Read the RGB VC shares and the original image
  2. Consider Red color share and calculate its PSNR, MSE value
  3. Initialization
         Initialize no. of ants, no. of iteration, initialize all the parameters
         Initialize the position of all ants with pixel intensity values
  4. Loop until maximum iteration
  5. For each ant k (currently in state t) do
       repeat
         compute the probability of the state to move into using equation (1)
         calculate the new position (value) of each ant to move into
           randomly from the current position
         calculate new MSE value based on the new position of the
           ants and save it in the array. (1/MSE is considered as
              pheromone deposit)
       until ant k has completed its solution.
     end for
  6. Trail update
         For each ant do
              compute Δτij using equation (2) and (3)
              update the trailing matrix.
         end for
  7. Check terminating condition (No. of Iteration) If not end of test go to
       step 4
  8. Array element with maximum pheromone is considered as the optimal
       solution
  9. Go to step 2 and repeat for green and blue shares
  10. Combine the first set of RGB shares to create VC share1 and the second
       set to create VC share2 and send it to the receiver on two different
       channels over the network to get a decrypted and enhanced VC Image
END
```

6.4 Results and Discussion

The Nature-Inspired Ant Colony Optimization algorithm given here is tried for standard images of size 512 × 512 employing MATLAB code. Enhancement algorithm is used to improve the quality of the decoded VC image. The results of the proposed method are compared with different existing enhancement algorithms. The effectiveness of every one of these algorithms are evaluated subjectively as far as human visual recognition and quantitatively utilizing standard measurements like Discrete Entropy, Contrast Improvement Index. Histogram, Peak Signal-to-Noise Ratio, Number of Edges identified, Universal Image Quality Index, Absolute Mean Brightness Error, and Image Enhancement Factor, to help to understand the dominance of the ACO algorithm.

The imperative uniqueness of VC is that the image incorporates just two color values (minimum and maximum) for each color channel. The fundamental target of this technique is to hold the essentialness of VC and in the meantime to enhance the nature of the decrypted image by modifying the minimum and maximum color value. ACO is utilized to locate the ideal minimum and maximum color value for each color channel that will best speak to the original image.

In ACO, the pheromone saved during the iteration fix the position of the ant. In this calculation, the fitness function to decide pheromone deposit relies upon MSE value of the image. Based on this the intensity value of the pixels are estimated. The pheromone trail of ants in search space for different images for red color space is shown in the Fig. 6.3a–f. The PSNR increases in all cases except for Fig. 6.3f which is a binary text image. When we try to enhance a binary image which is made up of zeros and ones, it gives unconventional result.

The benefit of VC is exploited in this strategy to conceal the secret message as an image by creating VC shares and sent over the network. The secret message can be recovered by simply superimposing or performing OR task on the VC Shares and HVS unveils the secret. The decrypted secret image can be improved utilizing available MATLAB code for standard Histogram Equalization (HE), Contrast-Limited Adaptive Histogram Equalization (CLAHE), Recursive Mean Separate Histogram Equalization (RMSHE), and Adjust image intensity (AII) methods. The results of improvement are not powerful, as VC utilizes only two color intensities. Subsequently, in this proposed strategy, image is enhanced utilizing Nature-Inspired Algorithm ACO with VC shares. This gives extra benefit of improving decrypted messages and the receiver need not perform any computation. The quantitative results of image improvement of VC shares utilizing the proposed ACO are given in Table 6.1 and are examined in detail underneath.

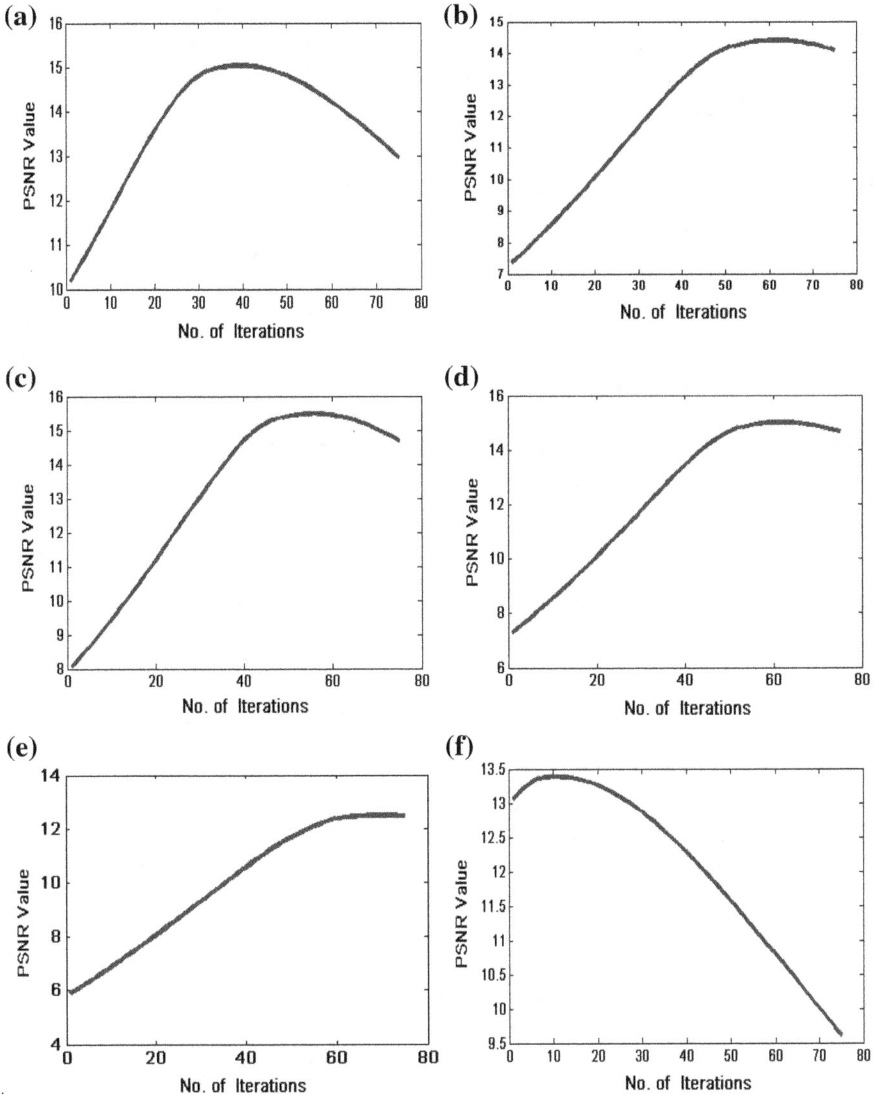

Fig. 6.3 Final pheromone trail of ants in search space for different images **a** Lenna **b** Baboon **c** Peppers **d** Barbara **e** Cameraman **f** Text image

6.4.1 Average Information Content (AIC)

Average Information Content (AIC) also known as the Entropy is simply the average amount of the information from the image. Higher value of Entropy signifies richness of the information in the output image. The AIC (Entropy) value of a color image (Baboon), Grayscale image (Cameraman) and a Binary image

(TextImage) for various enhancement techniques are compared in Fig. 6.4. From the chart below it is clear that the proposed method gives a higher value compared to other existing enhancement methods. Average Information Content values for different methods of test images are given in Table 6.1.

Table 6.1 Performance metrics compared for different enhanced VC images

Image	Enhanced image	PSNR	Entropy	CII	Q	No. of edges	AMBE	IEF
Baboon	HE	5.9502	0.7498	0.2721	0.4592	60,828	112.382	0.7643
	CLAHE	5.9507	0.7515	0.2722	0.4594	60,774	112.375	0.7644
	AII	6.9747	1.3690	0.8371	0.8947	46,386	70.0936	0.9666
	RMSHE	10.721	0.8460	0.2549	0.3504	60,928	59.562	2.3600
	ACO	**15.255**	**3.1005**	**0.8814**	**0.7439**	**59,738**	**5.9151**	**10.834**
Lenna	HE	6.5495	0.9581	0.3510	0.5356	54,542	105.886	0.7020
	CLAHE	6.5497	0.9588	0.3510	0.5357	54,499	105.883	0.7020
	AII	7.4979	1.3987	0.6348	0.7647	27,055	80.2050	0.9164
	RMSHE	10.348	0.8575	0.2980	0.3726	51,169	64.4322	2.0254
	ACO	**15.794**	**2.7818**	**0.8835**	**0.9179**	**50,030**	**6.3929**	**6.6962**
Peppers	HE	5.4698	1.1258	0.4727	0.6171	40,470	115.160	0.6780
	CLAHE	5.4702	1.1270	0.4728	0.6173	40,365	115.155	0.6781
	AII	6.4613	1.5615	0.7413	0.8537	20,379	89.745	0.8633
	RMSHE	11.118	0.8690	0.2784	0.3411	31,454	46.767	2.2897
	ACO	**14.709**	**2.9462**	**0.7527**	**0.8303**	**35,324**	**19.446**	**6.3520**
Barbara	HE	4.8546	0.6072	0.1711	0.3808	52,853	134.36	0.6680
	CLAHE	4.8547	0.6077	0.1712	0.3809	52,821	134.36	0.6681
	AII	6.3658	1.3285	0.8226	0.8688	36,370	78.765	0.9555
	RMSHE	11.962	0.8962	0.2588	0.3694	51,861	45.477	3.5685
	ACO	**15.456**	**2.8991**	**0.4470**	**0.7513**	**49,142**	**4.4403**	**10.642**
Castle	HE	5.1564	0.5535	0.1081	0.3559	32,890	126.829	0.6799
	CLAHE	5.1565	0.5537	0.1081	0.3559	32,896	126.829	0.6799
	AII	6.3409	1.1706	0.6121	0.7079	17,593	85.3532	0.8737
	RMSHE	11.115	0.8886	0.20	0.3120	24,484	50.5878	3.4575
	ACO	**15.147**	**2.6322**	**0.2981**	**0.3463**	**29,525**	**2.8068**	**8.1238**
Cameraman	HE	4.6173	0.7308	0.4574	0.9801	47,271	95.6083	1.0416
	CLAHE	4.6177	0.7641	0.5683	0.9998	47,268	85.5841	1.0057
	AII	6.4101	0.7308	0.5774	1	47,268	84.7711	1
	RMSHE	11.115	0.7308	0.2667	0.4592	48,214	46.992	2.9548
	ACO	**12.57**	**2.2050**	**0.3412**	**0.1361**	**5243**	**9.7490**	**4.1636**
TextImage	HE	10.179	0.3461	0	0.7402	0	29.670	0.5255
	CLAHE	10.179	0.3595	0	0.7402	0	29.670	0.5255
	AII	12.973	0.3461	1	1	36,437	13.151	1
	RMSHE	4.1826	0.3461	0.3098	0.5342	36,441	145.45	0.1321
	ACO	**12.246**	**1.5446**	**2.7490**	**0.9861**	**36,454**	**21.557**	**1.1046**

Bold letter indicates the result of the proposed method, which is compared with other existing methods

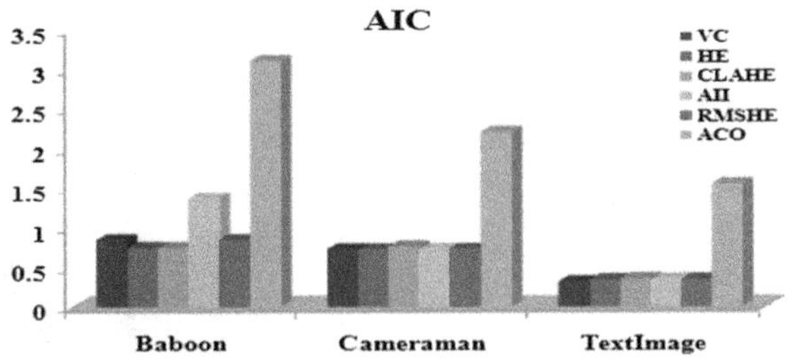

Fig. 6.4 Comparative analysis of AIC

6.4.2 *Contrast Improvement Index (CII)*

Contrast Improvement Index (CII) is a quantization measure of contrast enhancement. If the value of CII increases, then it shows improvement in contrast of an image. A comparative analysis of CII is shown in Fig. 6.5. The result confirms that the proposed ACO algorithm has an edge over other available methods. The findings in Table 6.1 confirm that the CII values of all color images are higher than other existing enhancement techniques.

6.4.3 *PSNR*

PSNR is an expression for the ratio between the maximum possible value of a signal and the value of distorting noise that affects the quality of its representation. The higher the PSNR value improved is the quality of the reconstructed image. PSNR value is higher for ACO enhanced images and is presented in Table 6.1 and Fig. 6.6. It is obvious that the obtained results are enhanced and lucid than the original one. The Baboon VC image with PSNR value 6.826 is enhanced to 15.255 using ACO which means it is enhanced to 223%.

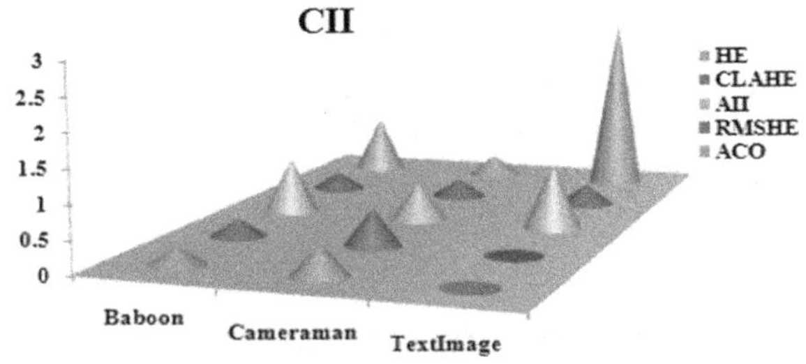

Fig. 6.5 Comparative analysis of CII

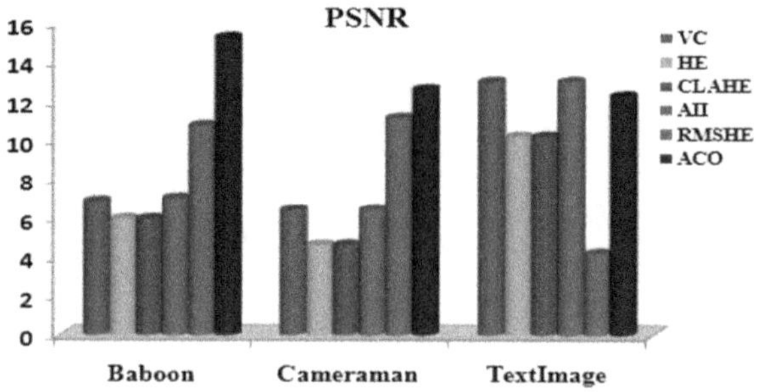

Fig. 6.6 PSNR values compared for different enhancement techniques

6.4.4 *Histogram*

Histogram refers to a histogram of the pixel intensity values. It confirms that the input image is a VC share with just 2 intensity pixels. The histogram comparison of the Red channel for various Lenna images is given in Fig. 6.7. The vertical axis represents the number of pixels in a particular color, whereas, the variations in color is represented by the horizontal axis. The right side of the horizontal axis represents the color pixels and the left side represents black pixels.

The color distribution of the VC image indicates that the pixel values are either 0 or 255. The proposed ACO optimization techniques determine the best lower and higher value for pixels for the red channel for share1 and share2 and on performing bitor operation resulting in four intensity values which are shown in the Fig. 6.7g. The results confirm that ACO enhanced VC shares of Lenna image individually consists of pixels with only two intensity values for each color channel, thus maintaining the property of VC.

6.4.5 *Universal Image Quality Index (Q)*

This quality index performs significantly better than the widely used distortion metric, mean squared error. The proposed ACO enhanced images has a higher value of Q compared to the original VC and enhanced VC images and is shown in Table 6.1 and Fig. 6.8. The value of Q is more for color images than gray-scale and binary image.

6.4.6 *Absolute Mean Brightness Error (AMBE)*

Absolute Mean Brightness Error (AMBE) is the absolute difference between original and enhanced image. Lesser the AMBE value, higher will be the clarity of the image. AMB error of ACO enhanced image is merely 1%, in contrast to VC

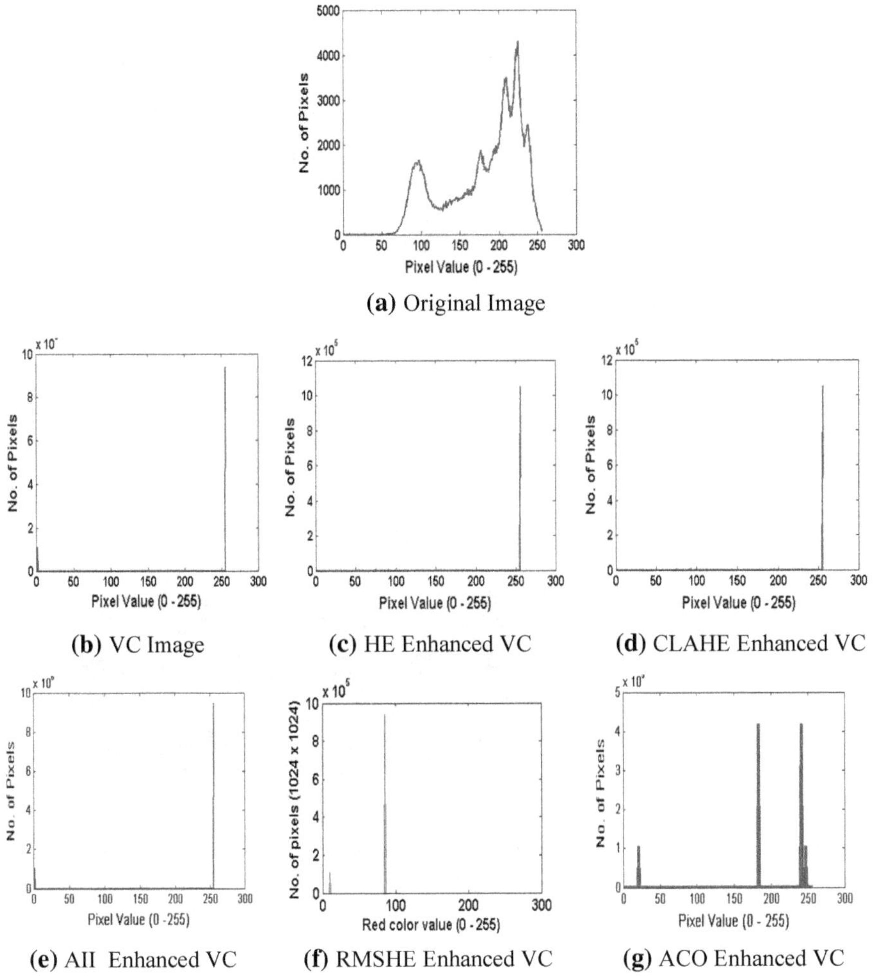

Fig. 6.7 Histogram comparison of original image, VC image, and various VC enhanced Lenna Images for red channel

image which is 14%. VC enhanced by HE and CLAHE has a maximum of 27%. The comparative analysis in the form of pie chart is given in Fig. 6.9.

6.4.7 Edge Detection

Edge detection is a fundamental tool in as image processing, computer vision, and machine vision Sobel operator is used for edge deduction using Matlab. The number of Edges detected for test images using the proposed enhancement method

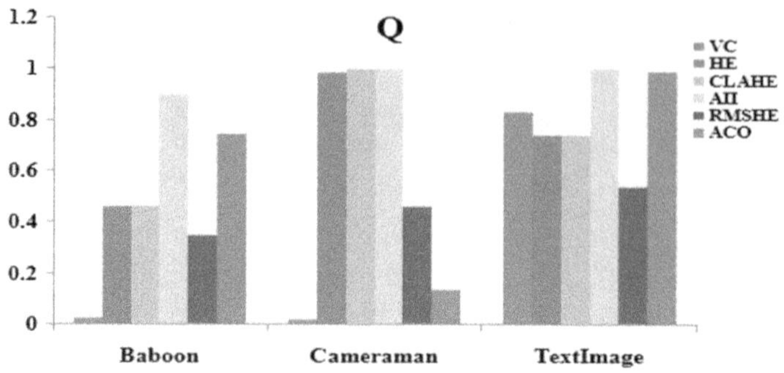

Fig. 6.8 Comparative analysis of *Q*

Fig. 6.9 Comparison of AMBE values

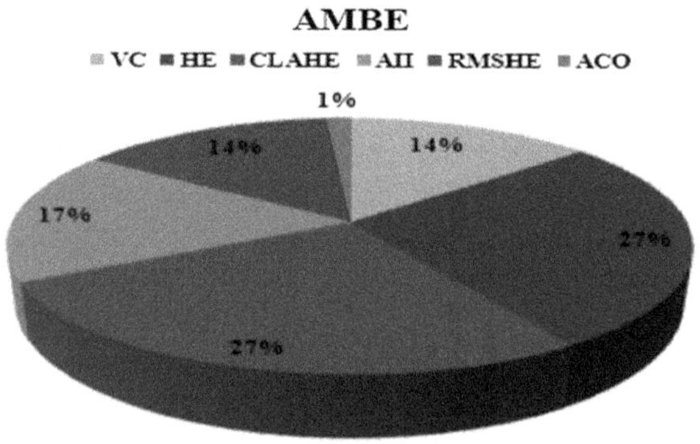

is presented in Table 6.1. A comparative analysis of the number of edges detected using existing GA, PSO and ACO is given in Table 6.2. The edges detected in the proposed method are with respect to VC Images in contrast to the original image used in other existing methods.

Edgel value refers to number of short linear edge segments in an image and can is calculated using Sobel Edge detection technique. The low edgel values are due to pixel expansion of VC images. In the case of Tire and Pout images, edgel value is increased from 0 to 152 and 36 respectively.

Table 6.2 gives the comparative analysis of the edges detected using various algorithms like Genetic Algorithm (GA), Particle Swam Optimization (PSO) and Ant Colony Optimization (ACO) for the same test images. The GA and PSO values are taken from reference [16] and ACO^{b} values are taken from reference [17]. The values obtained using the proposed algorithm is given in the last column of Table 6.2. The proposed method uses VC shares as input and enhances the image where as the original image is used as input in the references [16, 17].

Table 6.2 Comparative analysis of edges detected

Image	Original	VC	GA[a]	PSO[a]	ACO[b]	Proposed ACO (w. r. to VC)
Cameraman	2485	2292	2575	2674	5411	5243
Tire	1823	0	1917	2020	3962	152
Pout	1492	0	2040	2048	5841	36

[a]Braik et al. [16]
[b]Gupta and Gupta [17]

6.4.8 Image Enhancement Factor (IEF)

Image Enhancement Factor (IEF) is the ratio of mean square error before enhancement to the mean square error after enhancement. The comparative analysis of image enhancement factor is given in Table 6.1 and Fig. 6.10. The graph indicates maximum enhancement for color image, gray and binary image. There is no significant enhancement in binary image as it is made up of only 2 colors. The proposed algorithm is well suited for enhancing VC color images.

6.4.9 Qualitative Analysis of the Proposed Method

The performance analysis in terms of qualitative metrics namely Human Visual Perception of the tested images is depicted in Fig. 6.11. From the result, it is clear that image is enhanced better using ACO in comparison to other enhancement techniques.

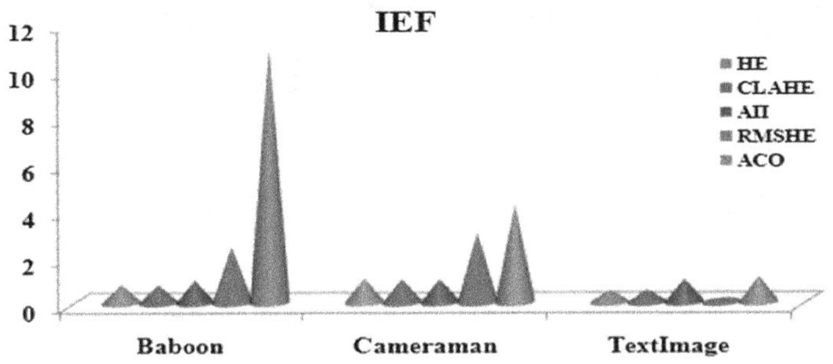

Fig. 6.10 Comparative analysis of IEF

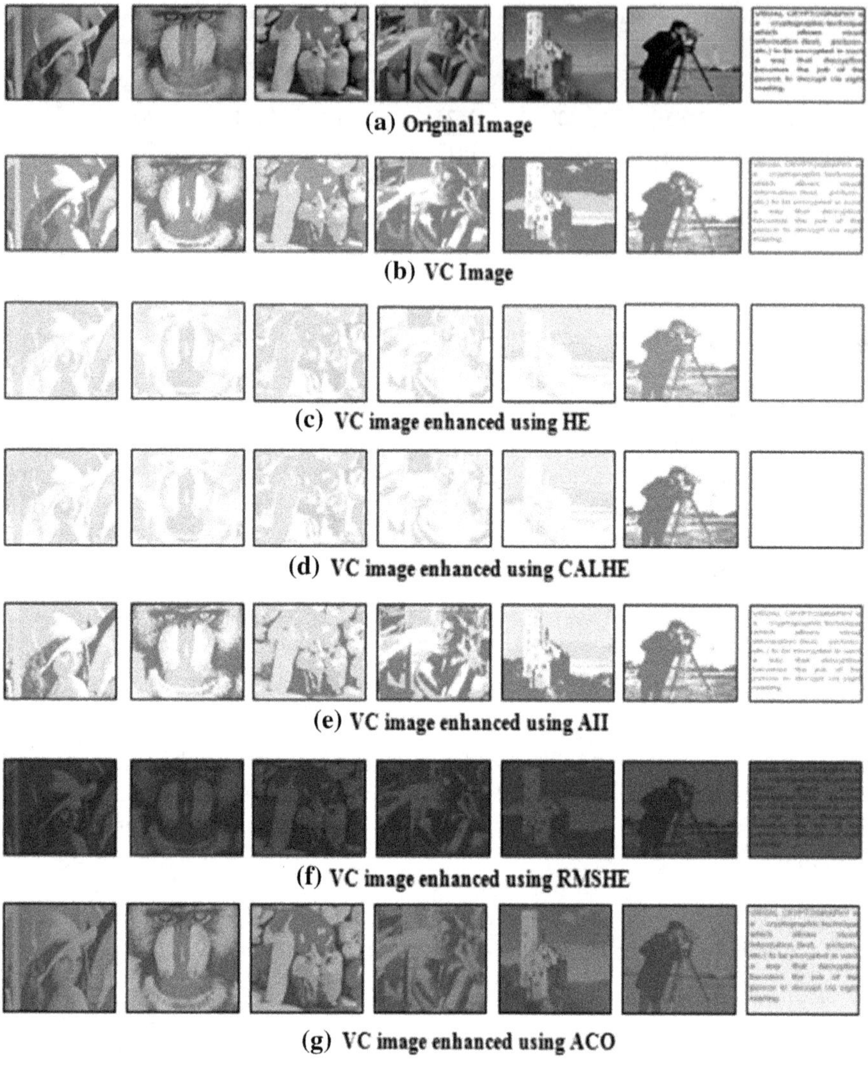

Fig. 6.11 Qualitative analysis of different enhancement methods

6.5 Conclusion

The proposed ACO based metaheuristic algorithm enhances the VC shares before it is sent over the network, and therefore, there is no computation involved at the receiving end. Simple fitness is applied to optimize problems of large dimensions, generating excellent outcome more rapidly. Tuning of input parameters and experimenting with different values for parameters α, β, ρ, and pheromone might create a different constructive outcome. ACO can be tried with different objective functions to calculate pheromone deposit to improve the enhancement quality of color in different channels. The proposed algorithm guarantees highly safe, secure, rapid and excellent quality transmission of the secret message in the form of the image, without the need for the mathematical operation, to reveal the secret. In future, a hybrid approach combining various Nature-Inspired Optimization Algorithms can be experimented to enhance VC shares.

References

1. Naor, A., Shamir, M., Santis, A. (eds): Visual cryptograph. In: Proceedings Advances in Cryptology__Eurocrypt '94, Lecture Notes in Computer Science, Vol. 950, pp. 1–12, Springer, Berlin (1995)
2. Ateniese, G., Blundo, C., De Santis, A., Stinson, D.R.: Extended capabilities for visual cryptography. Theoret. Comput. Sci. **250**, 143–161 (2001)
3. Bhattacharjee, T., Singh, J.P., Nag, A.: A novel (2,n) secret image sharing scheme. In: Procedia Technology, Second International Conference on Computer, Communication, Control and Information Technology, Vol. 4, pp. 619–623 (2012)
4. Verma, J., Khemchandani, V.: A visual cryptographic technique to secure image shares. Int. J. Eng. Res. Appl. **2**(1), 1121–1125 (2012)
5. Gonzalez, R.C., Woods, R.E.: Digital Image Processing, 3rd edn. Pearson Publications, London (2014)
6. Pourya, H., Shayesteh, M.G.: Efficient contrast enhancement of images using hybrid ant colony optimization, genetic algorithm, and simulated annealing. Digit. Signal Proc. **23**(3), 879–893 (2013)
7. Om Prakash, V., Kumar, P., Hanmandlu, M., Chhabra, S.: High dynamic range optimal fuzzy color image enhancement using artificial ant colony system. Appl. Soft Computing **12**(1), 394–404 (2012)
8. Katteda, S.R., Raju, C.N., Bai, M.L.: feature extraction for image classification and analysis with ant colony optimization using fuzzy logic approach. Signal Image Process. Int. J. (SIPIJ) **2**(4), 137–143 (2011)
9. Gupta, K., Gupta, A.: Image enhancement using ant colony optimization. IOSR J. VLSI Signal Process. (IOSR-JVSP) **1**(3), 38–45 (2012)
10. Rani, K., Kaur, G.: Image enhancement by adaptive filter with ant colony optimization. Int. J. Adv. Res. Ideas Innov. Technol. **2**(5), 1–6 (2016)
11. Kumar, D., Singh, S., Saini, V.: An efficient ant colony optimization based medical image enhancement. Int. J. Innov. Res. Sci. Eng. Technol. **5**(8), 15053–15063 (2016)
12. Pan, B.: Application of ant colony mixed algorithm in image enhancement. Comput. Model. New Technol. **18**(12B), 529–553 (2014)
13. Gonzalez, R.C., Woods, R.E., Eddins, S.L.: Digital Image Processing Using MATLAB, 2nd edn. McGraw Hill Education Publication, New York (2010)
14. Kaur, S., Agarwal, P., Rana, R.S.: Ant colony optimization: a technique used for image processing. Int. J. Comput. Sci. Technol. **2**(2), 173–175 (2011)
15. Pizzo, J.: Ant Colony Optimization, 1st edn. Clanrye International, New York (2015)
16. Braik, M., Sheta, A., Ayesh, A.: Image enhancement using particle swarm optimization. In: Proceedings of the World Congress on Engineering 2007, vol. I, pp. 1–6 (2007)
17. Gupta, K., Gupta, A.: Image enhancement using ant colony optimization. IOSR J. VLSI Signal Process. (IOSR-JVSP) **1**(3), 38–45 (2012)

Chapter 7
Plant Phenotyping Through Image Analysis Using Nature Inspired Optimization Techniques

S. Lakshmi and R. Sivakumar

Abstract It becomes mandatory to raise the crop and plant production for meeting the needs globally. Wheat is considered as a second food crop in India and it occupies nearly 30 million hectares. The estimated wheat production in 2030 is about 700 million tones. Since we are in technical era, we can make use of those techniques to utilize for extracting the valuable information easily and accurately. Plant phenotyping is the process of the assessment of a plant which is very important to estimate the growth of the plant. It is essential to measure the phenotype details of a crop like wheat for producing high throughputs and analyze the yields effectively. Background estimation and plant image segmentation are the first step in an automated phenotyping process. Swarm intelligence is a global optimization technique can be applied for segmenting the plant images to analyze the growth rate of the plant efficiently. Due to its simplicity, robustness and flexibility, Swarm intelligence acts as a backbone for extracting the phenotyping properties of the plants in designing computer vision systems which can help to raise the food production.

Keywords Phenotype · Nature inspired optimization techniques
Image segmentation · Computer vision and swarm intelligence

7.1 Introduction

In order to face the demanding needs of food and fuel for the rapidly growing population, it becomes necessary to breed high yielding crops. United Nations Food and Agriculture Organization stated that the cereal production should be doubled before 2050 to satisfy the demand of world requirements. Due to the shortage of

S. Lakshmi (✉)
Jeppiaar SRR Engineering College, Chennai, India
e-mail: lakshmi1503@gmail.com

R. Sivakumar
Divya Consultancy Services, Chennai, India

J. Hemanth and V. E. Balas (eds.), *Nature Inspired Optimization Techniques for Image Processing Applications*, Intelligent Systems Reference Library 150,
https://doi.org/10.1007/978-3-319-96002-9_7

fossil fuels, bio fuel and bio energy turns the attention of researchers drastically. To meet these global challenges, new techniques are required to do the quantitative analysis of plant phenotyping traits. The term phenotype is used to describe the characteristics of the plant such as height, biomass, growth, tolerance, resistance, architecture, yield, leaf shape and so on.

The environmental conditions such as frost and drought severely affects the phenotype of the plants. Severe winter affects the crop development and reduce the yield and drought permanently affects the soil. Moreover, salinity of the soil is another threat for crop production. Since the fertile land is used for expansion of cities which is significantly decrease the crop production. Agricultural scientists have been proposing novel techniques for extracting the phenotype traits of the plants for improving the nutrient values of the crops regularly.

Generally phenotyping techniques are described by using various image processing systems and methodologies to obtain high throughput phenotype. The following issues in the development of machine learning and computer vision algorithms with feature extraction for evaluating the plant phenotypes as a challenging task:

- The shape of the plant varies over time
- As the new leaves appear, make the changes in the shape of the plant
- The new leaves overlap and occlude other leaves
- Identification and detection of leaves and plants
- Counting of the leaves
- Estimation of the boundary of the leaves and plants
- Segmentation of leaves and plants for further processing
- Classification of the plants.

7.1.1 Wheat Production and Cultivation

Wheat cultivation takes place mainly in Utter Pradesh, Punjab, Haryana, Rajasthan, Madhya Pradesh, Gujarat and Bihar. There are six division in India to produce wheat commercially. They are

- NHZ (Northern Hlls Zone)
- NWPZ (North Western Plain Zone)
- NEPZ (North Eastern Plain Zone)
- CZ (Central Zone)
- PZ (Peninsular Zone) and
- SHZ (Southern Hills Zone).

Basically, all these zones are different soil type, different temperature, rainfall, biotic and a biotic stress. The Fig. 7.1 shows the several stages of wheat cultivation images.

Fig. 7.1 Several stages of wheat production

The nutritional value of wheat is mainly rich in fiber and protein. It forms the basic ingredients of bread, pasta and other bakery items. The detailed nutritional value is given in Table 7.1 per 100 g of wheat. It forms the basic ingredients of bread, pasta and other bakery items.

The quality of wheat is defined by grain color, grain size, weight, protein and flour color. The growth stage of wheat consists of the following stages and these stages are represented diagrammatically in Fig. 7.2.

(A) Tillering: one shoot, begins, formed, leaf strengthen, leaf erected
(B) Stem extension: first node of Stem visible, second node of stem visible and last leaf visible
(C) Heading—flowering
(D) Ripening.

Table 7.1 The nutritional value of Wheat

Nutrient	Availability
Carbohdrate	72 g
Protein	13.6 g
Saturated fatty acid	0.3 g
Omega-3-fatty acid	38 mg
Omega-6-fatty acid	738 mg
Vitamin A	91 IU
Riboflavin (B2)	0.25 mg
Sodium	5 mg
Zinc	2.9 mg
Copper	0.4 mg

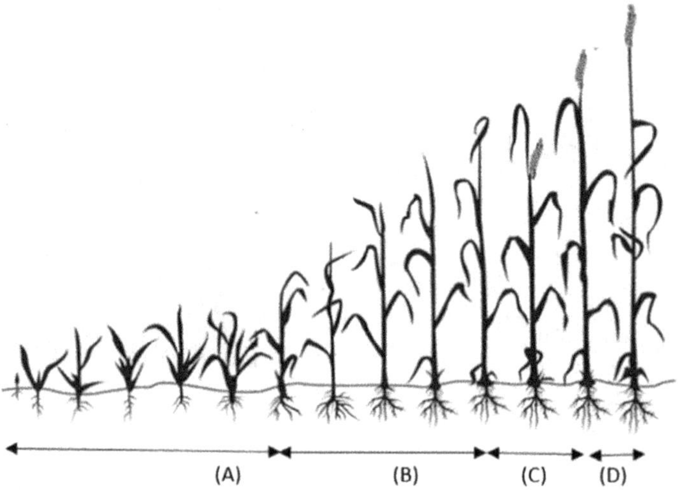

Fig. 7.2 Schematic diagram of wheat cultivation in various stages

Wheat Benefits

Wheat is a rich food in nutrients. The following health benefits are listed when we are consuming wheat regularly. They are

- Since it is low in calorie, it reduces the risk of heart diseases and regulating the blood glucose
- Used to reduce the cholesterol level and blood pressure
- Regular wheat consumption reduces the risk of cancer amongst people
- Daily consumption of wheat will provide a feeling of fullness and reducing the risk of overeating

- It decreases the obesity and high blood pressure
- Promotes healthy heart and it contains antioxidant and anti-aging properties that are good for skin and hair
- It improves the functioning of nervous system.

There are many plant image analysis tools have been introduced by [1–3] for plant phenotyping. Houle et al. [4] and Grobkinsky et al. [5] were discussed the gap between genotype to phenotype as the main problems in plant breeding. Image based phenotyping techniques with new technologies like robotic and conveyer belt systems in greenhouse environment automatically produces and transforms the images with reliable phenotypic measurements. Deep learning methods such as convolutional neural networks are used for extraction of phenotype details effectively. The Deep Plan Phenomics is an open source designed for the purpose of plant phenotyping by Ubbens et al. [6]. Convolutional Neural Networks were used for plant disease diagnosis and detection [7] and also used for classification of fruits [8].

7.1.2 *Wheat Phenotyping*

Generally, plant phenotype plays a vital role in plant breeding. Sustainable intensification is defined as the ability to increase the food production while still utilising existing farmland and minimising negative impact on the environment.

A lack of detailed phenotypic data impedes plant biologists from gaining a complete understanding of the interactions between the genotypes and the environment [9]. In [10], the authors manually trace the leaf of multiple tree and shrub species against graph paper and cut out a stencil to obtain a non-destructive estimate of leaf area. Estimating fruit and vegetable volume is important for measuring yield and studying the effects of diseases and physiological defects on the fruit. Manual approaches include the water displacement method [11, 12] and taking manual measurements of parameters such as diameter and height, to approximate fruit volume using standard formulae such as the volume of an ellipsoid or sphere [13, 5]. In our case, two subsets would exist, plant regions and non-plant regions such as background, soil, pots and stakes, etc. The process can be thought of as defining every pixel in the image as a member of one of these two classes. In plant phenotyping experiments, image segmentation is a crucial pre-processing step. After segmenting the image, the user has restricted the data to plant pixels only, thus making it much easier to extract statistical and structural information about the plants. An automated image analysis based plant phenotyping technique provides a powerful alternate to normal naked eye evaluation for reaching the high throughput.

7.1.2.1 Role of Imaging Techniques

Different types of imaging methods are used for plant phenotyping such as Spectroscopy, Fluorescence, thermal infrared, visible light and so on. Since the modern imaging techniques have better resolution and produce multi-dimensional data, the quantification of phenotyping properties could be done easily and effectively. By capturing the images of the plants alone won't produce useful information about plants. The goal of the imaging technique is to measure the phenotype characteristics of the plants such as absorption, reflection and transmittance in plant cells. When capturing the plant image for analysis, it is important to choose the type of imaging technique ie., whether 2D or 3D. Suppose we plan to go for 3D, we need to make it very clear that how to get the third dimension. The depth information is captured using various ways.

Visible Light

It is a conventional method of capturing images in which Silicon sensors are used. Since it is affordable in nature, it is called as digital cameras for plant phenotyping applications. The photons are absorbed by chlorophyll in blue and red spectral regions. Plant phenotypes are measured by varying the wavelengths such as leaf area, number of fruits, leaves, shoot biomass and germination rates. It can also be used to estimate the stress in the plants in a controlled environment like green houses. The separation of yellow and green areas in the plants and leaves are done to detect the salt accumulation. The root systems are analyzed by means of segmentation through plant image processing. By applying image analysis techniques, the following activities are performed easily.

1. The physiological information of plants
2. Speed and accuracy of the extracted information
3. salinity stresses and water stresses are identified and
4. Extraction of plant's metabolic information.

Fluorescents Light

It is a relevant and suitable imaging system to describe the phenotypes of plants. The fluorescence is the light emitted in the shorter wavelength during the absorption of radiation. The reemission of the absorbed light by the chlorophyll is observed by irradiation. The multicolor fluorescence imaging is about UV illumination which generates blue to green region and red to far-red region type of fluorescence. It can be utilized for identifying the diseases in plants at the early stages so that we can take some actions to raise the production. Plant metabolism could be affected first during the disease infection. In a controlled environment, the fluorescence imaging can be used for monitoring the metabolism. Other plant phenotyping is also captured using fluorescence imaging.

Thermal Infrared

These sensors include near-infrared, multispectral line scanning cameras. It is used to characterize the plant temperature responses. The phenotypic parameters Leaf area index, leaf temperature, Surface temperature and severity of diseases are extracted.

Hyperspectral Imaging

Here the hyper spectral thermal cameras are used to produce spectral data for processing and the spatiotemporal growth patterns are extracted easily. The phenotype like water content, leaf growth and grain quality are measured.

MRI

The Magnetic Resonance Imaging technique is used to visualize metabolism of the plants.

CT

This Computed Tomography is used to assess the tissue density and grain quality and it is based on X-ray digital radiography images.

Some of the open source of automated and semi-automated phenotyping tools are listed below:

1. ImageJ—It is an application used to measure the phenotypic traits.
2. IAP—It is used to analyse large scale plant phenotype for different plants from various spectrum proposed by Klukas et al. 2014 [14].
3. HTPheno—It is a high throughput phenotyping image analysis as a plugin for ImageJ introduced by Hattmann et al. 2011 [1].
4. Phenophyte—It is a web based application which is used to measure the area related phenotypic traits introduced by Green et al. 2012 [15].

Each software tool has its own limitation. Hence designing a tool to perform an image analysis to improve the production of wheat is a challenging task.

7.1.2.2 Datasets

We have collected this Wheat image data set from University of Nottingham for doing our work. The dataset consists of several wheat image sets with top view, front view and side view options. For the experiments described here, we focus on the images of individual plants. All images are in RGB. The important phenotype properties extraction from the dataset is a challenging job.

7.1.3 Nature Inspired Optimization Techniques

Some complex real-world problems are effectively solved by nature. This is the inspiration for researchers to find algorithms to solve problems based on the natural world phenomena. It works on two principles.

1. Exploration—Generates diverse solutions and explores the search space
2. Exploitation—Strives to improve the quality of the solution.

These two steps are applied iteratively to find the optimum solution. It is very simple and flexible in nature, Moreover, finding the locally maximum or minimum concept is avoided. These algorithms are referred as nature inspired algorithms (NIA) or Clever algorithms (CA). Some of the natural inspired algorithms are listed below:

- Genetic Algorithm (GA): Darwin's theory of evolution is an inspiration for the researchers to develop genetic algorithm. It is a population based algorithm with the operators such as Selection, Crossover and Mutation [4, 9, 16].
- Ant Colony Optimization (ACO): Generally, the ants are having the ability to find the shortest path between the food and the nest. Hence ACO can be used to find the optimum solution [17, 18].
- Swarm Intelligence (SI): It is inspired by the flocking behavior of birds. Here, in each iteration, the candidate solution is moved around the search space based on its position and velocity [19].
- Bacterial Evolutionary Algorithm (BEA): here the gene transfer operation allows the chromosomes directly to other chromosomes in population [20].
- Simulated Annealing (SA): It is based on the cooling process of a molten metal [21, 22].
- RRO is an Raven Roosting Optimization algorithm which is based on the social roosting and foraging behavior of the bird raven [23].
- Fish Swarm Algorithm is based on the schooling behavior of fish. It uses the social behaviors of fish to search food and face the dangerous situation [24].

Conventional methods are not able to solve the real-world problems efficiently, this NCA can act as an alternate option for finding solutions significantly. In this work plant phenotyping traits are extracted from the plant images for knowing the growth of plants and improve its production. The Particle Swarm Optimization techniques are used to segment the given plant images for identifying the phenotypes. The detailed explanation about PSO for segmenting the images are given in Sect. 1.3.

7.1.4 Social Intelligence

It is a new computation method based on the cooperation of individuals to find the best solution using other individual information. There are two familiar swarm models exists which are inspired by the social insects. They are

1. Ant Colony optimization and
2. Particle Swarm Optimization (PSO).

This Particle Swarm Optimization is a global optimization technique for solving the problems with the best solution. Birds are naturally attracted by food and they have the ability in flocking for food searching. Generally, the birds have the social interaction that enables them

- to fly without collision during changes in the direction
- to regroup quickly when facing threats.

The local interactions among the birds (particles) shows the shared motion direction of the swarm as in the Fig. 7.3. which is based on the nearest neighborhood principle. Particles are moved in the solution space and evaluated according to the fitness function. Each particle positions are stored in memory to attain the best solution in the search space so far is called experience. Moreover, the aim of PSO algorithm is sharing of experience in which each particle is communicated through the neighborhood properly.

In this work, an attempt is made to review some of the existing techniques for image segmentation, plant phenotyping and a brief explanation about wheat. Section 7.2 provides the proposed work for segmenting the wheat images for further processing. and describes the steps for calculating the various phenotyping measurements in detail. Section 7.3 describes the results and discussion about the proposed work and Sect. 7.4 provides the conclusion of this chapter.

Fig. 7.3 Birds flocking behavior

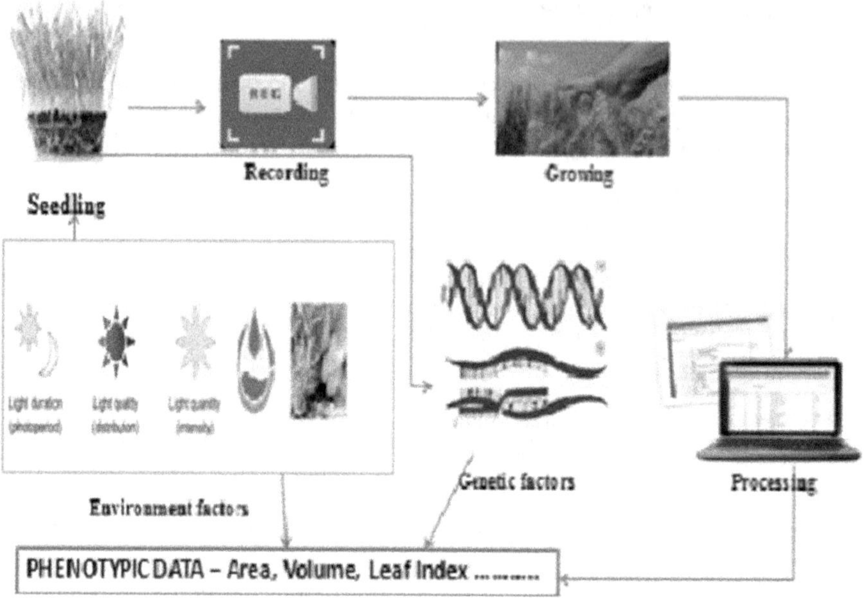

Fig. 7.4 Generalized wheat phenotyping extraction

7.2 Proposed Work

Automatic identification of phenotyping of plants are necessary for estimating the growth of plants to meet the challenges globally. The generalized wheat phenotyping details extraction procedure is given in Fig. 7.4. from seeding stage to growing stage.

Images are captured using digital cameras or mobile cameras and the captured images are stored in memory for further processing. The growth rate of the wheat as well as the phenotype properties are measured from the acquired images.

Every day at least twenty to thirty images are captured from the plants from the seeding stage to growing stage and the images are converted into frames. The influences of environmental factors such as sun light, moon and other genetical factors decide phenotypic data of wheat.

The following block diagram in Fig. 7.5 explains the various steps involved in the proposed work. The detailed explantation of the block diagram is as follows:

1. Image acquisition: The images are acquired from the wheat plants using mobile cameras or normal digital cameras and then the images are stored in a system to prepare training set and test data.
2. Pre-processing: When the images are acquired from the real world, there could be some unwanted information is added due to the image acquisition process. i.e., noisy pixels are added automatically when the images are captured. Those

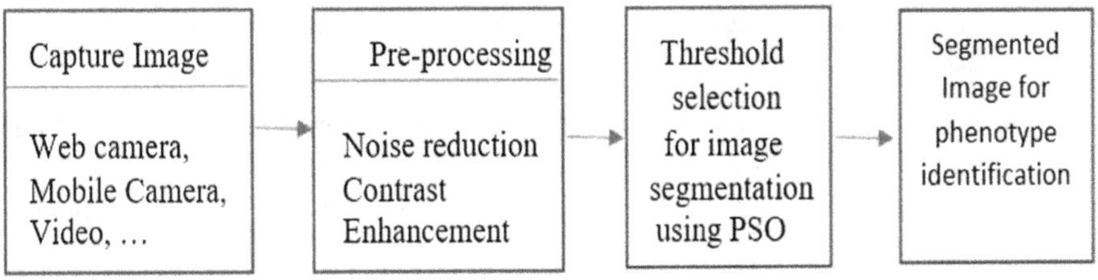

Fig. 7.5 Block diagram of proposed work

noisy pixels are eliminated by using the filtering techniques. The median filter can be applied to remove the noisy pixels from the captured images so that the mean square error is reduced and the signal to noise ratio is raised. The following median filter syntax is used to remove the noisy pixels from the images in MATLAB.

$$\mathrm{M} = \mathrm{medfilt2(I)};$$

where I is an noisy input image. The adaptive histogram equalization technique is used to enhance the contrast of the denoised image.

3. Image Segmentation: Since the segmentation process played a major role in almost all image processing applications, it is mainly used to locate objects in agriculture. There are a lot of image segmentation algorithms available in the literature, threshold based segmentation is very simple and it produces accurate results. In this work, the threshold selection is important to produce better results and the selection of threshold is done through the particle swarm optimization technique. Here, the foreground objects are segmented and separated from the background information. Normally, threshold is denoted by T and it can be calculated by using the Eq. (7.1).

$$f(x,y) = \begin{cases} 1, & \textit{if}\ f(x,y) > T \\ 0, & \textit{if}\ f(x,y) \le T \end{cases} \quad (7.1)$$

Since the threshold value can separate the image and the background, it is always a challenging task to choose the optimum threshold value. Thresholding can be done by using histogram, entropy based methods, clustering and so on. The threshold will be chosen by using the following steps:

1. The input image is divided into sub images.
2. Threshold value is fixed based on the pixel values of sub images first.
3. Sub image pixel values are compared with the threshold value and do the segmentation.

4. The segmentation process will be stopped when all the sub images are processed.

To provide the best result for finding the optimum threshold value for image segmentation is the key issue in this work. This work aims to use particle swarm optimization technique for selecting the best threshold value to extract the phenotype properties of wheat.

Swarm Intelligence Models: These models are also called as computational models which are naturally inspired by swarm flocking. These techniques are heuristic search in nature. The boundaries of the objects are extracted effectively using particle swarm intelligence algorithms. Any one observes the nature and try to imitate by creating models to solve problems. Those models are frequently tested and refined by checking the results for developing nature inspired algorithms to solve real world problems efficiently. The common flow of swarm intelligence technique is depicted in the Fig. 7.6.

Particle swarm intelligence algorithm: It was developed by Kennedy and Eberhart in 1995 [25] as an evolutionary technique based on the social behavior of birds flocking. PSO is a population based search algorithm.

Each individual particle $X_i = (i = 1, 2, 3, \ldots n)$
The initial velocity $V = [V_1, V_2, \ldots V_N]^T$
The velocity of each particle $V_i = V_{i,1}, V_{i,2} \ldots V_{i,D}$.

The Eq. (7.2) is used to calculate the new particle velocity based on the previous velocity and c1 and c2 are the learning rates in social influence.

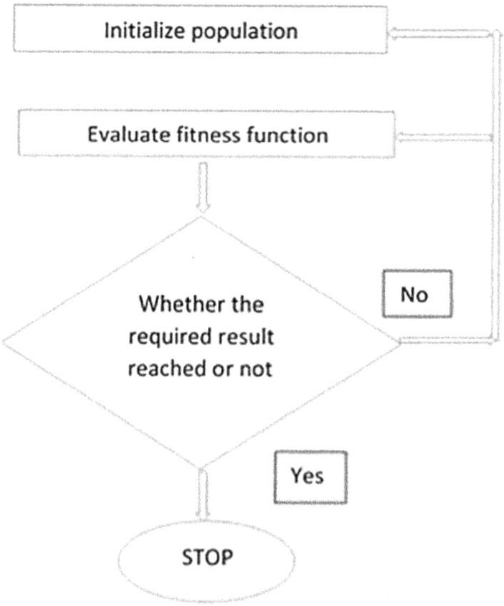

Fig. 7.6 Swarm intelligence technique

$$V_{i,j}^{k+1} = \mathrm{w} \times V_{i,j}^{k} + c_1 * r_1 * (Pbest_{i,j}^{k} - X_{i,j}^{k}) + c_2 * r_2 * (Gbest_{j}^{k} - Xk_{i,j}^{k}) \quad (7.2)$$

$$X_{i,j}^{k+1} = X_{i,j}^{k} + V_{i,j}^{k+1} \quad (7.3)$$

In Eq. (7.2) $\mathrm{Pbest}_{i,j}$ represents personal best *j*th component of *i*th individual, whereas Gbest_j represents *j*th component of the best individual of population upto iteration k.

The following steps are involved in PSO algorithm.

Step 1: Set the parameter of c1 and c2 of PSO
Step 2: Set k = 1
Step 3: Calculate the fitness of the particle and find the index of the best particle
Step 4: Select Pbest and Gbest
Step 5: Update Velocity and position of particles

$$V_{i,j}^{k+1} = \mathrm{w} \times V_{i,j}^{k} + c_1 * r_1 * (Pbest_{i,j}^{k} - X_{i,j}^{k}) + c_2 * r_2 * (Gbest_{j}^{k} - Xk_{i,j}^{k})$$
$$X_{i,j}^{k+1} = X_{i,j}^{k} + V_{i,j}^{k+1}$$

Step 6: Evaluate fitness function and find the index of the best particle
Step 7: Update Pbest of population and the Gbest of population
Step 8: If k < Maxiteration then k1 and goto step 6
Step 9: Print the optimum solution ie., Gbest^{k}.

The flow chart of particle swarm algorithm is depicted in Fig. 7.7

The syntax of the particle swarm optimization function in MATLAB is as follows:

$$\mathrm{x = particleswaram(fun, nvars)}$$
$$\mathrm{x = particleswaram(fun, nvars, lb, ub)}$$
$$\mathrm{x = particleswaram(fun, nvars, lb, ub, options)}$$
$$\mathrm{x = particleswaram(problem)}$$

where

fun	function
nvars	number of design variables
lb	lower bound
ub	upper bound
options	default optimization parameters are replaced by the values of options.
problem	is a structure

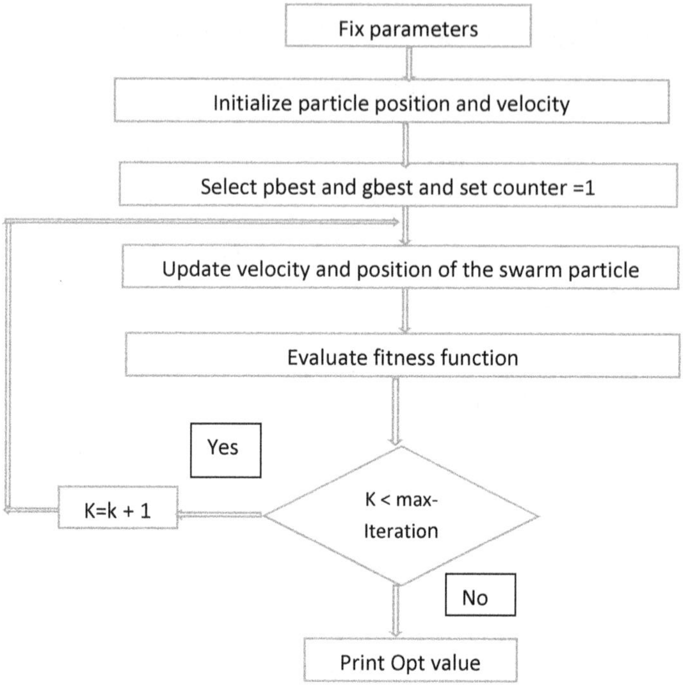

Fig. 7.7 Flowchart of PSO

The following parameters are used in this work

Initial weight: 0.45 to 0.9

Acceleration factors: 2 to 3

Population size: 50 to 200

Maximum number of iterations: 1000

Initial velocity = 10

The sample MATLAB program for using the particle swarm optimization function is given as follows:

```
fun = @(x)x(1)*exp(-norm(x)^2); % objective function
lb = [−10, −15];
ub = [15, 20];
options = optimoptions('particleswarm', 'SwarmSize', 100, 'Hybrid Fcn', @fmincon);
nvars = 2;
x = particleswarm(fun, nvars, lb, ub, options)
```

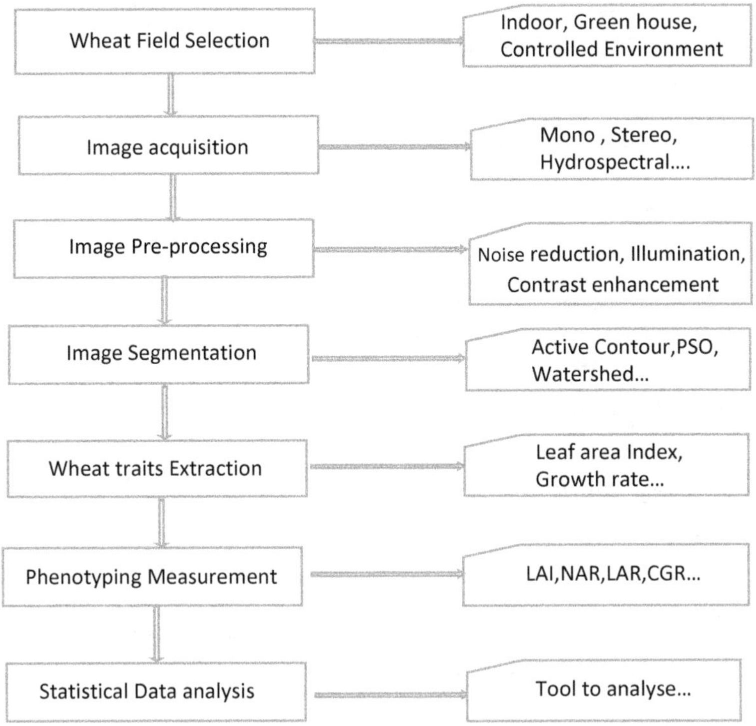

Fig. 7.8 Work flow of wheat phenotyping

Thus, the particle swarm function can be used in matlab successfully.

4. Phenotype measurement: In order to raise the growth of the wheat the following steps are taken and the Fig. 7.8 shows the workflow of wheat phenotyping.

The growth analysis of plants depends on the parameters like plant area, Leaf area index, Net Assimilation ratio, Relative growth rate, Leaf area ratio and so on. Using particle swarm intelligence, the given plant images are segmented and then the growth parameters are extracted to analyze the features of the plants in detail for raising the yields. Instead of using a single parameter for measuring the growth, combination of multiple parameters will produce better results. The growth stage can be divided based on the samples taken from the field and the statistical analysis are performed for measuring the phenotyping. The duration from sowing stage to leaf erected as Tillering stage and the duration from first sample to second sample is called Stem extension, third stage is called flowering stage and the last stage is called ripening stage [14, 26 and 27].

Plant samples were taken and the results are analyzed. Since the measurement of wheat phenotyping or growth traits are difficult, we can build the system like this for estimating the growth in the field.

Roderick in his paper discussed the basic growth analysis of plants in detail [28]. He used the following equations to calculate the different growth indices as follows:

1. LAI = L/ P,

   where LAI—Leaf Area Index
   L—Initial leaf area
   P—Ground area

2. NAR = 1/L*dw/dt,

   where NAR—Net Assimilation Rate
   dw—dry weight production in t days
   dt—Number of days

3. RGR = 1/W * dw/dt,

   where RGR—Relative Growth Rate
   W—initial dry weight

4. LAR = L/W,

   where LAR—Leaf area ratio

5. LAD = LAI * number of days from beginning of flowering to maturity,

   where LAD—Leaf Area Duration

6. CGR = 1/P * dw/dt

   where CGR—Crop Growth Rate,
   P—Ground area.
   Hence, the following traits can be taken for estimation of growth in the field.

7.3 Results and Discussions

The segmentation process was evaluated by means of visual comparison with the original images. The following sample images are taken from the dataset to perform the phenotype properties extraction (Figs. 7.9, 7.10, 7.11).

The ground area of the wheat plant can be calculated by using this segmented image and it can be used to calculate the phenotyping traits of wheat such as leaf area index and so on.

These growth stages are depicted graphically by using the tabulated data. The precision and recall methods can also be used to evaluate the performance of the segmented image and they are defined as follows:

Fig. 7.9 Front view of wheat in database and the segmented image

$$\text{Precision} = \text{TP}/\text{TP} + \text{FP}$$
$$\text{Recall} = \text{TP}/\text{TP} + \text{FN}$$

where

TP true positive
FP False positive and
FN False negative.

The method was evaluated using different plants. The wheat plant dataset is selected for evaluation. We have totally 100 leaves,

TP = 66 true positive leaves were correctly detected.
FP = 29 false positive values were correctly detected and
FN = 35 false negative samples., 35 leaves could not be detected.

Precision and recall was used to evaluated as follows:

$$\begin{aligned} \text{Precision} &= \text{TP}/\text{TP} + \text{FP} \\ &= 66/66 + 29 \\ &= 66/95 = 0.69 \\ \text{Recall} &= 66/66 + 35 \\ &= 66/101 \\ &= 65.34 \end{aligned}$$

The growth traits of wheat plants were recorded and tabulated and ten sample details for the growth traits are shown in the Table 7.2.

Fig. 7.10 Wheat image of Side view, the segmented result and the contrast enhanced image using histogram equalization

Using these growth sample traits of wheat the average values of the traits are calculated and tabulated as follows:

$$\begin{aligned}\text{The LAI(Leaf Area Index)of SG} &= (\text{Min value} + \text{Max Value})/2\\ &= 0.24 + 0.30/2 = 0.27\end{aligned}$$

$$\begin{aligned}\text{The LAI(Leaf Area Index)of LG} &= (\text{Min value} + \text{Max Value})/2\\ &= 1.61 + 2.56/2 = 2.085\end{aligned}$$

$$\text{The LAR of SG} = 53.55 + 78.06/2 = 65.805$$

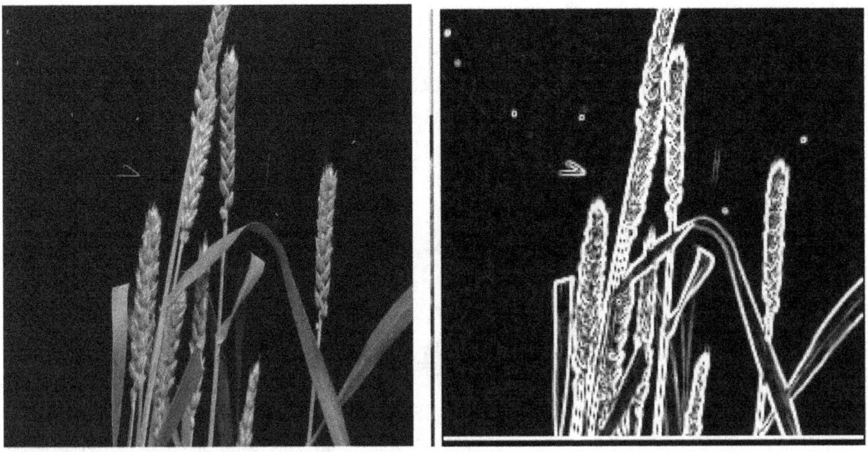

Fig. 7.11 Sample wheat images from the database (Top view) and the segmented image

Table 7.2 Sample wheat growth parameters

Leaf area index (LAI)	Samples	SG	LG	FS	GF
	1	0.24	1.61	−0.94	−1.63
	2	0.26	1.65	−0.99	−1.55
	3	0.28	2.76	−0.87	−1.49
	4	0.3	1.87	−0.86	−1.37
	5	0.25	1.98	−0.83	−1.28
	6	0.24	2.09	−0.79	−1.11
	7	0.27	2.76	−0.76	−0.91
	8	0.22	2.31	−0.75	−0.65
	9	0.28	2.42	−0.63	−0.87
	10	0.24	2.56	−0.61	−0.75
RGR	1	0.25	0.3	0.39	0.56
	2	0.27	0.33	0.4	0.58
	3	0.29	0.36	0.42	0.61
	4	0.31	0.39	0.38	0.62
	5	0.33	0.42	0.44	0.64
	6	0.28	0.45	0.48	0.63
	7	0.32	0.46	0.43	0.68
	8	0.23	0.45	0.51	0.7
	9	0.24	0.43	0.54	0.72
	10	0.25	0.41	0.56	0.74

(continued)

Table 7.2 (continued)

Leaf area index (LAI)	Samples	SG	LG	FS	GF
NAR	1	3.88	7.34	20.22	12.56
	2	3.95	7.92	21.43	14.78
	3	4.02	8.44	21.78	17.74
	4	4.75	9.02	22.23	19.32
	5	5.23	9.34	22.89	23.22
	6	5.95	10.11	23.67	28.87
	7	6.32	10.78	24.66	31.88
	8	7.11	11.12	27.11	32.65
	9	8.02	11.78	29.12	39.87
	10	8.78	12.22	32.34	45.22
LAR	1	53.55	45.86	2.88	0.12
	2	57.67	43.21	2.65	0.34
	3	58.98	42.23	2.11	0.56
	4	63.56	41.56	3.43	0.65
	5	69.34	40.98	3.65	0.33
	6	72.34	39.67	3.78	0.31
	7	74.35	38.65	3.92	0.29
	8	76.55	32.12	4.12	0.27
	9	77.98	29.43	4.54	0.23
	10	78.06	26.45	5.23	0.16
LAD	1	75.55	98.51	27.3	7.34
	2	78.16	99.23	28.4	8.6
	3	86.44	99.78	28.9	9.34
	4	89.32	101.12	29.5	8.56
	5	90.12	111.34	30.12	10.33
	6	92.34	114.76	31.36	15.32
	7	97.34	117.9	36.78	17.4
	8	99.67	119.32	39.46	19.56
	9	102.33	123.5	46.44	26.5
	10	110.45	128.78	47.21	28.33

Table 7.3 Mean of wheat plant growth parameters

Growth parameters	SG (mean)	LG (mean)	FS (mean)	GF (mean)
LAI (m^2)	0.27	2.09	−0.24	−1.08
LAR (cm^2)	65.8	36	2.55	0.34
LAD (m^2)	98	111	37	21
NAR (m^2)	4.56	9.22	33.56	21.5
RGR (gm)	0.22	0.25	0.06	0.05

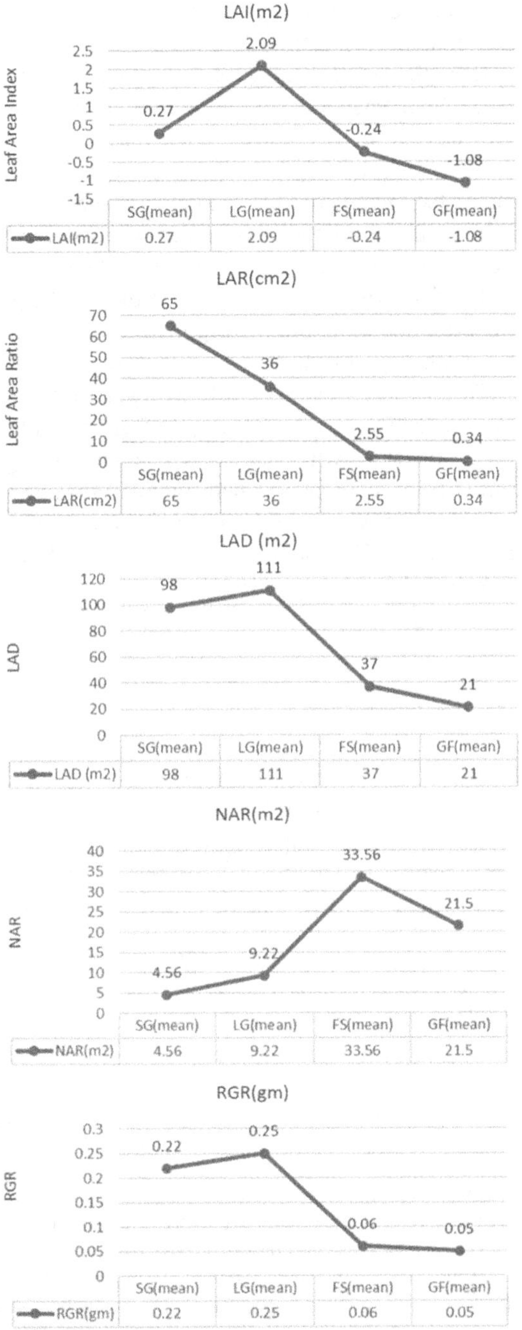

Fig. 7.12 Dynamic changes of LAI, LAR, RGR, NAR and LAD in various stages of wheat growth

Using the above formula mean values are calculated for each and every trait and the value are calculated as in Table 7.3.

Here the number of false positive and number of false negatives are similar, hence the precision and recall are also similar. The growth parameters mean values are recorded for LAI, RGR, NAR, LAD, LAR and CGR.

According to Ghost and Singh report, if there is any variation for growth parameters, there could be variation in yields. Leaf area index was recorded during the second and third stage and there was a decrease in leaf area and also in Leaf area index. The flowering stage is an important to show the highest yield.

The following graph shows that during the growth stage LAI values were very high and in grain filling stage there was a decrease. The leaf area during the flowering stage of plant is more important factor to produce high yield (Fig. 7.12).

The LAR i.e., Leaf Area Ratio is very high during Slow vegetative growth stage and it reached very low during the grain filling stage. The LAD value touch the peak at linear growth stage and low at grain filling stage. The RGR and NAR values started from low and touched the high at grain filling stage.

LAI and LAD are played a major role to show the growth rate of wheat effectively. For higher yield, the high LAI and LAD should be selected during linear growth and flowering stage.

7.4 Conclusion

In this work, we used particle swarm optimization technique for wheat image segmentation and from the segmented images the phenotyping traits are extracted and measured. We have tested this method in wheat image dataset. Our approach shows the better performance for image segmentation and we are able to extract the phenotyping of wheat like leaf index, leaf area easily and effectively.

References

1. Hartmann, A., Czauderna, T., Hoffmann, R., Stein, N., Schreiber, F.: HTPheno: an image analysis pipeline for high-throughput plant phenotyping. BMC Bioinform. **12**, 148 (2011). https://doi.org/10.1186/1471-2105-12-148
2. Fahlgren, N., Feldman, M., Gehan, M.A., Wilson, M.S., Shyu, C., Bryant. D.W., Hill, S.T., McEntee, C.J., Warnasooriya, S.N., Kumar, I., Ficor, T., Turnipseed, S., Gilbert, K.B., Brutnell, T.P., Carrington, J.C., Mockler, T.C., and Baxter, I.: A versatile phenotyping system and analytics platform reveals diverse temporal responses to water availability in Setaria. Mol. Plant (2015). https://doi.org/10.1016/j.molp.2015.06.005
3. Knecht, A., et al.: Image harvest: an open-source platform for high-throughput plant image processing and analysis. J. Exp. Bot. **67**, 3587–3599 (2016)
4. Houle, D., Govindaraju, D.R., Omholt, S.: Phenomics: the next challenge. Nat. Rev. Genet. **11**, 855–866 (2010)

5. Grobkinsky, D.K., Svensgaard, J., Christensen, S., and Roitsch, T.: Plant phenomics and the need for physiological phenotyping across scales to narrow the genotype-to-phenotype knowledge gap. J. Exp. Bot. **66**(18), 5429–5440 (2015). https://doi.org/10.1093/jxb/erv345 Advance Access publication 10 July 2015
6. Ubbens, J.R., Stavness, I.: Deep plant phenomics: a deep learning platform for complex plant phenotyping tasks. Front. Plant Sci. **8**, 1190 (2017). https://doi.org/10.3389/fpls.2017.01190]
7. Mohanty, et al.: Enotypic and phenotypic diversity of *Bacillus* spp. isolated from freshwater ecosystems (2011)
8. Pawara, P., Okafor, E., Surinta, O., Schomaker, L., Wiering, M.: Comparing local descriptors and bags of visual words to deep convolutional neural networks for plant recognition. ICPRAM, Porto (2017)
9. Sureja, N., Chawda, B.: Random traveling salesman problem using genetic algorithms. IFRSA's Int. J. Comput. **2**(2) (2012)
10. James, K., Russell, E.: Particle swarm optimization. In: Proceedings of 1995 IEEE International Conference on Neural Networks 1995. pp. 1942–1948
11. Akhilendra, V., Singh, S.P.: Studies on physico-chemical attributes of guava (*psidium guajava*) cultivars. Progress. Hortic. **47**, 53–56 (2015)
12. Wetzstein, H.Y., Zhang, Z., Ravid, N., Wetzstein, M.E.: Characterization of attributes related to fruit size in pomegranate. HortScience **46**, 908–912 (2011)
13. Martnez-Espl, A., Zapata, P.J., Castillo, S., Guilln, F.: Preharvest application of methyl jasmonate (meja) in two plum cultivars. 1. Improvement of fruit growth and quality attributes at harvest. Postharvest Biol. Technol. **98**, 98–105 (2014)
14. Klukas, et al.: High throughput phenotyping of maize. Plant Physiol. Preview, published on January 30, 2017 (2014). https://doi.org/10.1104/pp.16.01516
15. Green, J.M., Appel, H., MacNeal Rehrig, E., Harnsomburana, J., Chang, J.-F., Balint-Kurti, P., Shyu, C.-R.: PhenoPhyte: a flexible affordable method to quantify 2D phenotypes from imagery. Plant Methods **8**, 45 (2012). https://doi.org/10.1186/1746-4811-8-45
16. Holland, J.H.: Adaptation in Natural and Artificial Systems. Cambridge. MIT Press, MA, USA (1992); Sastry, K., Goldberg, D.E., Kendall, G.: Genetic algorithms. In: Search Methodologies, pp. 93–117. Springer, Berlin (2014)
17. Colorni, A., Dorigo, M., Maniezzo, V., et al.: Distributed optimization by ant colonies. In: Proceedings of the First European Conference on Artificial Life, vol. 142, pp. 134–142 (1991)
18. Chawda, B.V., Sureja, N.M.: An ACO approach to solve a variant of TSP. Int. J. Adv. Res. Comput. Eng. Technol. IJARCET **1**(5), 222 (2012)
19. Kennedy, J., Eberhart, R.: Particle swarm optimization. In: Proceedings of the IEEE International Conference on Neural Networks, 1995. Proceedings, vol. 4, pp. 1942–1948 (1995)
20. Niu, B., Wang, H.: Bacterial colony optimization. Discrete Dyn. Nat. Soc. 2012 (2012)
21. Kirkpatrick, S., Gelatt, C.D., Vecchi, M.P.: Optimization by simulated annealing. Science **220** (4598), 671–680 (1983)
22. Bookstaber, D.: Simulated annealing for traveling salesman problem. Spring (1997)
23. Brabazon, A., Cui, W., O'Neill, M.: The raven roosting optimisation algorithm. Soft. Comput. **20**(2), 525–545 (2015)
24. Li, X., Qian, J.: Studies on artificial fish swarm optimization algorithm based on decomposition and coordination techniques. J. Circuits Syst. **1**, 1–6 (2003)
25. Pandey, S.K., Singh, H.: A simple, cost-effective method for leaf area estimation. J. Bot. **2011**, 1–6 (2011)
26. El-Din, A., Omar, K., Ahmed, M.A., Al-Obeed, R.: Improving fruit set, yield and fruit quality of date palm (phoenix dactylifera, l. cv. mnifi) through bunch spray with boron and zinc. J. Test. Eval. **43**, 1–6 (2014)
27. Holland, J.H.: Genetic algorithms and the optimal allocation of trials. SIAM J. Comput. **2**(2), 88–105 (1973)
28. Roderick, H.: Basic Growth Analysis. Unwin Hyman Ltd., London. 112pp (1990)

Chapter 8
Cuckoo Optimization Algorithm (COA) for Image Processing

Noor A. Jebril and Qasem Abu Al-Haija

Abstract Image optimization is the process of enhancing the image quality and visual appearance in order to provide preferable transfer representation for many future automated images such as medical images, satellite and aerial images which might suffer from poor and bad contrast and noise. There are many state-of-art algorithms in the literature that have been used for image optimization process in which they were inspired from the nature such as the Particle Swarm Optimization (PSO), Differential Evolution (DE) and more recently, the Cuckoo Optimization Algorithm (COA). COA is very efficient optimization technique developed by Yang and Deb through applying a special versions of gauss distribution for solving optimization problems. COA is differentiated from the life-style and the characteristics of Cuckoo sparrow clique. Cuckoo sparrow society initiates with an elementary inhabitance that is classified into two portions: cuckoos and eggs. The cuckoo societies then start to change their environment to better one and start reproducing and putting eggs. Such endeavor of Cuckoos to enhance their life's environment is the Cuckoo Optimization Algorithm. In this chapter, a comprehensive discussion about one of the metaheuristic algorithms called Cuckoo Search Optimization (CSO) has been carried out. Also, the usefulness of CSO for solving image optimization problems is discussed in detail. Moreover, to support the theoretical discussion of CSO algorithm, the performance evaluation of CSO algorithm is provided in the results and comparisons section which compare and benchmark the execution of CSO algorithm with other genetic algorithms and particle swarm optimization. Finally, the analysis of comparison results illustrated the superior capability of Cuckoo search algorithm in optimizing the enhancement functions for digital image processing.

N. A. Jebril
Computer Sciences Department, King Faisal University,
P.O. Box 380, Ahsa 31982, Saudi Arabia
e-mail: njebril@kfu.edu.sa

Q. Abu Al-Haija (✉)
Electrical Engineering Department, King Faisal University,
P.O. Box 380, Ahsa 31982, Saudi Arabia
e-mail: qalhaija@kfu.edu.sa

J. Hemanth and V. E. Balas (eds.), *Nature Inspired Optimization Techniques for Image Processing Applications*, Intelligent Systems Reference Library 150,
https://doi.org/10.1007/978-3-319-96002-9_8

Keywords Genetic-Algorithm · Particle swarm optimization · Cuckoo optimization algorithm · Image enhancement · Histogram-Equalization · Linear contrast stretch

8.1 Introduction

In the last decades, the image processing and optimization have played the key role in many areas such as computer science, medical and astronomy fields as well as many other fields of engineering and technology. Most of these disciplines need to increase the capabilities of digital cameras and emerging devices—like Google Glass and autonomous driving cars- to Magnetic Resonance Imaging (MRI) and analyzing astronomical data that require many processing stages such as image transformation, correction of distortion effects, noise purge, histogram equalization, Image contrast enhancement, noise elimination, thresholding, edge detection and image segmentation. These processing steps are the key stages in the digital image processing techniques.

The image is vulnerable to be influenced by the noise process resultant from the image transmission which may produce harmful effects on the image during the image processing operations. Thus, to eliminate all these effects, the noise must be passed through removing or diminishing stage keeping all information of the image as possible.

The image information should be kept clean and safe as its considered crucial for helping the image processing specialized in making decisions depending on the information provided from the processed image. Therefore, high-performance image enhancement mechanisms are required for processing and analyzing the data generated from imaging systems. This is needed due to the huge amount of data doubled by high resolution and frame rates and the requirements of interactive and mission-critical applications. Consequently, the active domain of the nominated image features of the in-process-image must be increased as it is deemed an essential operation of the image-enhancement systems. Whenever the image is transformed from its specific sort to another sort after external operations such as photography, image scanning, or transferring, this processing will produce output results with more quality-dependent image over the authentic inputted-image.

There is much relevant information that can be obtained from an image such as the value of intensity from an image or the edge of the image. In similar cases, the transformed image from one form to another is required to get the required features since the image processing generates these result as needed. The enhancement of an image is one strategy under image processing [1].

The most essential property of any image is that it's usually stored and processed in a discrete (i.e. digital) form. In the digital image, the pixel concept is a physical point in a raster image which is the smallest controllable element of an image that appears on the screen. The pixels are the function of spatial coordinates of an image represented as [1]:

$$p = (x, y) \tag{8.1}$$

where: x and y are the image spatial coordinates and p is the resultant pixel value which might be any of the image-intensity at that point to the gray-scale value.

All the image processing mechanisms are essentially performed as applications of the mathematical function in which they are applied over each pixel with its encirclement pixels [1] where all mechanisms strive to get information from a certain pixel along with its encirclement pixels and then obtain new value for the targeted pixel.

In the meantime, image enhancement is a mechanism of image processing which strives to generate a more visually-appropriate digital image as the appearance of its visional inspection by individuals or devices.

The prime dilemma of applying conventional image enhancement mechanisms is the need for human intervention to inspect and decide whether the processed image became convenient for the desired mission or not. This is due to the absence of pre-defined or specific criteria to measure the amount of image enhancement. Therefore, it can only be drawn and then the human can inspect and evaluate the processed image to judge if the resultant image is appropriate or not [1]. Thus, huge efforts where consumed later to replace human inspection and construing of the processed images by the introduction of metaheuristic algorithms and genetic approaches [2].

The metaheuristic algorithms and genetic approaches would get better image-enhancement neutralizations and equations in which can be elaborated in the topmost level of effectiveness or as confined by the user to generate the desired outcomes. For instance, Munteanu and Rosa [3] were the first researchers who have proven the employment of genetic algorithms to beat the human-intervention dilemma and have proven the upgraded capability of the image-enhancement functions in the field of image transformation mechanisms. These consequences have a key role in raising the use of image enhancement-based metaheuristic algorithm operations.

Continuing the enormous successive developments which are later implemented in metaheuristic algorithms, it was discovered that as the algorithm became more capable, then the capacity of the function is increased. The genetic algorithm (GA) was followed by particle swarm optimization (PSO) and then cuckoo search optimization (CSO) has been recently shown up as an efficient solution for the aforementioned issues.

This chapter discusses the strengths of CSO based optimization for image-enhancement functions and formulas and discusses the results of several classical grayscale image enhancement mechanisms by employing several enhancement techniques such as Cuckoo Search (CS), Histogram equalization (HE) and Linear Contrast Stretch (CS) techniques [4]. In comparison with other metaheuristic algorithms, CSO has proved to be more efficient in image enhancement process. The coming text will thoroughly debate the capability of CSO and the comparison if its performance with other stated approaches.

8.2 Image Enhancement Functions

Image enhancement is the most important image processing techniques that convert the image to another enhanced image in which it can be used for perception or explanation of information for human viewers, or to get the best input for other image processing techniques (i.e. in serial image processing system that contains many stages). A Genetic Algorithm (GA) was suggested in [5] to enhance the digital images by implementing the contrast enhancement of multi-objective function that contains four non-linear mapping functions. It is used to find the optimal mapping of the grey levels of the input image to new grey levels that can give better contrast for the image.

Lately, the quality of the image is measured and used for grey-level and color image enhancement. Particle Swarm Optimization (PSO) has been proposed in [6] which is primarily used for preserving the mean value of image intensity for improving the contrast levels of digital grayscale images.

Generally, image enhancement approaches have four basic classifications: point operations, spatial operations, transformations, and pseudo-coloring methods. Contrast stretching, window slicing, modeling of the histogram are zero memory operations that assign a given input grayscale image into output grayscale image. The most popular between the classification is the linear contrast stretch and histogram equalization. In spatial operations, each value for all original pixels is replaced with the neighborhood pixel value, this process provides enhancement of the noise in the input image, but it might result with some degree of smoothing the image which can affect the accuracy of image details. The Linear, homomorphic and root filtering employed under transform operations depends on the inverse transformation of the transformed image.

In pseudo coloring methods, the grayscale image is colored by using a suitable color map, and due to non-singularity of the color maps, a lot of trails have been required to choose a suitable mapping. The function which is used to define the amount of the quality for the enhanced image for all the methods is known as the evaluation function or criterion.

8.2.1 Image Transformation Formulas

In this operation, the image can be enhanced by employing the image transformation function that uses the intensity value of each pixel of a $(P x Q)$ image. The process of manipulating the gray level distribution for each pixel in the neighborhood of the input image by the implemented image transformation function is called local enhancement technique. The traditional local enhancement transformation function is given in Eq. (8.2) below [7]:

$$g(x,y) = \frac{G}{\sigma(x,y)}(f(x,y) - m(x,y)) \tag{8.2}$$

where, $m(x,y)$ is the gray level mean and $\sigma(x,y)$ is the standard deviation which both calculated in a neighborhood centered at (x, y) contain $M \times N$ pixels, G is the global mean of the input image and $f(x,y)$ as well as $g(x,y)$ are the gray level intensities for the input and output images' pixel at location (x, y).

The other method of local enhancement technique is adaptive histogram equalization which is used in medical image processing. Also, one of the simplest and most popular methods to achieve the mission contrast enhancement is the global intensity transformation in which its function is derived from Eq. (8.2) and applied to each pixel at location (x, y) of the given image as in the following equation [7]:

$$g(x,y) = \frac{k.G}{\sigma(i,j)+b}[f(x,y) - c \times m(x,y)] + m(x,y)^a \tag{8.3}$$

$$Here: 5 < k < 1.5, a \in \psi_1, b \in \psi_2, c \in \psi_3 \quad \text{with } \psi_1, \psi_2, \psi_3 \subset R$$

where, b $\neq$ 0 allows for zero standard deviation in the neighborhood, c $\neq$ 0 allows for the only fraction of the mean value m (x, y) to be subtracted from original pixel gray level. The last term might have brightened and smoothened the effects on the image. G is the global mean, m (x, y) is the local mean and σ (x, y) is defined as the local standard deviation of the pixel (x, y) of the input image over n × n window, that can be debriefed as [7]:

$$\begin{gathered} m(x,y) = \frac{1}{n \times n}\sum_{x=0}^{n-1}\sum_{y=0}^{n-1} f(x,y), \quad and\, G = \frac{1}{M \times N}\sum_{x=0}^{M-1}\sum_{y=0}^{N-1} f(x,y), \\ \sigma(x,y) = \sqrt{\frac{1}{n \times n}\sum_{x=0}^{n-1}\sum_{y=0}^{n-1}((f(x,y) - m(x,y))^2} \end{gathered} \tag{8.4}$$

The accurate value of a, b, c and k parameters in the Eq. (8.3), will give a large variation in the processed output image by keeping the original version of it.

8.2.2 Objective Function

To evaluate the quality of the enhanced image without human intervention, a specific pre-defined function can be used to measure the image performance in terms of the number and intensity of edge pixels and entropy of the whole image.

Image entropy is the amount of information which must be coded by a compression algorithm. Low entropy images, like the black sky, have very little contrast and large runs of pixels with the same DN values. An image that is completely flat contains an entropy of zero. Therefore, this can minimize the size to the lowest, on the contrary of the image that has high entropy such as the image of cratered areas on the moon that cannot be compressed. Image entropy is calculated with the same formula used by Galileo imaging team [8] as follows:

$$Entropy = -\sum_{i} P_i \log_2 P_J \tag{8.5}$$

where P_i is the probability that the difference between two adjacent pixels is equal to I, and Log 2 is the base 2 logarithms.

Image quality metrics is mentioned in [7] which describes the final quality of the enhanced image. Hence, Peak Signal-to-Noise Ratio (PSNR) is one of the goals in the objective function. Thus, we put the aggregated weight based on objective function as follows:

$$OF = W_1 \times OF_1 + W_2 \times OF_2 \tag{8.6}$$

where OF is the objective function, W_1 and W_2 are weight factors such that $W_1 = 0.5$ and $W_2 = 0.5$ (equal weight age).

$$\begin{array}{c} OF_1 = F(I_e) = \log(\log(E(I_s))) \times \frac{n_{edgels(I_S)}}{M \times N} \times H(I_e), \\ OF_2 = PSNRI(I_e) \end{array} \tag{8.7}$$

- $F(I_e)$ is the objective function that describe the quality of the output image with transformation function Eq. (8.7).
- $F(I_s)$ is the sum of edge pixel intensities that calculated by using Sobel edge detector.
- $n_{edgels(I_s)}$ is the number of edge pixels.
- $H(I_e)$ is the entropy value.
- M and N are the number of pixels.
- $PSNRI$ is peak signal to noise ratio of the enhanced image, where is used the mean squared error [9]:

$$MSE = \frac{1}{mn} \sum_{i=0}^{m-1} \sum_{j=0}^{n-1} [I(i,j) - K(i,j)]^2 \tag{8.8}$$

The PSNR is defined as:

$$PSNR = 10.\log_{10}\left(\frac{MAX_I^2}{MSE}\right) = 20.\log_{10}\left(\frac{MAX_I}{\sqrt{MSE}}\right) \tag{8.9}$$

$$Therfore, PSNR = 20.\log_{10}(MAX_I) - 10.\log_{10}(MSE) \quad (8.10)$$

where MAX_I is the most (largest) conceivable value of the image pixel. Similarly, many other Sobel operators can be used for the same purposes such as Canny Operators in which are proposed in [3].

The automatic threshold detector needs the edge detector where the summation of the intensity of the edges is calculated using the following equation [7]:

$$E(I(J)) = \sqrt{\partial u(x,y)^2 + \partial v(x,y)^2} \quad (8.11)$$

where:

$$\begin{aligned}
\partial u(x,y) &= g(x+1,y-1) + 2g(x+1,y) + g(x+1,y+1) \\
&\quad - g(x-1,y-1) - 2g(x-1,y) - g(x-1,y+1) \\
\partial v(x,y) &= g(x-1,y+1) + 2g(x,y+1) + g(x+1,y+1) \\
&\quad - g(x-1,y-1) - 2g(x,y-1) - g(x+1,y-1)
\end{aligned}$$

8.2.3 Parameter Setting

Parameters a, b, c and k are determined by the real positive numbers in comparing of Eqs. (8.2)–(8.3). The values of the parameters are used as constants as follows:

$$b = 0, \quad c = 1 \quad k = 1 \quad m(x,y) = 0$$

In Eq. (8.3): $b = 0$ prohibits the *Not A Number* (*NAN*) values, $c = 1$ allows for only a fraction of the mean to be subtracted from the pixel's input gray-level intensity value, while the last term may have brightened and smoothened the effects on the image. Accordingly, Eq. (8.2) broadened the spectrum of the transformation output range by modifying the original equation.

The optimization algorithm drives the solutions of the image enhancement problem by setting the four parameters $(a, b, c\, and\, k)$ to obtain the maximum possible mix based on the objective standards that mention the contrast in the image. The selected variable is given as in [10]:

$$a \in [0, 1.5], b \in [0, 0.5], c \in [0, 1] \quad and\, k \in [0.5, 1.5]$$

However, they failed to get a perfect output that provides the domain of b since the difference value of b will have a significant effect on the intensity stretch even if the difference is small. Therefore, the original image might get lost by the intensity normalization value. Thus, to overcome this problem, we need to modify b to [1, G/2] where G is the global mean of the input image [11].

8.3 Related Work

Recently, the image enhancement has been heavily researched by collaborators and contributors as a world-wide critical research topic. There are many recent state-of-art works that focus on this issue. For instance, Weigel et al. 2013 [12] provided a mechanism to associate Image Inversion Microscopy (IIM) with digital holography by applying additional computations to prove their proposed essence by means of the utilization of Point Spread Function (PSF). The PSF function describes the reply to point object from an imaging system. The system's impulse response is a general expression of PSF in which it focused on optical systems as they also presented an explanation of how to reduce the distance between the first zeros by a factor of about two. Figure 8.1 shows image formation in a confocal microscope: central longitudinal (XZ) slice that distribution arises from the convolution of the real light sources with the PSF, as used general form, as follows:

$$Image\,(Object1 + Object2) = Image\,(Object1) + Image\,(Object2) \tag{8.12}$$

In addition, they listed images of 10 gratings to clarify and comprehend the boosted resolutions and to calculate a portion of the optical transfer functions of the coherent, the incoherent and the image-inverted case.

In a related work, Lin (2011) [13] proposed another scheme for enhancing the image quality by a means of Infrared images (IR) for an extended-domain surveillance system. The Infrared images which were taken at long-domain that encompass reduced contrast and brightness levels. The most important feature of the proposed method is that no need for pre-knowledge about the IR image and no parameters must be preset. Briefly, two main objectives for this research: enhancement for adaptive contrast by using Adaptive Histogram-Based Equalization (AHBE), and enhancement of the strength of elevated spatial-frequency of infrared images to maintain the datum of original inputted- images.

Another noticeable technique has been proposed by Zhao (2011) [14] who suggested to employing the Gravitational Search Algorithm (GSA) for image enhancement method. Gravitational Search Algorithm (GSA) strives to make the best and most effective use of the normalized incomplete Beta function parameters by employing the greyscale image generalized convention. This function used to enhance the damaged image which gives effectively enhanced results, as follows:

$$f\left(i_{xy}, \alpha, \beta\right) = \frac{\int_0^{i_{xy}} t^{a-1}(1-t)^{\beta-1}dt}{B(\alpha, \beta)} \tag{8.13}$$

The fitness function is defined as:

$$Fintness(f) = \frac{1}{n}\sum_{x=1}^{M}\sum_{y=1}^{N} f^2(x, y) - \left(\frac{1}{n}\sum_{x=1}^{M}\sum_{y=1}^{N} f(x, y)\right)^2 \tag{8.14}$$

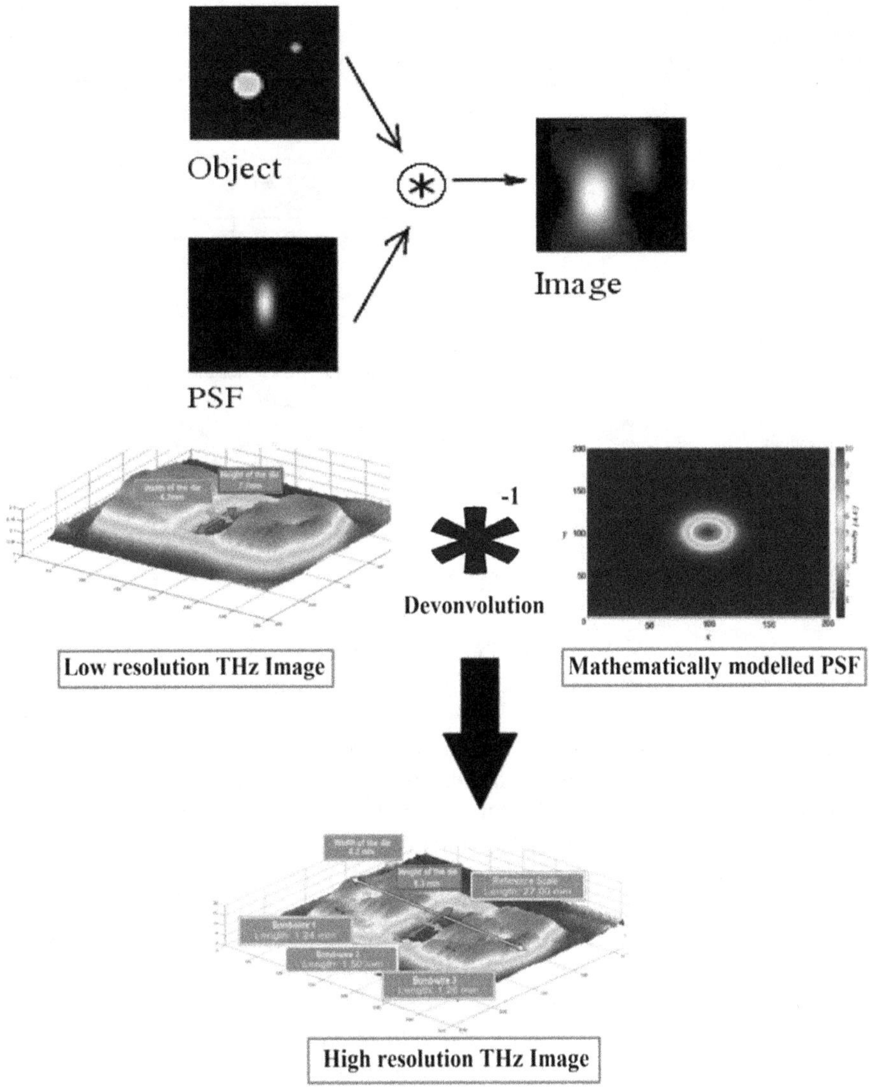

Fig. 8.1 The point spread function (PSF)

where M and N are the coordinates of the image to be enhanced with $n = M \times N$, and $f(x, y)$ is the enhanced pixel value of the input image. Thus, the higher value of the fitness function, the image will be more enhanced. Figure 8.2 shows the flowchart of adaptive enhancement for greyscale image based on GSA and it also explains the gravitational search algorithm [6].

The histogram improvement-based solutions were also valid and practiced in many state-of-the-art works such as the unprecedented formulation of image

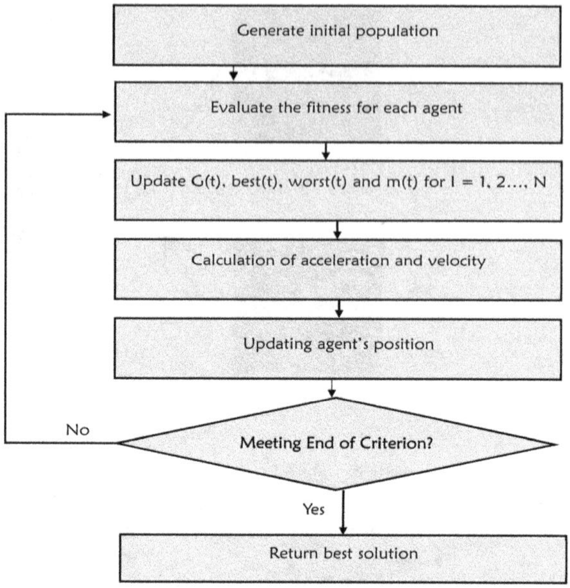

Fig. 8.2 Procedural diagram of Gravitational Search Algorithm of enhancing the grey images

histogram operation to ameliorate the contrast process of the mage which was formally proposed by Zeng et al. (2012) [15]. Zeng method encompasses several steps started by dividing the image into several equal-sized regions by using the values of the intensities of gradients, followed by adjusting their values of gray levels and finally, the histogram for the whole image can be obtained by the summation of all weighted values of the divided regions. This method improved the enhancement by testing X-Ray images that support this method. Figure 8.3 shows the final content of adaptive contrast enhancement algorithm.

Furthermore, recently, Santhanam and Radhika (2010) [16] have implemented a mechanism in which they employed the noise identification predominant operation in the processing stages of any digital image operation to distinguish the image smoothing filters. The noise detection was performed by using Artificial Neural Network (ANN) to isolate the noise and extracts needed features [8]. Eventually they concluded that ANNs provided a better solution in identifying the noise which provide the precise filter that can be used for enhancing the given image.

Finally, several enhancement algorithms for enhancing the digital image have been implemented by Garg et al. (2011) [17] through the manipulation of the greyscale Histogram-Equalization and greyscale Filtering. The most noticeable issue of the set of proposed frameworks is that they preserve the brightness level of the inputted-image on the resultant outputted-image by applying considerable enhancement in the image-contrast levels. Histogram-Equalization (HE) scheme can be effectively used to re-modulate the image gray grades. Figure 8.4 shows the effect of HE method on the entered (pre-processed) image [9].

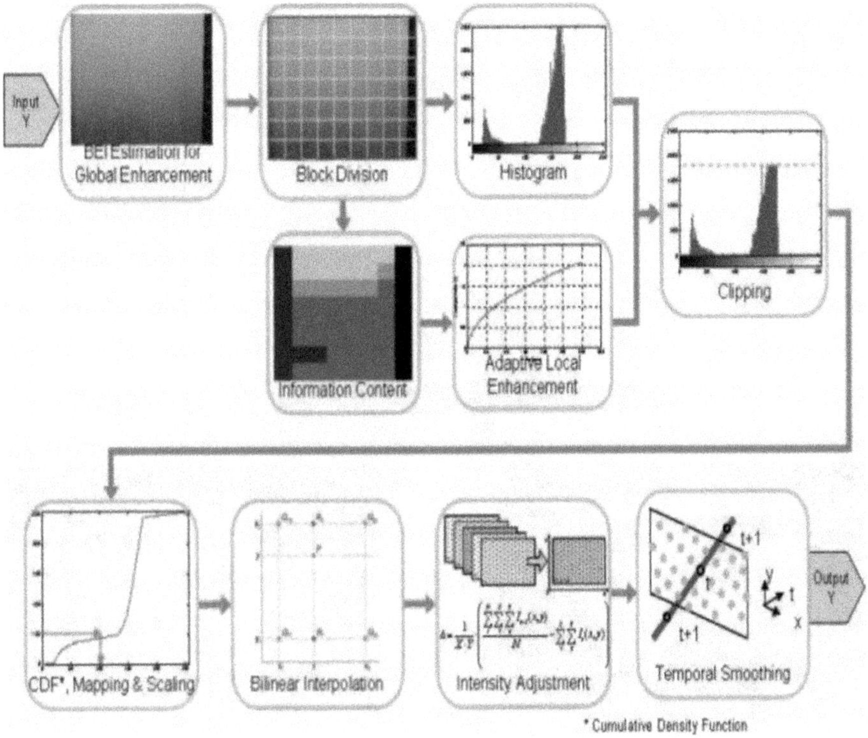

Fig. 8.3 Flowchart of the adaptive contrast enhancement algorithm

8.4 An Introduction to Cuckoo Search Optimization Algorithm

Cuckoo search (CS) [5] is a meta-heuristic algorithm inspired by the cuckoo bird, these are the "Brood parasites" birds. It is impossible to create its nest and place its eggs in the nests of other birds. Some steward fowls can engage directly with the other cuckoo that comes to own nest. The host bird recognizes the eggs if they belong to its nest or not. If the eggs are not belonging to it, it will throw the eggs away or remove the nest and build another one. According to this phenomenon, suppose that each existing egg is a solution and cuckoo egg is a new and workable solution. Thus, for every nest, there has one egg of cuckoo in which every nest shall contain various eggs that appears as a group of solutions.

Practically, any new egg put by cuckoo acts as an unprecedented settling to the search algorithm and prior the implementation of the next step, a distribution process formula defines the number of remaining eggs. The new number of eggs can be represented as the populace for the next iteration; therefore, increasing the

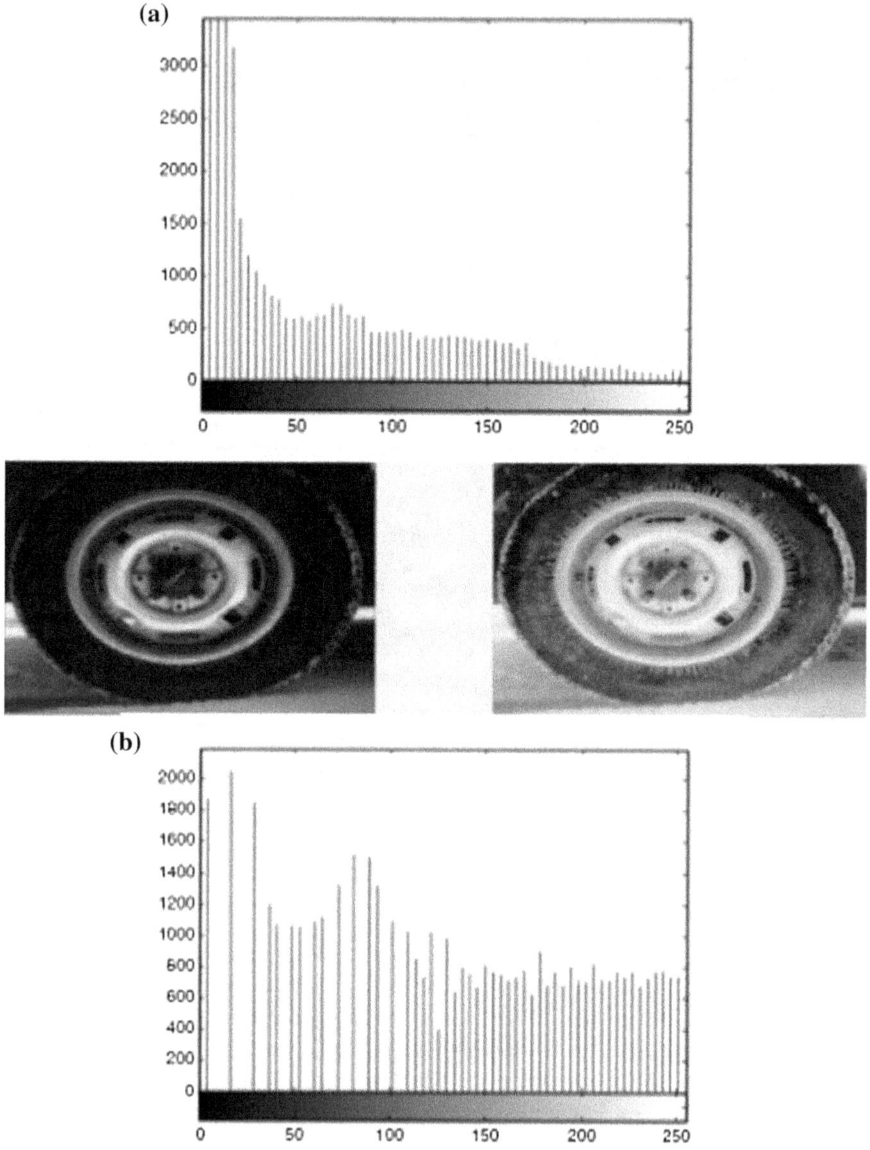

Fig. 8.4 **a** The original picture of the tire. **b** HE's picture of the tire

number of iterations is better to obtain enhanced results. The iterations carry on until satisfying the desired optimization. Shortly, CS algorithms and morphological operations are efficiently used to enhance the image by modifying its contrast and intensity. Figure 8.5 clarifies the general steps of CS algorithm. Eventually, CS

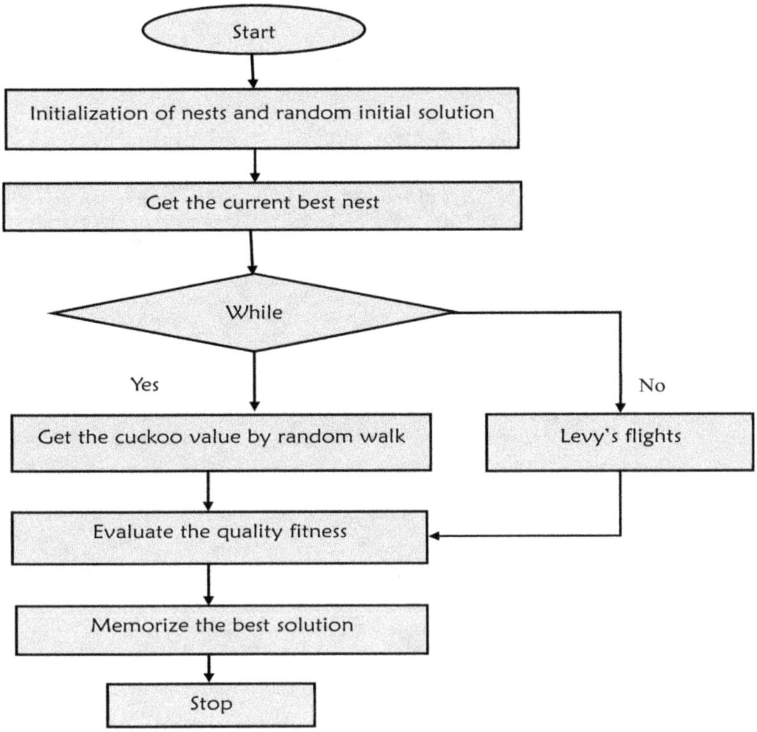

Fig. 8.5 General CS Algorithm Steps

algorithm became very applicable in many optimization fields and its exemplary for how the upbringing behavior [7] since it has been successfully implemented to solve the scheduling issues and design optimization problems such as speech recognition, job scheduling, and global optimization.

8.5 Image Enhancement via Cuckoo Search Methodology

Cuckoo search algorithm focuses mainly on the repeated calculation of the Mean Square Error (MSE) and Minimum Fitness Function (MFF) until the optimum requested results are satisfied (reaching the threshold values). Figure 8.6 explains the exact steps of CS algorithm in which it contains mainly two iteration steps: one for checking the population (less than the max value) and the other iteration step for checking if the condition is met or not [18].

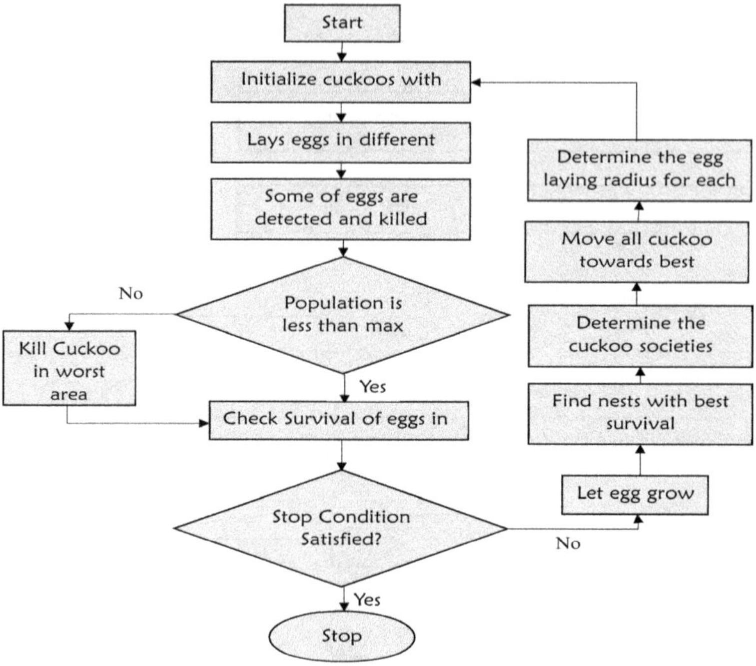

Fig. 8.6 The flowchart of Cuckoo Search (CS) Method

The fitness function is defined as:

$$Fitness\,Function = W1 \times MSE + W2 \times Iteration \tag{8.14}$$

The main idea of the image enhancement techniques is to get rid of noise and its process can be summarized as follow: Consider I is the original inputted image with dimensions $P \times Q$. The first step is the transforming of the original image (I) from its RGB coloring system into gray-scale coloring system, by applying the next formula [19]:

$$I_{gy} = 0.289 \times r + 0.5870 \times g + 0.1140 \times b \tag{8.15}$$

where: R, G, B are the levels of image-coloring components and I_{gy} is the transformed grayscale image. Also, the mean value (μ) of the grayscale image can be obtained by applying the following law [19]:

$$\mu = \frac{\sum_{p=0}^{p-1} \sum_{q=0}^{Q-1} Iqp}{P \times Q} \tag{8.16}$$

At this stage, the image is ready to be applied to Cuckoo Search algorithm via Levy flight to obtain an enhanced image and the process works as follows:

(1) *Every cuckoo place only single egg at a time and into any of the selected nests.*
(2) *The top-quality eggs of the superior nest will be renewed to the next descent.*
(3) *The number of ready host nest doesn't change and when a host bird recognizes the cuckoo egg with the probability of* $p \in (0, 1)$ [20], *the host bird can either remove the egg away or leave them and build a new nest.*

In the Lévy flight, the third method is selected to implement the cuckoo search and could be converged as the fraction '*p*' of the '*n*' nests. Many researchers proved that the flight behavior of many animals and insects has the typical features of Lévy flights which is a random walk; that the stages are determined according to the step-lengths, which contain a specific probability distribution with the directions being random. This random walk can be seen in animals and insects. The next motion depends on the current position which produces new settlings $x(t + 1)$ for a cuckoo is considered as integrating between Levy flight with controls of the search capability. This can be implemented by the following equation [19]:

$$X_i(t+1) = X_i(t) + \alpha \oplus \mathrm{Levy}(\lambda) \tag{8.17}$$

where: $\alpha > 0$ is the stage size and in most case, α is supposed to equal to one, and the product $\oplus$ is entry-wise multiplication; in other words, it is an Exclusive-OR operation. Levy-flight is a random walk with random stage size and it is considered as step lengths that are distributed according to the following probability distribution equation (PDE) [20]:

$$levy \sim u = t^{-\lambda}(1 < \lambda \leq 3) \tag{8.18}$$

The produced image can be acquired by utilizing the structuring-element of the genuine image. Thereafter, the adjusted pixels of the entered-image are calculated to build up the numerical value for each pixel into the produced image. The image I is transformed to the binary digital form I_b through the adjustment process of both factors; the intensity and the contrast of the image. Finally, the enhanced image is earned by applying the morphological operation in which it depends on the use of the structuring element '*se*' as follows [19]:

$$I_b \oplus Se = Se \oplus I_b \tag{8.19}$$

8.6 Pseudo-code of Cuckoo Search and Algorithm Implementation

The Pseudo-code of Cuckoo Search is shown below [18]:

```
Algorithm CSA:
Begin
Objective function f(x), x= (x1, x2..., xd) T;
Generate initial population of n host nests xi i=1, 2...n
While (t< Max Generation) or (stop criterion)
        Get a cuckoo randomly by Levy Flights
        Evaluate its fitness Fi
        Choose a nest among n (say j) randomly
        If (Fi>Fj)
        Replace j by the new solution;
        End If
        A fraction (pa) of worse nests is abandoned and new ones
        are built;
        Keep the best solutions (or nest with quality solutions)
        Rank the solution and find the solution and find the current
        best
End while
Post process results and visualization
End Begin
```

By using the connotation of Cuckoo searching and Levy flight, the developed algorithm is giving an efficient image optimization results. The implementation steps of the algorithm are shown in the flowchart of Fig. 8.7 and explained below in details [21]:

Stage 1: Reading the colored-image and converts it into its grayscale equivalent image. The function that is used for the conversion process is:

```
rgb2rray function. grayI = rgb2gray(imread(fn));
```

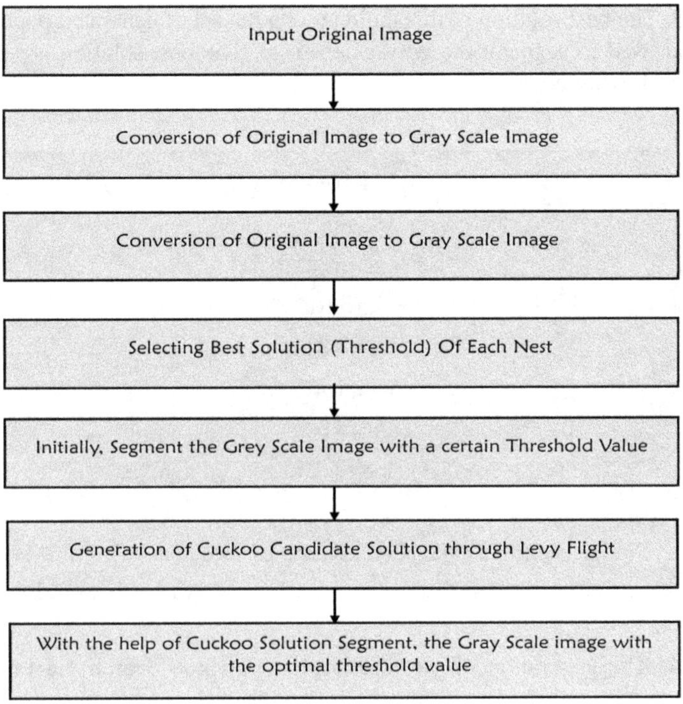

Fig. 8.7 A flowchart explaining the steps of the algorithm

Stage 2: Threshold will be created, and its upper and lower bounds are defined as:

```
Lb=0; Ub=255;
for i=1:10, for j=1: l
nest (i, j) = round((Lb+(Ub-Lb). * rand (size (Lb))));
```

Stage 3: The best solution of threshold among the set of generated populations is chosen and used to segment the grayscale image. The best solution is selected by using the following code:

```
a=0;
b=0;
c = zeros;(10 ,1)
d = zeros;(10 ,1)
for k=1:10
a=0; b=0;
for j=1: k
[c (k), d (k)] = calbest (gi, nest (k, j), a, b, j);
a =c (k);
b = d (k); %best solution of each nest
```

Stage 4: The generation of the threshold by Cuckoo Search via Levy Flight Algorithm is accomplished by using the following piece of code:

```
c1=0;
d1=0;
for k=1:10
for j=1: k
cuck = levii (nest (k, j), d1, j);
if cuck ~=z
[c2 d2] = calbest (gi, cuck, c1, d1,1);
c1=c2;
d1=d2;
```

Stage 5: The Levy flights concepts will be used in the generation the optimized threshold values as in the following code:

```
beta=3/2;
sigma=(gamma(1+beta) *sin(pi*beta/2)/(gamma((1+beta)/2)
*beta*2 ^ ((beta-1)/2))) ^ (1/beta);
for j=1: k,
s=n;
u = randn (size(s)) * sigma;
v = randn (size(s));
step = u./abs (v) ^ (1/beta);
best = curthresh;
stepsize = 0.01*step * (s-best);
s = s + stepsize * randn (size (s));
s=round(s);
```

Stage 6: This step is used if you need to segment the image by using all the obtained solutions. The code for segmenting an image and calculating the correlation of segmented image as follows:

```
function [gi1] = seg (gi, bestthresh)
[o p] = size (gi);
for i =1: o
for j =1: p
if gi (i, j) < bestthresh
gi1 (i, j) = 0;
else gi1 (i, j) = 255;
end; end; end;
message = sprintf ('Best Threshold Value: %d', bestthresh');
msgbox (message);
end;
```

8.7 MSE and PSNR Value Calculations

After getting the segmented image, the mean squared error (MSE) of the image is created using the grayscale image and the final segmented image to get the PSNR value. The higher value of PSNR the better the result [21].

The term 'signal' in the PSNR value formula shows that the original image has some 'noise' that will produce an error in the image segmentation process. Thus, if high PSNR value is found, it gives an idea that decent quality segmentation and success. The term "MSE" will be used in the formulas to give the cumulative squared error between the segmented image and the original image. The code for getting the MSE and PSNR values is as shown below:

```
function [sei] = psnr1 (gi, gi1)
format long g;
format compact;
fontSize = 20;
ii = gi;
[rows columns] = size (ii);
si = gi1;
sei = (double (ii) – double (si)) ^ 2;
mse = sum (sum (sei)) / (rows * columns);
PSNR = 10 * log10 (256^2 / mse);
```

8.8 Results and Discussion

The performance of CS algorithm and morphological operation [19] which were used in the image enhancement process has been tested with different images and the upcoming figures and values show the execution of the suggested methodology. Here, independent stages are performed sequentially: firstly, we processed the cuckoo search algorithm, then we applied a morphological operation to improve the image and finally, the enhanced image was obtained. In order to adjust the contrast value of the processed image, the CS algorithm was iteratively applied until we obtained the optimal contrast value.

Table 8.1 Performance analysis of GA, PSO, ABC, and CS

GA	PSO	ABC	CS
16.8841	18.11698	18.65656	18.61738
17.7551	18.06440	18.07211	18.71658
15.0671	15.56377	15.76718	15.71828
14.208	15.10659	15.41357	16.23400

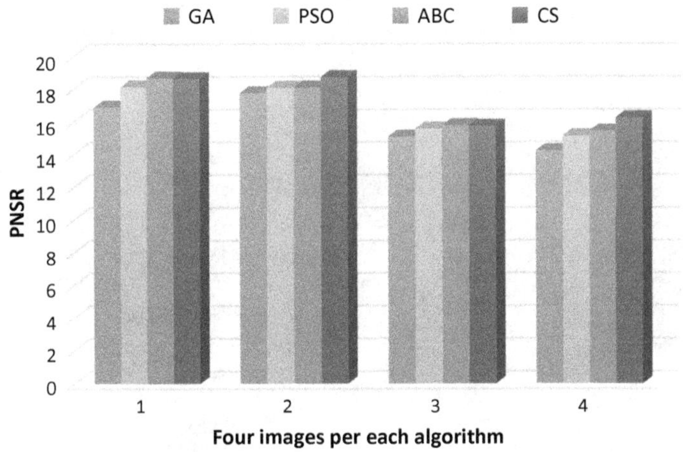

Fig. 8.8 Performance benchmarking of GA, PSO, ABC, and CS for four images

Table 8.1 benchmarks the performance of CS algorithm with other three well-known existing algorithms namely: genetic algorithm (GA), particle swarm optimization algorithm (PSO), and Artificial Bee Colony (ABC) respectively. Also, Fig. 8.8 illustrates the graphical representation of the comparison result between GA, PSO, ABC, and CS in terms PNSR for the images provided in the figure set 8.9a–d that provides the input images and their enhanced images. It's clearly seen that GA, PSO, ABC, and CS have recorded PNSR values as 17.627926, 16.0594, 17.113572, and 17.334228 respectively. The performance figures show that the grey images are effectively improved with image enhancement technique by using CS algorithms and morphological operations.

Another noticeable system implementation has been found in [21] where they have used the MATLAB package to utilize the identical combination of algorithms [7] as shown the figure set 8.10. It is clearly seen that the resulted image contains a very little amount of noise due to the high PSNR value with significantly less size but the same quality as the original image, because of the algorithm.

Fig. 8.9 Four original images each with its enhanced image

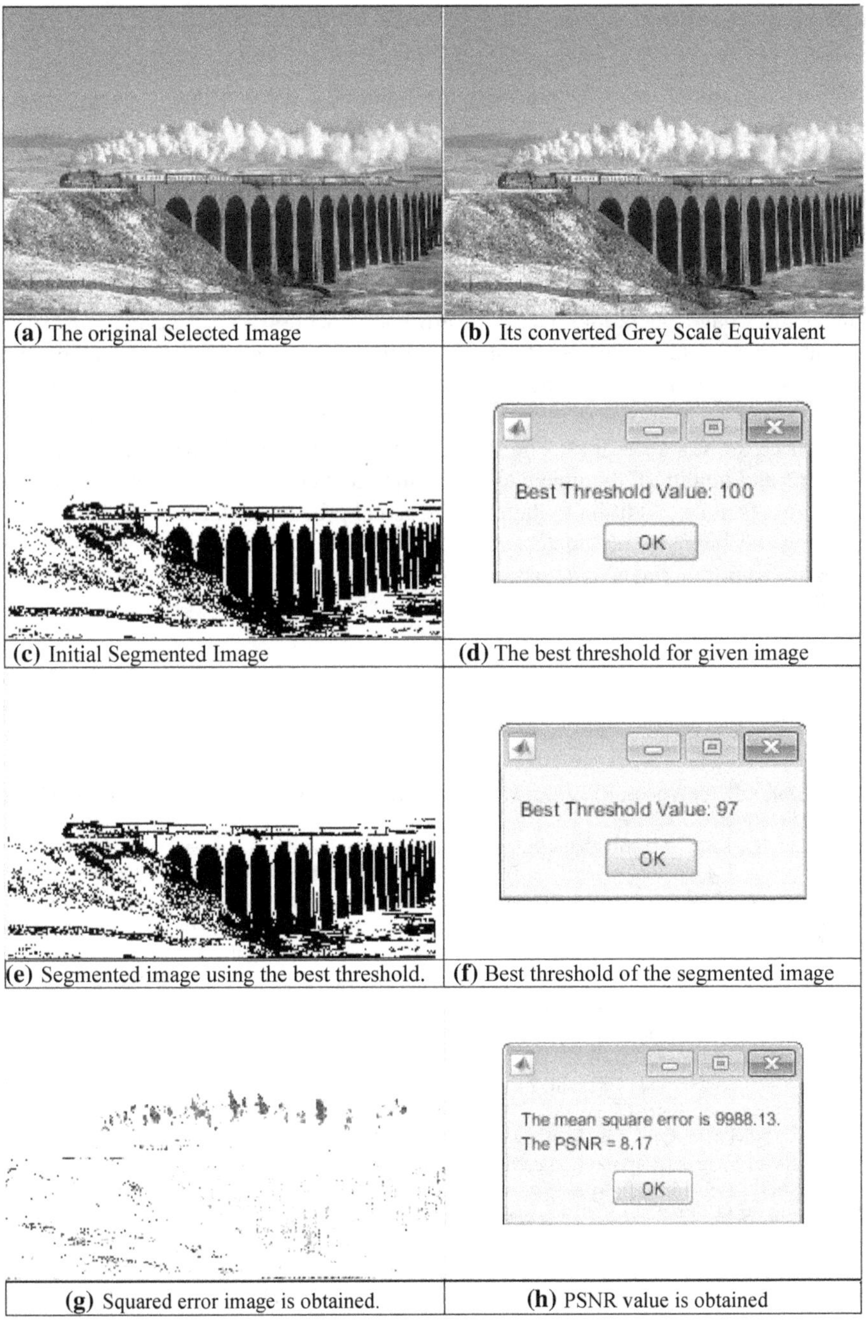

Fig. 8.10 The sample results of several processing stages at MATLAB

8.9 Conclusions

The development of image enhancement techniques was a worldwide research for a long time ago. Such issue formed a pool of research to avoid or mitigate image noise to clarify several image features which positively contribute to many image processing operations such as the image segmentation, image detection, feature extraction, edge detection and others. The milestone discovery of Cuckoo Optimization Algorithm (COA) in the image processing world has maximally improved the image enhancement techniques by minimizing the human intervention in the image operations using an artificial selection of image parameters to optimize the inline image processing algorithm. To sum up, the CO algorithm works by firstly transforming the inputted full-colored image into gray-scale image and followed by computing the image-contrast parameter by applying the fitness function of the CS algorithm. The main purpose of using CS algorithm to elevate the merit and quality of the image to obtain the most effective and desirable amount of contrast factor in addition to the morphological processes which were fulfilled by adjusting the intensity parameters. CS algorithm showed the best results of noise removal from the image and choose the best parameter and contrast value to enhance the image.

References

1. Gonzalez, R.C., Woods, R.E.: Digital Image Processing, 3rd edn. Pearson Publications, Upper Saddle River, NJ (2007)
2. Singh, N., Kaur, M., Singh, K.V.P.: Parameter optimization in image enhancement using PSO. Am. J. Eng. Res. (AJER) **2**(5), 84–90. e-ISSN: 2320-0847, p-ISSN: 2320-0936
3. Munteanu, C., Rosa, A.: Towards automatic image enhancement using Genetic Algorithms. In: Proceedings of the 2000 Congress on Evolutionary Computation, vol. 2, pp. 1535–1542. Instituto Superior Tecnico, University Tecnica de Lisboa, Portugal (2000)
4. Gupta, A., Tripathi, A., Bhateja, V.: De-speckling of SAR images via an improved anisotropic diffusion algorithm. In: Proceedings of (Springer) International Conference on Frontiers in Intelligent Computing Theory and Applications (FICTA 2012), Bhubaneswar, India. AISC, vol. 199, pp. 747–754 (2012)
5. Pal, S.K., Bhandari, D., Kundu, M.K.: Genetic algorithms for optimal image enhancement. Pattern Recogn. Lett. **15**(3), 261–271 (1994)
6. Kwok, N.M., Ha, Q.P., Liu, D., Fang, G.: Contrast enhancement and intensity preservation for gray-level images using multiobjective particle swarm optimization. IEEE Trans. Autom. Sci. Eng. **6**(1), 145–155 (2009)
7. Thampi, S.M., Gelbukh, A., Mukhopadhyay, J. (eds.): Advances in signal processing and intelligent recognition systems. Adv. Intell. Syst. Comput. (2014). https://doi.org/10.1007/978-3-319-04960-1_25
8. Shannon, C.E.: A mathematical theory of communication. Bell Syst. Tech. J. **27**, 379–423 (1948)
9. Jaime, M., Beatriz, J., Salvador, S.: Towards no-reference of peak signal to noise ratio. (IJACSA) Int. J. Adv. Comput. Sci. Appl. **4**(1) (2013)

10. Lei, X., Hu, Q., Kong, X., Xiong, T.: Image enhancement using hybrid intelligent optimization. Opt. & Optoelectron. Technol. 341–344 (2014)
11. Gorai, A., Ghosh, A.: Gray-level image enhancement by particle swarm optimization, pp. 72–77 (2009)
12. Weigel, D., Elsmann, T., Babovsky, H., Kiessling, A., Kowarschik, R.: Combination of the resolution enhancing image inversion microscopy with digital holography. Opt. Commun. **291**, 110–115 (2013). https://doi.org/10.1016/j.optcom.2012.10.072
13. Lin, C.L.: An approach to adaptive infrared image enhancement for long-range surveillance. Infrared Phys. Technol. **54**, 84–91 (2011). https://doi.org/10.1016/j.infrared.2011.01.001
14. Zhao, W.: Adaptive image enhancement based on gravitational search algorithm. Procedia Eng. **15**, 3288–3292 (2011). https://doi.org/10.1016/j.proeng.2011.08.617
15. Zeng, M., Li, Y., Menga, Q., Yang, T., Liu, J.: Improving histogram-based image contrast enhancement using gray-level information histogram with application to X-ray images. Optik Int. J. Light Electron Opt. **123**, 511–520 (2012). https://doi.org/10.1016/j.ijleo.2011.05.017
16. Santhanam, T., Radhika, S.: A novel approach to classify noises in images using artificial neural network. J. Comput. Sci. **6**, 506–510 (2010). https://doi.org/10.3844/jcssp.2010.506.510
17. Garg, R., Mittal, B., Garg, S.: Histogram meequalization techniques for image enhancement. Int. J. Electron. Commun. Technol. **2**, 107–111 (2011)
18. Pentapalli, V.V.G., Varma, R.K.P: Cuckoo Search Optimization and its Applications: A Review. CSE Department, MVGR College of Engineering, Vizianagaram, India1 Associate. Professor, CSE Department, MVGR College of Engineering, Vizianagaram, India
19. Babu, R.K., Sunitha, K.V.N.: Original Research Paper Enhancing Digital Images Through Cuckoo Search Algorithm in Combination with Morphological Operation (2014)
20. Yang, X.-S., Deb, S.: Cuckoo search via Lévy flights. In: Proceedings of World Congress on Nature and Biologically Inspired Computing (NaBIC 2009), India, pp. 210–214. IEEE Publications, USA (2009)
21. Prashar, P., Jain, N., Mahna, S.: Image optimization using Cuckoo search and levy flight algorithms. Int. J. Comput. Appl. (0975–8887) **178**(4) (2017)

Noor A. Jebril is a lecturer of Computer Sciences at King Faisal University. She is a Jordanian resident (Married) proficient in both languages Arabic and English. Eng. Noor received his B.S. in software engineering from Philadelphia University in Aug/2009. She joined the graduate program at Yarmouk University of Science & Technology in Aug/2011 where she received her M.Sc. degree in computer engineering before she joins the faculty staff at King Faisal University as a lecturer of computer science department. Her research interests include Digital Image processing, Biomedical hardware, and software, Computer programming, algorithms, and security.

Qasem Abu Al-Haija is a senior lecturer in Electrical and Computer Engineering at King Faisal University. Eng. Abu Al-Haija was born on July-1982 and received his B.S. in ECE from Mu'tah University in Feb/2005. Then he worked as a network engineer in a leading institute at KSA as well as a lecturer before he joined the graduate program at Jordan University of Science & Technology in Sept/2007 where received his M.Sc. degree in Computer engineering in Dec/2009. His research Interests: Information Security, Cryptography, coprocessor & FPGA design, Computer arithmetic and algorithms, Wireless sensor networks.

Chapter 9
Artificial Bee Colony Based Feature Selection for Automatic Skin Disease Identification of Mango Fruit

A. Diana Andrushia and A. Trephena Patricia

Abstract Food is very important in Human's life and one of the best natural foods are fruits. These fruits have lot of nutrients that are needed for our day-to-day life, but there are diseases which affects the fruits and makes them non-edible or waste. These diseases in fruits also cause a huge problem in agricultural industries, worldwide. This research work presents the automatic skin disease identification system for mango fruit. Initially, the features of color, texture and shape are extracted from the input mango images. The optimal feature set is selected from the original feature set. Artificial Bee Colony (ABC) optimization is used to get the optimal feature set. The novelty of the proposed method lies on the usage of metaheuristic algorithm in the feature selection of mango skin disease identification. Support Vector Machine (SVM) classifier is used to train and test the feature set. Stem end rot and anthracnose are the two identified mango skin diseases. The performance of the proposed method yields reliable results interms of classification accuracy and receiver operating characteristics which are compared to the state-of-the art methods.

Keywords Feature selection · Artificial bee colony optimization
Skin disease · Mango fruit

9.1 Introduction

In India the most cultivated fruit is mango and many types of mangoes are cultivated in various places of India. The season for mango is in the month of May. These mangoes are even exported to other countries like UAE, Saudi Arabia,

A. Diana Andrushia (✉)
Department of ECE, Karunya Universtiy, Coimbatore, India
e-mail: andrushia@gmail.com

A. Trephena Patricia
Department of ECE, Panimalar Engineering College, Chennai, India
e-mail: trephyece@gmail.com

J. Hemanth and V. E. Balas (eds.), *Nature Inspired Optimization Techniques for Image Processing Applications*, Intelligent Systems Reference Library 150,
https://doi.org/10.1007/978-3-319-96002-9_9

Kuwait, Qatar and Nepal. According to the statistics taken in the year 2014–2015 India exported around 42,998.31 million tons of mangoes. However this production is greatly reduced because of the fruits affected by diseases and pests. In India monitoring the health and detection of diseases in fruits are done manually by experts but it is expensive and requires more time. The diseases of fruits affect the quality of the fruit in the pre harvesting and post harvesting time.

These diseases do not affect the fruit alone but also the twigs, leaves, branches and other parts of the tree. It affects overall yield and also results in deterioration of that particular variety from cultivation. If the diseases are found earlier then the fruits and trees can be protected by proper fertilizers.

Every process around us is automated in our day to day life. Therefore an attempt has been made to detect the fruit diseases using image processing techniques. This technique will be helpful to identify the diseases accurately at different stages. So that the farmers can use right amount of fertilizers and chemicals to cure the diseases. This method can reduce time, man power and also the money spent for consulting experts in identifying the diseases.

The automatic skin disease identification system plays a major role in the food processing industry. Once the fruit diseases are classified according to the severity of the disease the fruits may be contaminated or discarded from the system. Number of image processing techniques are used in the field of agriculture to identify the diseases in fruit items. It reduces the time and manpower consumption of food packaging industry [1]. Due to this many research findings are available in the area of automatic disease identification in citrus, mango, apple and pomegranates [2].

Fruit grading [3], fruit disease identification, fruit classification, fruit quality checking are some of the areas in which lot of image processing methods are applied. All these methods are automated with the help of computer vision methods. The defect detection in the automated system involves image processing algorithms. In order to identify the diseases in the fruits the image processing techniques are applied. Threshold methods, morphological operators, segmentation methods, neural network methods were also used by various researchers.

The proposed method deals with automatic detection of skin disease identification using artificial bee colony optimized feature set. The main stages of this method are: Pre-processing, Feature extraction, optimized feature selection and classification. Initially the backgrounds of the fruits are removed. The features of color, shape, texture are extracted. The features are selected based on the meta-heuristic approach. The optimized features are given to the classifier. The classifier classifies the healthy fruit and the diseased fruit. Stem end rot and anthracnose are the serious diseases in the mango fruit. The symptoms of both the diseases are same. Among post harvesting diseases, stem end rot is more serious disease than anthracnose. These diseases are common for mango, citrus and avocado. They are controlled by using appropriate chemicals in the post harvesting stages. So in this study also these two skin diseases are classified along with the healthy mangoes.

The organization of this research paper as follows: Sect. 9.2 explains about the backgrounds of the skin disease identification and classification. Section 9.3 presents the details of proposed method. The subsection clearly explains the artificial bee colony based feature selection and SVM classification. Section 9.4 shows the experimental results and performance analysis of proposed method. Last section gives the conclusion of the proposed research.

9.2 Backgrounds

9.2.1 Fruit Analysis

The analysis of fruit images include defect detection, fruit grading, fruit quality checking etc., Many researchers reordered their research findings by considering single fruit and some researchers have not done for specific fruits. Dubey [4, 5] proposed a method for segmenting disease region by adopting K-means clustering approach. Initially the images are divided into different regions. Based on the different clusters the defected regions are identified and segmented.

Zhang et al. [6] proposed a method for fruit classification. Eighteen different types of fruits are classified which includes berries, grapes and pine apples. Color histogram, shape and texture features are extracted from input image dataset. Feed forwarded neural network was used to classify the different types of berries, grapes and pine apples. The different fruit varieties are classified with classification accuracy of 89%. The improved method of Zhang et al. [6] is proposed in Zhang et al. [7]. It includes biogeography based optimization and wavelet based feature selection. So the accuracy of this system was improved to 90%.

Dubey et al. [8] presented a method for apple disease classification. The major features of color, shape and textures are used. Global color histogram for color features, completed local binary patterns for texture features, zernike moments for shape features are extracted. Three different types of apple diseases are classified namely scab, rot and blotch. K means clustering used initially for finding the infected apple fruits. Support vector machine with multiclass is used to identify the three diseases of apple. However for the apple disease classification, the effect of shape features are not suited which is highlighted in this research. The classification accuracy of this method is given by considering individual features and feature combinations. While classifying three types of apple diseases, it is hard to classify apple blotch. So this category yields poor accuracy when compared to the other classes of diseases.

Orange grading system was proposed by Thendral et al. [9]. In order to identify the good quality of oranges, a computer vision method was proposed. For the post harvesting time, it is mandatory to examine the skin defects of each fruit. The parameters which are considered for orange grading system are size, color, shape and texture. Genetic algorithm based feature selection is involved in this orange grading system. It empowers the classification tasks and accuracy of the overall system. The optimized features through genetic algorithm is given to various types of classifiers. Among the various classifiers auto associative neural network produced peak accuracy of 94.5%.

Bhange et al. [2] presented the method for defect detection on pomegranate fruits. A web based method is launched in this method which produced smart farm scheme. It is based on modern agricultural scenario. One type of skin disease in pomegranate is bacterial blight which is detected in this paper. The basic features of color and morphology are used to classify the defective pomegranate fruits and non defective pomegranate fruits. Based on the image quality various accuracy levels are obtained. The overall accuracy of this system is 82%. It has created smart farming approach for farmers to take decisions and preventive steps during initial stages of the disease.

Sabzi et al. [1] proposed a method for classifying orange varieties. Three varieties of oranges are classified namely bam, Thomson and paybandi. Three hundred color orange images are used for the classification. Hundred images are used under each category. Large number of features are extracted from each category of orange images. Three different metaheuristic algorithms are used to find the reduced feature set which are harmony search, artificial bee colony, particle swarm optimization. After getting the optimal feature set, the classifiers are used to classify three different types of oranges. The classification accuracy of the system is 96.7%.

Sa'ad et al. [3] proposed a method for mango grading using shape features. Harumanis mangoes are only used for this grading system analysis. A method of fourier descriptor is used to grade the different type of mangoes. The cylinder approximation method is used to find the weights of each mangoes. Initially all the input images are done with image enhancement, feature extraction and image restoration. Discriminant analysis and support vector machines are the two classifiers which are used in the grading system. Both the classifiers are used to analyse the shape features only. SVM method produced 100% classification accuracy in the mango grading system. Where as discriminant analysis classifier achieved 95% of classification accuracy.

Fernando et al. [10] and Li et al. [11] presented research paper in citrus defect detection. The proposed method [10, 11] detected fungal decay in citrus fruit. Citrus fruit affected by fungal infection in the post harvesting stage. The hyperspectral imaging is used in this method. At first mean normalization is used in the input images. The principal component analysis is used to reduce the dataset. Image segmentation method is applied in the last stages in order to segment the defected portions.

Chaugule et al. [12] presented a method for classification of paddy. The shape and texture features are the main features which are used in the four varieties of paddy. The four paddy varieties are ratnagiri-2, ratnagiri-24, ratnagiri-4 and karjat-6. Many texture and shape features are taken to train the classifiers. Among all these features, central moments of shape features play a major role in the classification task. It helps the overall system to get high classification accuracy. The average accuracy of the system is 88%. In this study the texture features are given lower accuracy compared to the shape features.

Dutta et al. [13] presented a method for classification of grapes using image processing techniques. This method was used to separate the fresh grapes among the pesticide affected grapes. All the input images are analysed in the frequency domain. The discriminatory features are extracted in transform domain by applying Haar transform. The discriminatory pattern is discovered by applying three level wavelet decomposition. The difference in the features which are related to fresh grapes and pesticide affected grapes are obtained. These feature sets are given to the training and classification method. SVM is used to classify both the cases of grapes. This method accurately classified the pesticide affected grapes than fresh grapes.

All the fruit analysis literatures are involving four main stages, which are image collection, preprocessing, feature extraction and classification. If the automatic system involves with fruit grading, defect detection then the color, texture and shape features are only considered. The first order and second order statistical features are calculated for texture feature extraction.

9.2.2 *Classification Methods*

Classification is the major step in the automatic disease detection and identification methods. There are many methods proposed to classify the types of fruits and classify their defects. Single stage classifiers and multi stage classifiers are used in many research papers. Auto associative classifiers, back propagation method, harmony search classifier, K nearest neighbour classifier, Support Vector machines are the important classifiers which are used in the fruit disease classification. Support vector machine [14] classifier yields good results in many research findings which are related to the fruit grading and fruit disease classification. In recent past, very few research papers used optimization methods for feature selection. The research papers which used optimization techniques for feature selection produced very good outcomes compared to the other methods.

The biological inspired optimization methods are used in many research findings in order to find the important features for training and classification of fruits. Genetic algorithm, ant colony optimization, artificial bee colony optimization, particle swarm optimization and simulated annealing are some of the optimization methods which are used in the feature selection of fruit grading and defect classification methods. The proposed method deals with mango fruit skin disease identification using optimized feature set which are derived from artificial bee colony method.

9.3 Proposed Method

The aim of this research paper is to develop a method to classify the skin diseases of mango based on optimized feature set. Color, texture and shape features are used in the proposed method. Artificial Bee Colony optimization (ABC) is used to identify

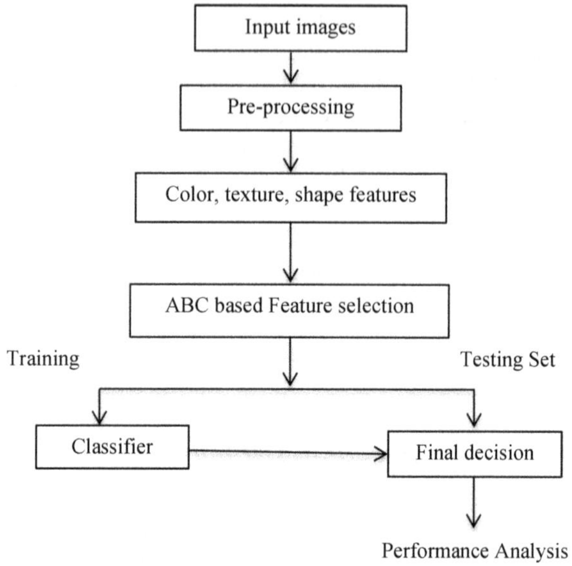

Fig. 9.1 Block diagram of the proposed method

the optimal feature set. In order to nullify the redundant features, the ABC optimization is used. SVM classifier is used to classify the healthy fruits and diseased fruits (Fig. 9.1).

9.3.1 Input Dataset

The mango samples are collected from the different agricultural farms and online database [http://www.cofilab.com/portfolio/mangoesdb/] [15]. In this present study totally 150 images are collected. Where in 70 images are healthy mango images and 80 images are defected mango images. 40 images from anthracnose disease and 40 images from stem end rot disease are taken for the proposed work.

9.3.2 Preprocessing

The initial step of any image processing approach is preprocessing the input image. It aims to remove the noise in the image. In this method the background of the fruit images are removed using appropriate thresholds. The input images are represented with different color spaces namely RGB, YCbCr, HSV, CMY. In this method RGB color space is chosen in order to remove the backgrounds. It is the optimal color space. The appropriate threshold is applied to segment the background from the fruit foreground. It is applied to all the three color channels R, G, B. Equation 9.1 is applied to the pixel (x,y) in the input image in order to identify the background.

$$\text{if}((R(x,y)+G(x,y)+B(x,y))/3) < 35\ (x,y)\ \text{is the background} \tag{9.1}$$

9.3.3 *Feature Extraction—Color, Shape, Texture Features*

Feature extraction is the key step in all the methods of the image processing. This step is used to represent the input images in efficient way. Three types of features namely color, shape and texture are used in this proposed method. Totally 50 features are taken under three categories.

9.3.3.1 Color Features

Most of the machine vision applications, Color is considered as important parameter for representing the input images. The methods which consider the color feature produced better results compared to the other features. In general three components are representing color and intensity. To determine the efficient color component is a big task. In this proposed method 12 different color components are considered which are RGB, HSI, YCbCr, CMY.

9.3.3.2 Shape Features

The shape features are representing the original fruit properties in terms of regularity, unevenness and elongation. The shape features are very much important to describe the foreground object. The shape features are usually represented in terms of geometric feature and shape related features [16]. The geometric features include area, length, width, convexity, major axis length, minor axis length and axis ratio. The shape related features are represented by second order statistical descriptors. The standard moments and normalized central moments are taken in this method. The zernike moment is the rotation invariant moment. Moments are used to find the skinniness of the fruit [17]. Ten standard moments, seven normalized central moments, seven invariant moments and seven geometric features are used in this proposed method. Totally 31 shape features are used.

9.3.3.3 Texture Features

In general texture features are representing the distribution of intensity in various level. These features are exhibiting the smoothness, uniformity, brightness and flatness. The statistical features are calculated to extract texture features. Usually the statistical texture features are well represented in the area of image retrieval, medical image analysis and image classification. In this proposed method, first order and second order statistical texture features are used.

The first order statistical features are mean, variance, standard deviation, correlation, skewness, kurtosis, entropy, homogeneity, energy. These features are calculated from image pixels. The second order texture features are taken from Grey level cooccurrence matrix (GLCM) which specifies the spatial relation tween the pixels. GLCM is calculated for the pixels (x,y) by considering the distance between (x,y) and orientation. GLCM is calculated for four different directions 0°, 45°, 90° and 135°. For four different directions all the 9 features are calculated. Totally 36 texture features are used in the proposed method. $P(x,y \mid d,\theta)$ is the matrix component which is having second order statistical probability with grey levels x and y at particular orientation and particular distance [12].

The local intensity variation/contrast is calculated by the following equation.

$$CR = \sum_{n=0}^{N-1} n^2 \left(\sum_{x=1}^{N} \sum_{y=1}^{N} p(x,y) \right) \quad |x-y| = n \tag{9.2}$$

Correlation specifies the linear relationship between pixels kin the particular direction which is calculated by

$$CRR = \sum_{x=0}^{N-1} \sum_{y=0}^{N-1} \frac{(x-\mu_x)(y-\mu_y)p(x,y)}{\sigma_x \sigma_y} \tag{9.3}$$

where

$$\mu_x = \sum_{x=0}^{N-1} \sum_{y=0}^{N-1} xp(x,y) \quad \mu_y = \sum_{x=0}^{N-1} \sum_{y=0}^{N-1} yp(x,y) \tag{9.4}$$

$$\sigma_x = \sum_{x=0}^{N-1} \sum_{y=0}^{N-1} (x-\mu_x)^2 p(x,y) \quad \sigma_y = \sum_{x=0}^{N-1} \sum_{y=0}^{N-1} (y-\mu_y)^2 p(x,y) \tag{9.5}$$

$$\text{Energy} = \sum_{x=0}^{N-1} \sum_{y=0}^{N-1} (p(x,y))^2 \tag{9.6}$$

$$\text{Homogeneity} = \sum_{x=0}^{N-1} \sum_{y=0}^{N-1} \frac{p(x,y)}{1+|x-y|} \tag{9.7}$$

$$Entropy = -\sum_{x=0}^{N-1} \sum_{y=0}^{N-1} p(x,y) \log(P(x,y)) \tag{9.8}$$

9.3.4 Feature Selection

In this proposed study 12 color features, 31 shape features and 36 texture features are extracted. So initial feature set contains 79 features for 150 images. The initial feature set consists of large number of features which includes some redundant features also. The high number of features may reduce the accuracy of system and affect the classifier accuracy. In the real time scenario it would affect the computational efficiency of the overall system. So it is necessary to remove the redundant or unwanted features [9].

Feature selection is the process of selecting suitable subset in feature space. The random search provides the random subset in the search space. So that many bio-inspired algorithms are developed to choose optimal subset [18]. In order to select or minimize the feature set a metaheuristic approach is followed in the proposed method. The optimal feature set will give the effective features to the classifiers.

9.3.4.1 Artificial Bee Colony (ABC) Algorithm

Artificial bee colony method is used in several domains in terms of wrapper and forward strategies. It is used in the optimization problem. ABC optimization method is used by very few researchers for feature selection.

Artificial bee colony algorithm highlights the brainy behaviour [19] of bees which are scheduled by their tasks: hired, observers and guides. Frisch et al. [20] and Seeley et al. [21] found the hunting fashion of bees, interior and exterior information. The information of about food location is expressed in terms of waggle dance. The ultimate model of ABC shows the intelligent behaviour of bees with respect to the following parameters: food location, engaged bees and unengaged bees.

ABC algorithm consists of three groups of bees: engaged bees, unengaged bees and scout bees. In this method number of bees in the engaged and unengaged are same. The unengaged bees make the decision of food selection in the dance area. The unengaged bee is named as engaged bee when it drives to the food sources. If the engaged bee has the food then it returns as scout bee and it will find new food source by random search [22].

The following are the steps of the ABC based feature selection process.

In this proposed method M is taken as total number of features. Each source is connected with the feature vector of size M. The position of the each feature vector represents the number of features to be checked. If the position is 1 then it indicates the feature is part of the optimal set to be checked. If the position is 0 then it indicates the feature is not a part of the optimal set to be checked. In addition, each food source have quality in terms of fitness function [18].

Step 1: Generate food sources: Let M is the number of food sources (i.e) total number of features. Assume each feature subset is filled with one feature which is having the single position vector

Step 2: Take accuracy as fitness function and give the feature subset to the classifier. So the accuracy is the fitness of the food source

Step 3: Find the neighbours using changing rate (CR) parameter. The neighbour is found from feature vector of food source. The neighbour is defined by taking perturbation for Eq. 9.9.

$$A_{bc} = X_{ij} + \rho_{ij}\left(X_{ij} - X_{kj}\right) \tag{9.9}$$

For the individual food source A_{bc} is determined through an optimization parameter j. The subset j and k is randomly selected. ρ_{ij} is the number in between -1 and 1. The fitness function of the food source is given by

$$fnss = \begin{cases} \frac{1}{1+v_i} \; if \; v_i \geq 0 \\ 1 + abs(v_i) \; if \; v_i < 0 \end{cases}$$

where v_i is the cost function. In the maximization problems the cost function is directly taken as fitness value

Step 4: Take the feature subset of neighbours to the classifier and assign accuracy as fitness

Step 5: Check whether the fitness of the neighbour is good or not? If the neighbour is having better fitness than the original then consider the neighbour source as the new one

Step 6: The unengaged bees collect data about the fitness of new food sources which are visited by the engaged bees. It will check the better fitness or better probability value. At this time the unengaged bees become engaged bees and run the step 3

Step 7: Remember the better food source. By distributing the unengaged bees to the food sources the best fitness is considered

Step 8: Find new scout bees. If the food source is uninhibited then, create a new scout bee which become engaged bees with new food source and run the step 3.

9.3.5 Support Vector Machine

Support Vector Machine (SVM) is found by Boser, Guyon and Vapnik. It is initiated by using statistical learning theory. SVM is a supervised learning method which is used to perform classification and regression analysis. It is also known as discriminative classifier. It consists of hyperplane which classifies the given input to different groups.

The hyperplane is capable of separating the data in the multi-dimensional space. Regularization parameter and kernel are the main parameters of this supervised learning method. Kernel parameter includes linear, polynomial, sigmoid gaussian, hardlim, sine and exponential. If the regularization parameter is large then it creates smaller margin hyperplane. If the regularization parameter is small then it generates larger margin hyperplane (Fig. 9.2). According to the training set single class and multiclass SVM classification are used in various applications. The SVM classification depends on the various parameter selection. It is one of the empirically best classifier.

SVM classifier is one of the most popular classifier which is used in many applications of image processing. It includes human brain tumor segmentation, salient region detection, line segmentation, image compression, image recognition and optical character recognition. It is not only used in image processing but also in bioinformatics and communication system for spectrum sensing.

The SVM algorithm separates the input by using the hyper plane. The training method determines the parameters for the hyperplane. For different class applications it is classified into different groups. The steps which are involving in the pattern recognition related to SVM is given below.

Consider F is the feature vector with features (a, b)

Where $a = (a_1, a_2, a_3, a_4, a_5 \ldots a_n) \quad a_i \in R \quad i = 1, 2 \ldots n$

Each feature belongs to the class $x_k \in \{-1, +1\}$

The training set

$$T = \{(a_1, x_1), (a_2, x_2), (a_3, x_3), (a_4, x_4) \ldots (a_n, x_n)\} \tag{9.10}$$

The linear classifier is denoted with the hyperplane with two regions which are class 1 and class-1 regions

The hyperplane in the search space S is given by

$$\{a \in S | w \cdot x + b = 0\}, \quad w \in S, \; b \in R$$

The product is given by

$$w \cdot x = \sum_{i=1}^{n} w_i x_i \tag{9.11}$$

Initially 79 features are extracted from different features for 150 images. After applying artificial bee colony optimization the feature set is reduced to 38. The feature selection step provides a set of features which are not redundant. The optimal feature set is having the features with important information. The optimal feature set is applied for the training set. The feature vector matrix size is 150 × 38. Training dataset contain 100 images and testing for 50 images which are separated from normal, disease1, disease2 category.

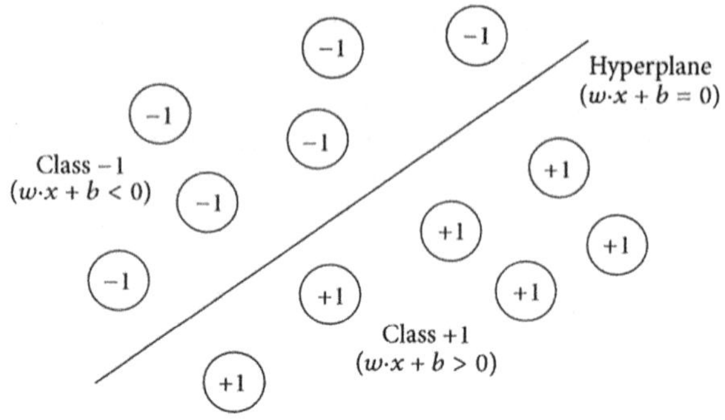

Fig. 9.2 SVM classifier with linear hyperplane—Serter et al. [23]

The training set is separated as linear if any one linear classifier is expressed by the pair (w,b). Once the training gets over, the classifier performs the prediction which is different from the training process. The different classes are obtained from the following expression,

$$class\,(x_k) = \begin{cases} +1 \; if \; w \cdot x_k + b > 0 \\ -1 \; if \; w \cdot x_k - b < 0 \end{cases} \tag{9.12}$$

SVM trains the machines with optimal hyperplace which segregate the input into different classes. It separates the input data into the maximum distance of hyperplane and nearest to the training data points. The results of SVM classifier is given interms of normal fruit, disease1 and disease2 categories (Fig. 9.3).

9.4 Experimental Results

As narrated earlier in the Sect. 9.3, the optimal feature set is selected and given to the SVM classifier. In the optimal feature set many texture and color features are selected by ABC method. Less number of shape features are only available in the reduced feature set. Figure 9.4 shows the images of healthy mangoes and defective mangoes. Two types of diseases are identified. They are stem end rot and anthracnose.

9.4.1 Background Removal

Initially the input images are subjected to the preprocessing stage in which the noises and backgrounds are removed. In order to separate the fruit from the backgrounds, segmentation is applied. Input images in RGB color space is transferred into HSV color space. The contrast in between the foreground and

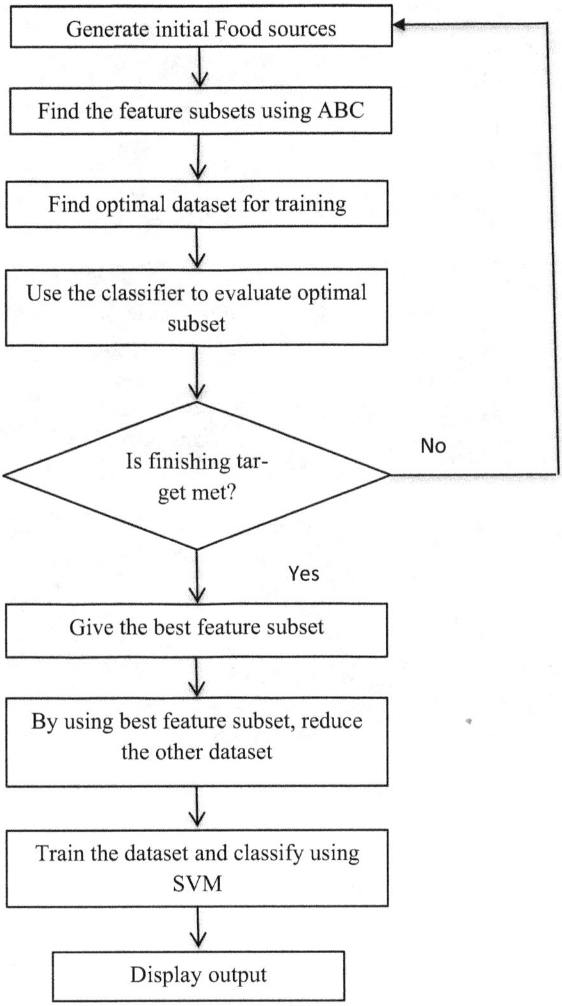

Fig. 9.3 Flow chart of ABC based feature selection and SVM based classification

Fig. 9.4 Images of healthy mangoes and defective mangoes

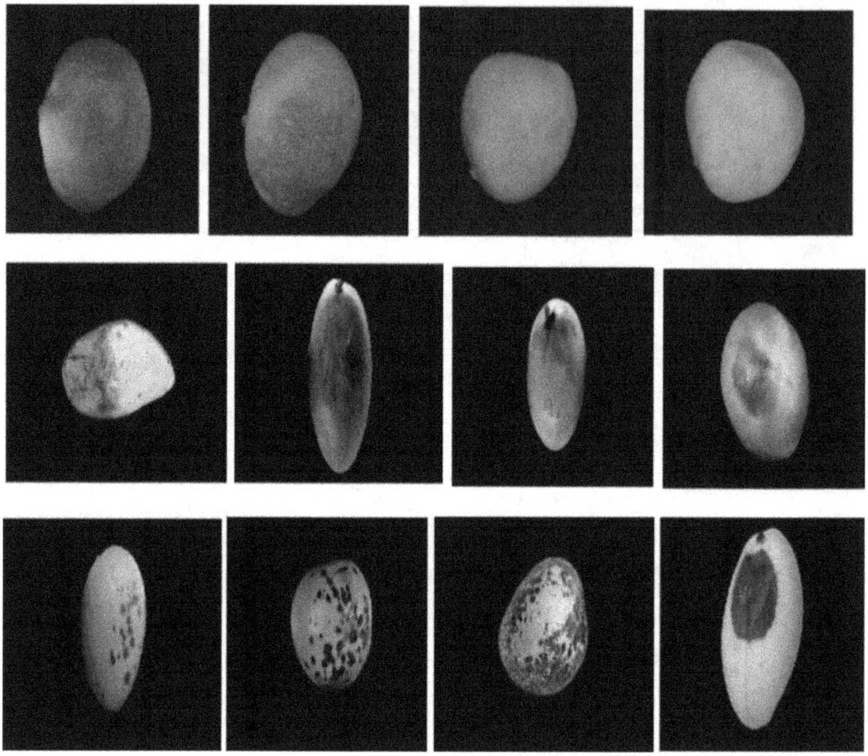

Fig. 9.5 Background removed images row 1: Healthy images, row 2: Anthracnose, row 3: stem end rot

background is not sufficient in RGB color space. So it is converted into HSV color space. In this color space the backgrounds are removed and it is shown in Fig. 9.5.

9.4.2 ABC Based Feature Selection

Artificial bee colony method is used to find the optimal features. Initial feature set is reduced to optimal feature set which includes the features from color, texture and shape. The feature value ranges for different features that are displayed in Table 9.1. It shows some of the important features and their feature values.

Figure 9.6 shows the example images of skin defect identified mangoes. The stem end rot and anthracnose are the commonly occur skin diseases in mango fruit.

Table 9.1 Selected GLCM features in the orientation of 0°

Features	Healthy images (low-high)	Defective images (low-high)
Energy	0.1261–0.467	0.512–0.785
Contrast	0.132–0.143	0.032–0.0867
Auto correlation	2.922–9.345	10.283–18.234
Average	4.121–9.342	11.23–15.333
Variance	1.222–4.534	5.876–8.887
Maximum probability	0.954–0.9712	0.981–1.023
Entropy	0.0812–0.154	0.023–0.0786
Correlation coefficient	0.791–0.982	0.235–0.678

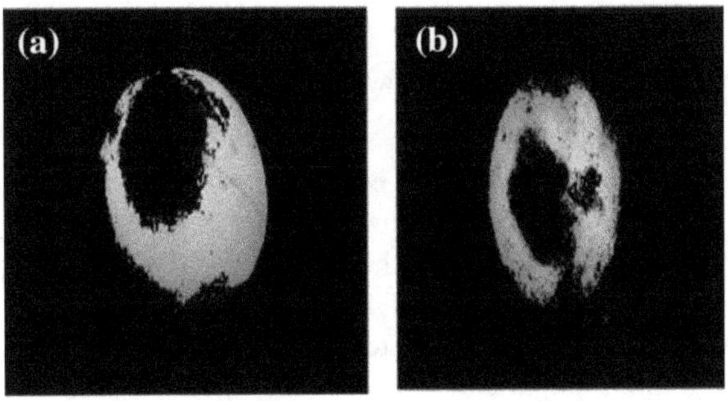

Fig. 9.6 Skin defect identified images **a** stem end rot **b** anthracnose

9.4.3 Performance Analysis

The performance of the proposed method is evaluated in terms of performance metrics. Accuracy of the automatic defect identification system is measured by classification accuracy parameter.

9.4.3.1 Classification Accuracy

The classification accuracy of the skin defect identification system of mango fruit is analyzed in terms of feature set which are color, shape and texture. It is a multi-class classification problem for classifying the mango fruit in three different categories which are healthy, stem end rot and anthracnose disease.

The Classification Accuracy (CA) is defined by

$$CA = \frac{Total\ images\ with\ correct\ classification}{Total\ testing\ images} \times 100 \tag{9.12}$$

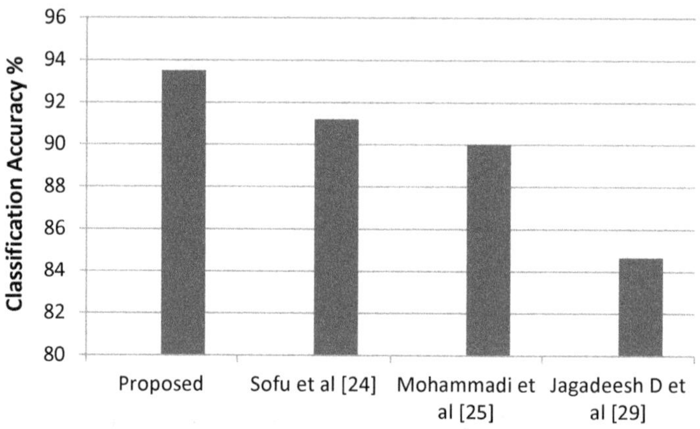

Fig. 9.7 Comparison of classification of accuracy parameter

The results of the proposed method is compared with state-of-the-art methods. Sofu et al. [24] presented a method for the identification of different types of apples. The decision trees are used to classify the types of apples. Mohammadi et al. [25] proposed a method of classification of fruits in three maturity levels. Quadratic Discriminant Analysis (QDA) is used to classify the maturity levels. Figure 9.7 shows the bar graph comparison of proposed method, Sofu et al. [24] and Mohammadi [25] methods.

Momin et al. [26] discussed about mango grading system. The mangoes are classified as small, medium and large level. Median filter based image processing algorithm is used. The parameters of ferret diameter, roundness, perimeter are used to classify the three types of mangoes. The system produced 97% classification accuracy. The mass estimation of mangoes are developed by Schulze et al. [27]. Mass estimation of a fruit is calculated by linear regression and artificial neural network. The system produced 96.7% accuracy rate. Pujitha et al. [28] presented a method for finding the external defects in mango fruit. The bacterial diseases of mango fruit are identified and classified. Input is applied as a video of mango fruit which is converted to 100 frames of image. Jagadeesh et al. [29] presented a survey on early fungal disease on mango and other fruits. The anthracnose affected mango fruit is identified with the classification of accuracy 84.65%.

In this present study the classification accuracy of the proposed method is 94.5%. The comparison methods are having the classification accuracy as 91.2 and 90%. In the proposed method optimal feature selection based on ABC is an important step. The best features are selected through the optimization technique and it yields better result in terms of classification accuracy. Whereas the comparison methods are missed to consider the optimal feature set, thus reduced the efficiency of the classifier. ABC based method reduces the irrelevant features in order to get the optimal feature set. So the proposed method produces reliable results in terms of classification accuracy.

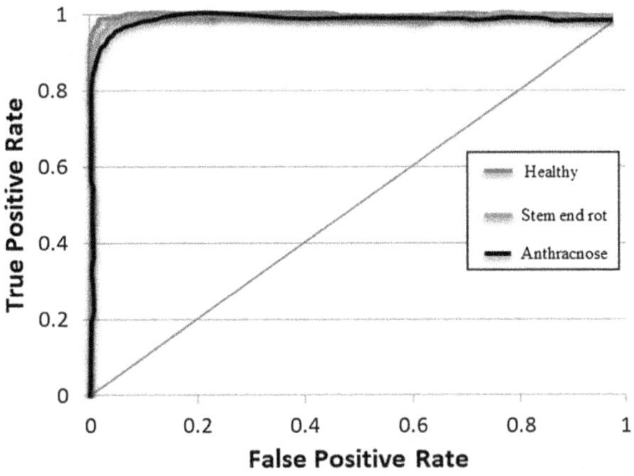

Fig. 9.8 ROC plot for the SVM classifier under three classes

The proposed method is experimented in personal computer with intel i7 dual core, 8 GB RAM, windows 10 and MATLAB 2013 version. The average computational efficiency to process an image is 0.5 s and for the feature extraction step it is 0.31 s.

9.4.3.2 Receiver Operating Characteristics

The proposed method is analyzed in terms of Receiver Operating Characteristics (ROC) as performance metric. ROC is a plot between true positive rate and false positive rate. Area under the curve is high, then the method produces good performance. If it is less than 0.5 then it produces wrong output. Figure 9.8 shows the ROC graph of the proposed method. These classes of output identification is given in Fig. 9.8. For the healthy mango fruit detection is one class. Stem end rot skin disease identification is second class and anthracnose skin disease identification is third class. The area under the curve for each class is 0.972, 0.910 and 0.873.

9.5 Conclusion

One of the favourite fruits of the world is mango and it is very important fruit in India. Identification of skin disease is the crucial problem in the post harvesting stages. This research paper presents a computer vision method to identify the skin diseases of mango fruit. Three set of features are evaluated from the input mango images. Among the initial feature set, the set of essential features are obtained from artificial bee colony optimization. This produces essential features for the training and classification. SVM classifier is used to classify the healthy mango fruit, stem end rot and anthracnose mangoes. The classification accuracy of the proposed method is 94.5% which is the reliable outcome of this research. The proposed method can be extended for the mango grading system with different set of mangoes. The future direction of this research involves, usage of deep learning neural networks to classify different mango diseases.

References

1. Sabzi, S., Abbaspour Gilandeh, Y., Garcia Mateos, F.: A new approach for visual identification of orange varieties using neural networks and metaheuristic algorithms. Inf. Process. Agric. (2017). https://doi.org/10.1016/j.inpa.2017.09.002
2. Bhange, M., Hingoliwala, H.A.: Smart farming: pomegranate disease detection using image processing. Proced. Comput. Sci. **58**, 280–288 (2015)
3. Sa'ad, F.S.A., Ibrahim, M.F., Md.Shakaff, A.Y., Zakaria, A., Abdullah, M.Z.: Shape and weight grading of mangoes using visible imaging. Comput. Elect. Agric. **115**, 51–56 (2015)
4. Dubey, S.R.: Automatic recognition of fruits and vegetables and detection of fruit diseases. Master's theses (2012)
5. Dubey, S.R., Jalal, A.S.: Adapted approach for fruit disease identification using images. Int. J. Comput. Vis. Image Process. **2**(3), 51–65 (2012)
6. Zhang, Y., Wang, S., Ji, G., Phillips, P.: Fruit classification using computer vision and feedforward neural network. J. Food Eng. **143**, 167–177 (2014)
7. Zhang, Y., Phillips, P., Wang, S., Ji, G., Yang, J., Wu, J.: Fruit classification by biogeography-based optimization and feedforward neural network. Exp. Syst. **33**(3), 239–253 (2016)
8. Dubey, S.R., Jalal, A.S.: Apple disease classification using color, texture and shape features from images. Signal, Image Video Process. **10**(5), 819–826 (2016)
9. Thendral, R., Suhasini, A.: Automated skin defect identification system for orange fruit grading based on genetic algorithm. Curr. Sci. **112**(8), 1704–1711 (2017)
10. Fernando, L.G., Gabriela, A.G., Blasco, J., Aleixos, N., Valiente, J.M.: Automatic detection of skin defects in citrus fruits using a multivariate image analysis approach. Comput. Elect. Agric. **71**(2), 189–197 (2010)
11. Li, J., Huang, W., Tian, X., Wang, C., Fan, S., Zhao, C.: Fast detection and visualization of early decay in citrus using Vis-NIR hperspectral imaging. Comput. Elect. Agric. **127**, 582–592 (2016)
12. Chaugule, A., Mali, S.N.: Evaluation of texture and shape features for classification of four paddy varieties. J. Eng. (2014)
13. Dutta, M.K., Sengar, N., Minhas, N., Sarkar, B., Goon, A., Banerjee, K.: Image processing based classification of grapes after pesticide exposure. LWT Food Sci. Technol. **72**, 368–376 (2016)
14. Zhang, Y., Wu, L.: Classification of fruits using computer vision and a multiclass support vector machine. Sensors **12**, 12489–12505 (2012)
15. Cubero, S., Diago, M.P., Blasco, J., Tardaguila, J., Millán, B., Aleixos, N.: A new method for pedicel/peduncle detection and size assessment of grapevine berries and other fruits by image analysis. Biosyst. Eng. Spec. Issue Image Process. Agric. **117**, 62–72 (2014)
16. Shouche, S.P., Rastogi, R., Bhagwat, S.G., Sainis, J.K.: Shape analysis of grains of Indian wheat varieties. Comput. Elect. Agric. **33**(1), 55–76 (2001)
17. Hu, M.K.: Visual pattern recognition by moment invariant. IRE Trans. Inf. Theory **8**, 179–187 (1962)
18. Schiezaro, M., Pedrini, H.: Data feature selection based on artificial bee colony algorithm. EURASIP J. Image Video Process. **47** (2013)

19. Karaboga, D.: An Idea Based on Honey Bee Swarm for Numerical Optimization. Technical Report-tr06, vol. 200, Erciyes University, Engineering Faculty, Computer Engineering Department (2005)
20. Frisch, K., Lindauer, M.: The language and orientation of the honey bee. Ann. Rev. Entomol. **1**, 45–58 (1956)
21. Seeley, T.: Honey bee Ecology: A Study of Adaptation in Social Life. Princeton University Press, Princeton (1985)
22. Zorarpaci, E., Ozel, S.A.: A hybrid approach of differential evolution and artificial bee colony for feature selection. Expert Syst. Appl. **62**, 91–103 (2016)
23. Uzer, M.S., Yilmaz, N., Inan, O.: Feature selection method based on artificial bee colony algorithm and support vector machines for medical datasets classification. The Sci. World J. (2013)
24. Sofu, M.M., Erb, O., Kayacan, M.C., Cetissli, B.: Design of an automatic apple sorting system using machine vision. Comp. Elect. Agric. **127**, 395–405 (2016)
25. Mohammadi, V., Kheiralipour, K., Ghasemi-Varnamkhasti, M.: Detecting maturity of persimmon fruit based on image processing technique. Sci Hortic-Amsterdam **184**, 123 (2015)
26. Momin, M.A., Rahman, M.T., Sultana, M.S., Igathinathane, C., Ziauddin, A.T.M., Grift, T. E.: Geometry based mass grading of mango fruits using image processing. Inf. Process. Agric. **4**, 150–160 (2017)
27. Schulze, K., Nagle, M., Spreer, W., Mahayothee, B., Müller, J.: Development and assessment of different modeling approaches for size-mass estimation of mango fruits (*Mangifera indica* L., cv. 'Nam Dokmai'). Comput. Elect. Agric. **114**, 269–276 (2015)
28. Pujitha, N., Swathi, C., Kanchana, V.: Detection Of External Defects On Mango. Int. J. Appl. Eng. Res. **11**(7), 4763–4769 (2016)
29. Pujari, J.D., Yakkundimath, R., Byadgi, A.S.: Image processing based detection of fungal diseases in plants. Proc. Comput. Sci. **46**, 1802–1808 (2015)

Chapter 10
Analyzing the Effect of Optimization Strategies in Deep Convolutional Neural Network

S. Akila Agnes and J. Anitha

Abstract Deep convolutional neural network (DCNN) is a powerful model for learning significant data at multiple levels of abstraction form an input image. However, training DCNN is often complicated because of parameter initialization, overfitting and convergence problems. Hence this work has been targeted to overcome the challenges of training DCNN with an optimized model. This chapter describes a deep learning framework for image classification with cifar-10 dataset. The model contains a set of convolutional layers with rectified linear unit activation function, max-pooling layers, and a fully-connected layer with softmax activation function. This model learns the features automatically and classifies the image without using the hand-crafted image based features. In this investigation, various optimizers have been applied in gradient descent technique for minimizing the loss function. Model with Adam optimizer constantly minimizes the objective function compared with other standard optimizers such as momentum, Rmsprop, and Adadelta. Dropout and batch normalization techniques are adapted to improve the model performance further by avoiding overfitting. Dropout function deactivates the insignificant node form the model after every epoch. The initialization of a large number of parameters in DCNN is regularized by batch normalization. Results obtained from the proposed model shows that batch normalization with dropout significantly improves the accuracy of the model with the tradeoff of computational complexity.

Keywords Batch normalization · Convolutional neural network
Dropout · Image classification · Optimization strategies

S. Akila Agnes · J. Anitha (✉)
Department of Computer Sciences Technology, Karunya Institute of Technology and Sciences, Coimbatore, India
e-mail: anitha_j@karunya.edu

S. Akila Agnes
e-mail: akilaagnes@karunya.edu

J. Hemanth and V. E. Balas (eds.), *Nature Inspired Optimization Techniques for Image Processing Applications*, Intelligent Systems Reference Library 150,
https://doi.org/10.1007/978-3-319-96002-9_10

10.1 Introduction

Object classification plays a significant role in the area of computer vision. The goal of this process is to classify the objects into different categories, in the field of robotics to any intelligent systems. It is applied in various application domains such as medical imaging, vehicle tracking, industrial visual inspection, robot tracking, biometric systems and image remote sensing. The classification system examines the numerical properties of different image features and classifies them into different categories. It consists of two stages including training and testing. In training stage, the significant features of the input images are used to train the classification system against the target class. In testing stage, the classifier predicts the class for the input image.

The plethora of image classification methods has been proposed in the literature [1]. Various Machine Learning (ML) approaches such as Artificial Neural Network (ANN), Decision Tree Classifier, Support Vector Machine (SVM), and Expert System have been employed in the field of computer vision that label the input images to the desired category. The supervised learning algorithms enable the computer to learn on its own from the available dataset with labels and make predictions for the given data.

The efficiency of the machine learning system relies on the design of several handcrafted features extracted from the images. Despite various object classification algorithms and systems are introduced, there lacks a general and complete solution for recent challenges. New computational models such as Deep Learning (DL) models motivate the researchers to move towards the Artificial Intelligence. Deep learning has been evolved in 2006 with Deep Belief Networks (DBNs) [2] as a part of a machine learning algorithm that exploits many layers of non-linear information processing for pattern analysis and classification [3].

Supervised deep networks employ with labeled information and classify the input data in these labels. They exemplify the most common form of ML, deep or not [4]. This network is more flexible to build, more appropriate for end-to-end learning of complex systems [5] and more capable to train and test. It can be categorized into linear supervised deep method (e.g. Deep Neural Networks with linear activation functions) and non-linear supervised method (e.g. Deep Stacking Networks, Recurrent Neural Networks and Convolutional Neural Networks).

Conventional Neural Network (CNN) is the kind of DL which has been used in various applications of computer vision [6–13], especially for the classification of large sets of images. The performance of deep CNN is highly associated with the number of layers. It also has millions of parameters to tune with, which requires a large number of training samples. First Convolutional Neural Network is introduced by LeCun et al., in 1998 [8], has been the mainstream architecture in the neuronal network family for image classification tasks. Naturally, a CNN is specialized to learn useful local correlations and associate features in low level layers that support higher order learning. Further using Fully Connected (FC) layers in a general feed-forward neural network, CNN also depends on several convolutional and pooling layers before FC layers.

AlexNet [14] and VGG [15] networks have achieved better performance on image classification using deeper convolutional neural networks. Recently many researches have been moving in the field of deep networks. The advantage of deep CNN in images classification is that the entire model is trained end-to-end, from raw pixels to specific categories, which removes the requirement of handcrafted feature extraction.

The popular deep CNN architecture [14] composed of five convolutional layers and three fully-connected layers with a final soft-max classifier, and contains more than 60 million parameters. Some deeper networks, such as models with 16 and 19 hidden layers [15], 22 hidden layers [16] have attained better performance with more number of parameters. However, training deep CNN has several difficulties including vanishing gradients and overfitting [17]. This can be resolved by training a deeper CNN with well-designed architecture, initialization strategies, better optimizers and transfer learning.

As the gradient is back propagated through the network only a few blocks that learn suitable representations and many blocks contribute very little information towards the final goal. This problem is called as diminishing feature. This is solved by a methodology called dropout that disables the corresponding residual blocks during training [18]. Dropout methodology has been first introduced by Srivastava et al. [19] and adopted in many successful architectures [14, 15]. Mostly this is applied to top layers that had a large number of parameters to prevent feature overfitting. Another methodology named batch normalization [20] has been introduced to reduce the internal covariate shift in neural network activations by normalizing them to have specific distribution. This can also works as a regularizer and the researchers experimentally show that a network with batch normalization achieves better accuracy than a network with dropout. Directly learning so many parameters from only thousands of training samples will result in serious overfitting even though the overfitting preventing technique is applied. Therefore, there is a challenge on how to make the deep CNN that fit small dataset while keeping the similar performance as on large-scale dataset.

As a popular benchmark in this field, the cifar-10 database [21] is frequently used to evaluate the performance of classification algorithms. Krizhevsky [22] has carried a classification task on the cifar-10 dataset using a multinomial regression model. This uses all the layers and a single hidden layer that resulted in an overall accuracy of 64.84%. Liu and Deng [23] has proposed a modified VGG-16 network and achieved 8.45% error rate on CIFAR-10 without severe overfitting.

This chapter presents a deep CNN (DCNN) architecture to classify the images in the cifar-10 dataset. The presented architecture overcomes the problems in gradient descents (such as vanishing gradients and overfitting) by integrating suitable layers, optimizers, drop out and batch normalization strategies. The architecture uses the Adam optimizer as an efficient optimizer for cifar-10 dataset classification. The suitable optimizer is selected based on the analysis of different optimization strategies, which aim to minimize the objective function. Further, the effect of dropout and batch normalization also evaluated in the presented architecture. The experimental results show that the presented architecture significantly decreases the loss function with improved validation accuracy.

The rest of the chapter is organized as follows: Sect. 10.2 describes the general CNN architecture and its specifications. Section 10.3 deals with the proposed deep CNN architecture for cifar-10 dataset classification. Results and discussions are reported in Sect. 10.4. Finally, Sect. 10.5 presents the conclusion.

10.2 CNN Architecture

Convolution neural network is a back propagation neural network that works on images. CNN architecture has set of convolutional layers followed by fully connected layers and a final softmax layer that makes predictions. CNN layers learn the parameters using backpropagation algorithm. Convolutional layer acquires the significant special representation from an image, which is essentially used for categorizing images. Generally, the performance of any classification technique depends on the features considered for grouping the data. Selecting interesting and discriminative features from images is the very tedious task. However these extracted features may not be appropriate for all classification problems.

Convolution neural network is able to learn these features automatically to make better predictions without human intervention. Almost every convolutional layer is followed by a non-linear activation function, which helps the network to learn discriminative representations of the image that improve the classification accuracy. Figure 10.1 shows the typical CNN architecture.

The layers involved in the architecture are:

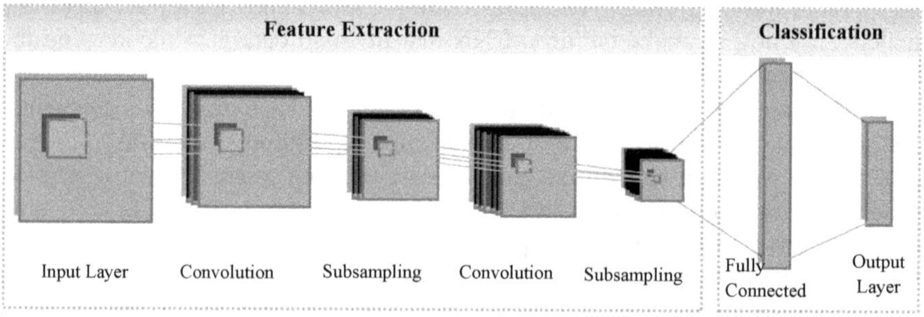

Fig. 10.1 The typical CNN architecture

10.2.1 Convolution Layer

Convolution layers are described by weights. This has multiple kernels per layer with fixed size, and each kernel is convolved over the entire image with a fixed stride that extracts a spatial or temporal features. The low-level features such as lines, edges, and corners are learned in the first convolution layer. More complex representations are learned in the consequent convolutional layers. As the network is deeper and deeper, the learned features contain higher-level information. The mathematical representation of the convolution operation is given in Eq. 1.

$$g(x,y) = h(x,y) * f(x,y) \tag{1}$$

where $f(x,y)$ is the convolution mask, $h(x,y)$ is the input image and $g(x,y)$ is the convoluted image.

In convolution operation, a filter slides over the input image to produce a feature map as shown in Fig. 10.2. Convolution operation captures different feature maps for the same input image with different filters. More features can be extracted by using more number of filters. In training, a CNN learns the values of these filters. The size of the feature map is determined by stride, padding and depth. Stride is the number of pixels that the filter jumps to slide over the input matrix. Larger stride will produce smaller feature maps. Affixing zeroes around the input matrix is called zero-padding or wide convolution. Padding allows the network to apply the filter to border elements of the input image matrix. Depth is the number of filters used in convolution operation.

10.2.2 Activation Layer

Activation layer uses activation functions that ignite a signal when a specific stimulus is presented. As compared to common activation functions such as tanh

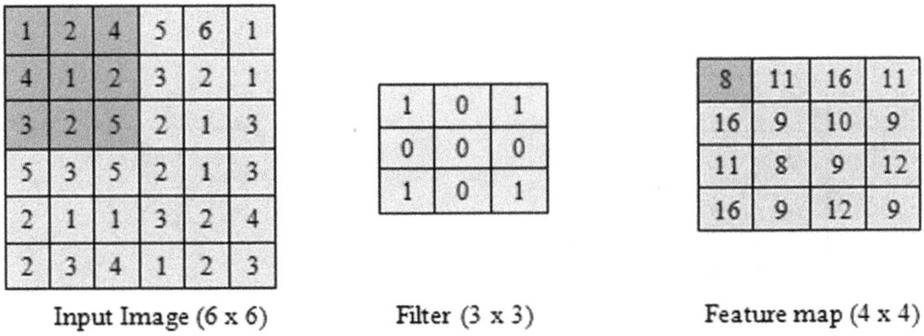

Fig. 10.2 Convolution of 5 × 5 image with 3 × 3 filter

and sigmoid, Rectified Linear Units (ReLU) is easy to compute and more robust to overfitting because of its sparse activation.

ReLU is the most common activation function used in convolution layer. Generally, activation function brings the non-linearity into DCNN. ReLU accelerates the convergence of the training procedure and leads to improved solutions. ReLU operation replaces all negative pixel values in the feature map by zero that is represented in Eq. 2.

$$relu(x) = \max(0, x) \tag{2}$$

where 'x' represents the input and $relu(x)$ represents the output function.

10.2.3 Pooling Layer

Pooling layer achieves a linear or non-linear downsampling. This layer reduces the computation complexity in terms of parameters reduction and alleviates overfitting. Pooling reduces the dimensionality of feature map but preserves the most important information. Various pooling methods are available for subsampling the feature map such as max, average, and sum. Max pooling operation takes the largest element from the rectified feature map within the window as shown in Fig. 10.3. As an alternative to taking the largest element, average or sum of all elements in that window can be taken.

10.2.4 Fully Connected Layer

All outputs of the preceding layer are attached to all inputs of the FC layer that predicts the image label. This layer uses activation functions such as softmax, sigmoid etc. for predicting the target class. Softmax function is used in the output layer for multi classification model, which return the probabilities of each class in that the target class has a higher probability. Softmax function provides a way of

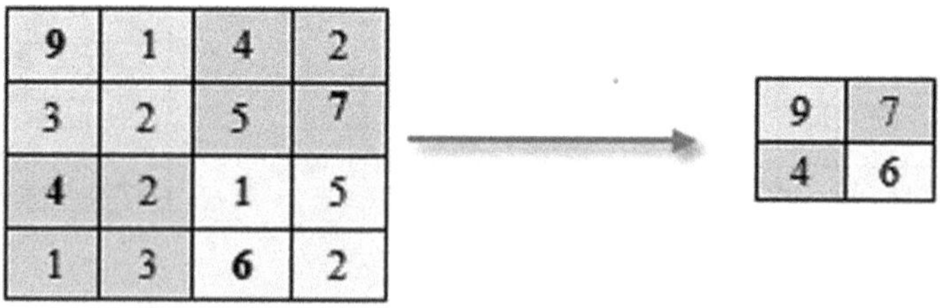

Fig. 10.3 Max pool with subsample 2 × 2

predicting discrete probability distribution over multiple classes and the sum of all the probabilities will be equal to one. Sigmoid function provides output in the range 0–1, which is mostly used for binary classification model.

10.3 Proposed DCNN Architecture

The DCNN architecture for classification of images in the cifar-10 dataset implemented in this work is shown in Fig. 10.4.

The DCNN model explored in this work consists of 6 consecutive convolutional layers and 3 fully connected layers. Each convolutional layer is followed by a subsampling layer. The input of the CNN model is a 32 × 32 × 3 image (i.e., the input has three channels of 32 × 32 pixels). The first convolutional stage consists of 48 kernels of size 3 × 3 with no subsampling. The second convolutional stage consists of 48 kernels of size 3 × 3 and a max pooling layer that subsamples the image by half. The third convolutional stage consists of 96 kernels of size 3 × 3 with no subsampling. The fourth convolutional stage consists of 96 kernels of size 3 × 3 and a max pooling layer that subsamples the image by half. The fifth convolutional stage consists of 192 kernels of size 3 × 3 with no subsampling. The sixth convolutional stage consists of 192 kernels of size 3 × 3 and a max pooling layer that subsamples the image by half.

Each kernel produces a 2-D image output (e.g., 48 of 32 × 32 images after the first convolutional layer), which is denoted as 48 @ 32 × 32 in Fig. 10.4. Kernels may contain different matrix values that are initialized randomly and updated during

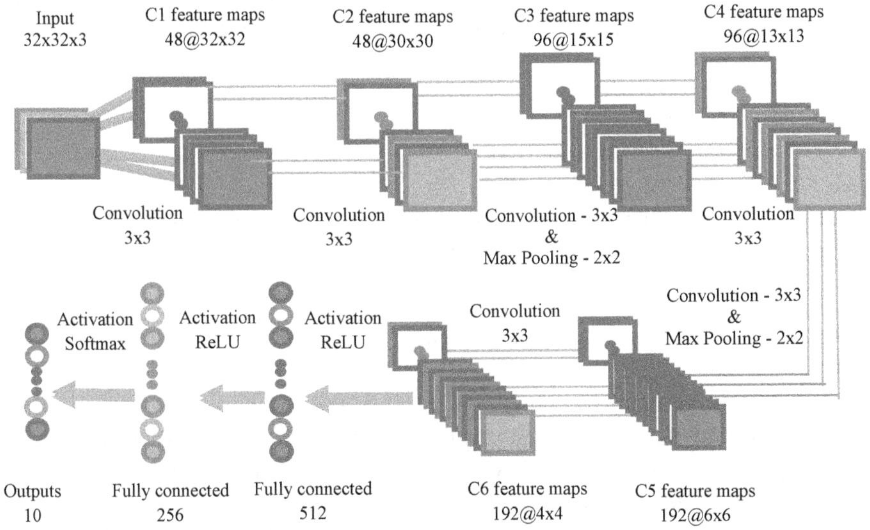

Fig. 10.4 Proposed DCNN architecture for cifar-10 dataset classification

training to optimize the classification accuracy. First, fully connected layer has 512 nodes and the second fully connected layer has 256 nodes, and the final stage has a softmax layer containing ten nodes. All convolutional and fully connected layers are equipped with the ReLU activation function. The last fully connected layer contains ten neurons which compute the classification probability for each class using softmax regression. To reduce overfitting, "dropout" and "batch normalization" is used after convolution layers. The effect of various optimizers in accelerating gradient descendent is analyzed in this work.

10.3.1 Dropout Layer

Dropout layer drops less contributed nodes in the forward pass by setting them to zero during training. Even some of the nodes are dropped out still the network can able to provide the correct classification for a given example, that makes sure that the network is not becoming too fitted to the training data and thus aids mitigate the overfitting problem. It is an optional layer in the architecture. The nodes to be dropped are randomly selected with a probability in each weight update cycle.

10.3.2 Batch Normalization

Normalization is simply a linear transformation applied to each activation. Batch normalization technique normalizes each input channel across a mini-batch as given in Eq. 3, which normalizes the activations of each channel with the mini-batch mean and mini-batch standard deviation i.e. applies a transformation that maintains the mean activation close to 0 and the activation standard deviation close to 1.

$$\hat{x} = \frac{x - E[x]}{\sqrt{Var[x]}} \tag{3}$$

where $E[x]$ is mini-batch mean and $Var[x]$ is mini-batch standard deviation. Activations y_i are computed with the following transformation function for all input neurons, x_i.

$$y_i = w\hat{x}_i + b \tag{4}$$

where 'w' is weight and 'b' is bias.

Figure 10.5 illustrates the transformation of inputs (x_i) into activations (y_i) with batch normalization technique. Batch normalization acts like a regulator between input layer and transformation function, which normalize the inputs intended for distributing activation values uniformly all through the training process. A batch normalization layer is used between the convolutional layer and activation layer that

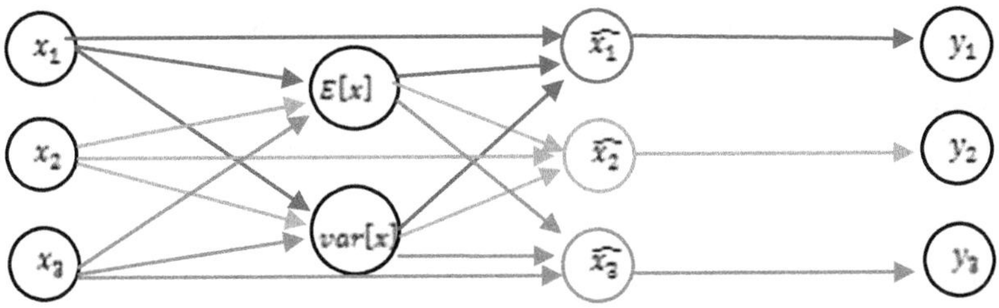

Fig. 10.5 Normalization of inputs with batch normalization

reduces the sensitivity to network initialization. Batch normalization significantly accelerates training speed by reducing vanishing gradient problems [24]. The presence of batch normalization has a benefit of optimizing the network training. Also, this has other benefits such as easy weight initialization, improvement in training speed, higher learning rate and regularization of values for the activation function.

10.3.3 Optimizing Gradient Descendant with Various Optimizer

The trainable parameters of CNN play a major role in efficiently and effectively training a model and produce accurate results. Optimization strategies have great influence on model's learning process and the prediction process. Optimization helps to minimize the error at training process and tune the model's internal learnable parameters such as weights (W) and the bias (b) values.

Gradient Descent is the most important technique used for training and optimizing Intelligent systems. Gradient descent works by iteratively performing updates based on the first derivative of a problem. For speedups, a technique called "momentum" is often used, which averages search steps over iterations. Gradient descent can be very effective, if the learning rate and momentum are well tuned. In order to achieve the objective, model learns appropriate model parameters in every iteration. Convergence of network depends on the internal structure of the model and optimizer [25]. The formula for updating the parameter in the model is given in Eq. 5.

$$\theta = \theta - \delta \cdot \nabla J(\theta) \tag{5}$$

where 'δ' represents the learning rate, '$\nabla J(\theta)$' represents the Gradient of Loss function $J(\theta)$ with respect to 'θ'.

Crossentropy loss function is the widely used cost function, which is used as an objective function to optimize the classification task. Crossentropy describes the loss between the predicted probability distributions and target probability distributions. Crossentropy is measured by Eq. 6.

$$H(p,q) = -\sum p_i \log q_i \tag{6}$$

where p_i is the target probability distribution and q_i is the predicted probability distribution of the current model.

Momentum: Momentum is a technique for quickening the Stochastic Gradient Descent (SGD) by changing the momentum largely towards the desired direction and minimally towards the fluctuating direction. When the objective function reaches local minima, the momentum is high. So the model is getting into local minima, is negligible. Moreover, this method performs larger updates frequently by which the model may miss the actual minima.
RmsProp: RmsProp is an optimizer that utilizes the magnitude of recent gradients to normalize the gradients. This method divides the current gradient by a moving average over the root mean squared gradients. RmsProp would boost the parameter multiple times and decrement it once by the current gradient. Also this has adaptable learning rate. This is a very robust optimizer which can deal stochastic objectives very nicely, making it applicable to mini-batch learning.
Adadelta: Adadelta is a method that uses the magnitude of recent gradients and steps to obtain an adaptive learning rate. This method stores an exponential moving average over the gradients and learning rate. The scale of learning rate for each individual parameter obtained by their ration.
Adam: Adaptive Moment Estimation (Adam) is another method, which determines the learning rates for each parameter. The scale of learning rate for each individual parameter obtained by their importance. Choosing a proper learning rate is a challenging task. Small learning rate leads to painfully slow convergence. Since adaptive algorithms dynamically adapt the learning rate and momentum, it supports network to converge quickly and discover the accurate parameter values. Whereas standard momentum techniques are deliberate in reaching the global minima. This method stores an exponentially moving average over the past squared gradients.

10.4 Results and Discussion

The proposed DCNN architecture has been trained and validated with images in the cifar-10 dataset. This dataset contains 60,000 images of size 32 $\times$ 32 with the following 10 categories such as airplane, automobile, bird, cat, deer, dog, frog, horse, ship, and truck. Some of the sample images from this dataset are depicted in Fig. 10.6. In the dataset 50,000 images are used as the training set and 10,000 images are used as the validation dataset. The experimental DCNN architecture is

Fig. 10.6 Sample images from the cifar-10 dataset

developed in Keras, written in Python. This is an open source high level library, used to build neural network models.

The model uses accuracy as a metrics that can be evaluated during training and testing. The performance of the model is measured by validation score. The efficient model which is trained with a part of the dataset could able to predict the new one, that has never used for training. A loss function or objective function used in this experiment is crossentropy which is commonly used for image classification tasks. The classifier tries to minimize this crossentropy between the target and the estimated class probabilities.

This section presents the experimental results obtained during training and testing stages on the cifar-10 image dataset. In order to speed up the experiments, the training of the network is stopped, if the validation accuracy is not improved for 5 consecutive epochs. The upper bound for the number of training epochs considered in this experiment is 25 epochs.

Figure 10.7 shows a training loss over time for DCNN with various optimizers. It is observed that the crossentropy loss over time is much higher throughout the training process for Rmsprop, momentum and Adadelta optimizers. Whereas the stepwise behavior of the entropy loss in Adam optimizer is significantly reduced over time.

The validation accuracy for DCNN with various optimizers for first 10 epochs is presented in Table 10.1.

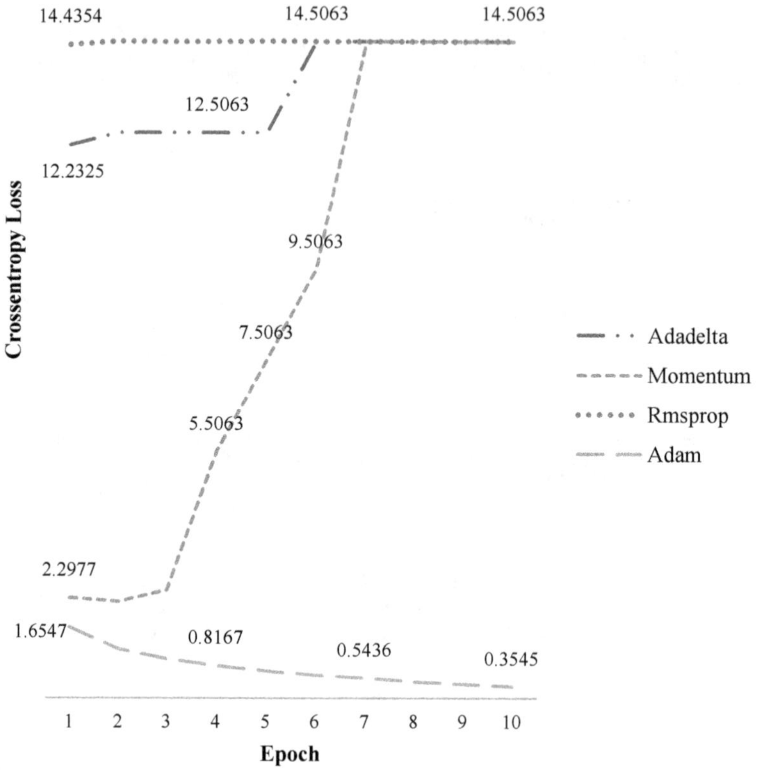

Fig. 10.7 Training loss over time for DCNN with various optimizers

Table 10.1 Performance comparison of various optimizers in DCNN in terms of validation accuracy (%)

Optimizers in DCNN				
Epochs	Adadelta	Momentum	Rmsprop	Adam
1	10	19.38	10	49.87
2	10	27.4	10	61.3
3	10	9.98	10	67.87
4	10	10	10	70.54
5	10	10	10	73.46
6	10	10	10	74.55
7	10	10	10	76.19
8	10	10	10	77.74
9	10	10	10	75.46
10	10	10	10	77.74

The back propagation network uses the batch gradient technique, which is a first-order optimization technique with favorable convergence properties. The performance of models with various optimizers in terms of validation accuracy is reported in Table 10.1. It is observed that Rmsprop, momentum, and Adadelta optimizers are not able to achieve validation accuracies whereas Adam optimizer achieves higher validation accuracy. From these observations, it is concluded that the Adam optimizer outperforms other optimizers in terms of entropy loss and accuracy due to its adaptive learning nature on the training set. The results show that the effect of optimizers also significantly changes the accuracy of the model, in addition to the structure of the architecture.

The model summary with a number of parameters used in the proposed DCNN architecture is shown in Table 10.2. It shows the number of parameters initialized at every layer in the DCNN architecture. Trainable parameters are initialized with minimum random numbers to avoid dead neurons, but not too small to avoid zero gradients. Uniform distribution is generally preferred for parameter initialization. Totally 1,172,410 parameters are tuned by DCNN training to classify the images in the cifar-10 dataset. The parameters in the model can be further re-tuned by introducing dropout and batch normalization.

Table 10.2 Model summary of the proposed DCNN architecture

Layer (type)	Output shape	Param #
conv2d_1 (Conv2D)	(None, 32, 32, 48)	1344
activation_1 (Activation)	(None, 32, 32, 48)	0
conv2d_2 (Conv2D)	(None, 30, 30, 48)	20,784
activation_2 (Activation)	(None, 30, 30, 48)	0
max_pooling2d_1 (MaxPooling2)	(None, 15, 15, 48)	0
conv2d_3 (Conv2D)	(None, 15, 15, 96)	41,568
activation_3 (Activation)	(None, 15, 15, 96)	0
conv2d_4 (Conv2D)	(None, 13, 13, 96)	83,040
activation_4 (Activation)	(None, 13, 13, 96)	0
max_pooling2d_2 (MaxPooling2)	(None, 6, 6, 96)	0
conv2d_5 (Conv2D)	(None, 6, 6, 192)	166,080
activation_5 (Activation)	(None, 6, 6, 192)	0
conv2d_6 (Conv2D)	(None, 4, 4, 192)	331,968
activation_6 (Activation)	(None, 4, 4, 192)	0
max_pooling2d_3 (MaxPooling2)	(None, 2, 2, 192)	0
flatten_1 (Flatten)	(None, 768)	0
dense_1 (Dense)	(None, 512)	393,728
activation_7 (Activation)	(None, 512)	0
dense_2 (Dense)	(None, 256)	131,328
activation_8 (Activation)	(None, 256)	0
dense_3 (Dense)	(None, 10)	2570
Total params: 1,172,410		
Trainable params: 1,172,410		
Non-trainable params: 0		

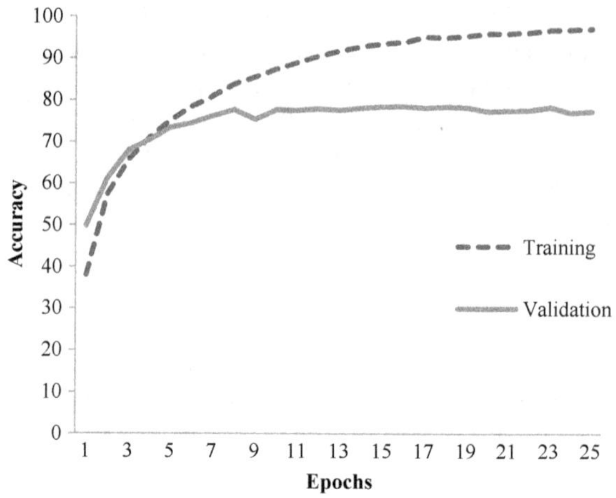

Fig. 10.8 Training and validation accuracy curve for DCNN with Adam optimizer

The training and validation accuracy for the proposed model with Adam optimizer is shown in Fig. 10.8. It is observed from the figure that in later epochs, there is no significant improvement in the validation accuracy as compared to the training accuracy. The best training accuracy and validation accuracy achieved in this model is 97.13 and 78.47% respectively. The validation accuracy is lower than the training accuracy due to overfitting of the model. This happens when the model learns the training data very detail and creates a negative impact on the performance of the model on new data. This issue can be solved by introducing dropout after the convolution layer.

Regularization is a very important technique to prevent over fitting in machine learning problems. In this model, the regularization technique called dropout is applied to avoid over fitting. Dropout does not rely on modifying the loss function but the network itself. Figure 10.9 shows the performance of the network model by introducing the dropout of 0.5 after the dense layer. It is noticed that the validation

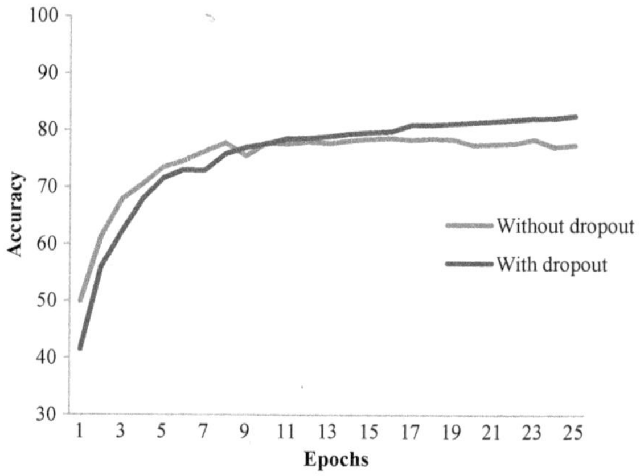

Fig. 10.9 Performance comparison of validation accuracy without dropout and with dropout

accuracy suddenly start to go up and oscillate on high values until the next learning rate drop.

The key idea of dropout is randomly droping the parts of neural network during training and thus preventing the over learning of features. It is observed from Fig. 10.9 that after 10 epochs, there is an improvement in the validation accuracy with dropout compared to a network without dropout. Dropout decreases the loss from 1.1423 to 0.6112 and improves validation accuracy from 77.4 to 81.32% on cifar-10 dataset. Also, time taken for training the neural network is minimized.

The model summary of the proposed DCNN architecture with batch normalization is shown in Table 10.3. To improve the efficiency of the DCNN model,

Table 10.3 Model summary of the proposed DCNN architecture with batch normalization

Layer (type)	Output shape	Param #
conv2d_1 (Conv2D)	(None, 32, 32, 48)	1344
batch_normalization_1 (Batch)	(None, 32, 32, 48)	192
activation_1 (Activation)	(None, 32, 32, 48)	0
conv2d_2 (Conv2D)	(None, 30, 30, 48)	20,784
batch_normalization_2 (Batch)	(None, 30, 30, 48)	192
activation_2 (Activation)	(None, 30, 30, 48)	0
max_pooling2d_1 (MaxPooling2)	(None, 15, 15, 48)	0
conv2d_3 (Conv2D)	(None, 15, 15, 96)	41,568
batch_normalization_3 (Batch)	(None, 15, 15, 96)	384
activation_3 (Activation)	(None, 15, 15, 96)	0
conv2d_4 (Conv2D)	(None, 13, 13, 96)	83,040
batch_normalization_4 (Batch)	(None, 13, 13, 96)	384
activation_4 (Activation)	(None, 13, 13, 96)	0
max_pooling2d_2 (MaxPooling2)	(None, 6, 6, 96)	0
conv2d_5 (Conv2D)	(None, 6, 6, 192)	166,080
batch_normalization_5 (Batch)	(None, 6, 6, 192)	768
activation_5 (Activation)	(None, 6, 6, 192)	0
conv2d_6 (Conv2D)	(None, 4, 4, 192)	331,968
batch_normalization_6 (Batch)	(None, 4, 4, 192)	768
activation_6 (Activation)	(None, 4, 4, 192)	0
max_pooling2d_3 (MaxPooling2)	(None, 2, 2, 192)	0
flatten_1 (Flatten)	(None, 768)	0
dense_1 (Dense)	(None, 512)	393,728
activation_7 (Activation)	(None, 512)	0
dense_2 (Dense)	(None, 256)	131,328
activation_8 (Activation)	(None, 256)	0
dense_3 (Dense)	(None, 10)	2570
Total params: 1,175,098		
Trainable params: 1,173,754		
Non-trainable params: 1344		

batch normalization layer is added after every convolution layer. In this experiment, a momentum of 0.99 is used in the batch normalization layer for moving the mean and variance. The presence of this layer improves the overall accuracy and learning rate. This layer performs a transformation at each batch by normalizing the previous layer's activations that in turn maintains activation mean towards to 0 and standard deviation towards to 1. The model re-tuned with batch normalization yields the number of trainable parameters as 1,175,098.

The effect of the batch normalization in the performance of the model in terms of validation accuracy is shown in Fig. 10.10. It shows that batch normalization really has positive effects on neural networks but it delays the convergence of network. By observing the loss over time, the regularizing effect of batch normalization becomes very prominent. The batch normalized network learns consistently. Overall, batch normalized models achieve higher validation and test accuracies on all datasets. Due to these results, the use of batch normalization is generally advised since it prevents model divergence and may increase convergence speeds through higher learning rates.

The performance of batch normalization with and without dropout is shown in Fig. 10.11. The batch normalization and dropout can be used at the same time for improving the accuracy in the validation dataset. The batch normalized model consistently achieves higher validation accuracy. Whereas it adds computational complexity that can be handled by keeping higher learning rate. It is recommended to keep the batch normalization between convolution and activation layers for getting best results. Also, the dropout layer introduced after dense layer reduces the over fitting issues. Figure 10.10 shows the accuracy improvement from 79.99 to 83.23% in first 25 epochs with the inclusion of dropout and batch normalization in the deep CNN for cifar-10 dataset.

Further, the performance of the classification system can be improved by changing the structure of architecture and tuning its parameters. Besides the number of layers and the layer density of the architecture, all tunable factors such as the

Fig. 10.10 Performance comparison of validation accuracy without batch normalization and with batch normalization

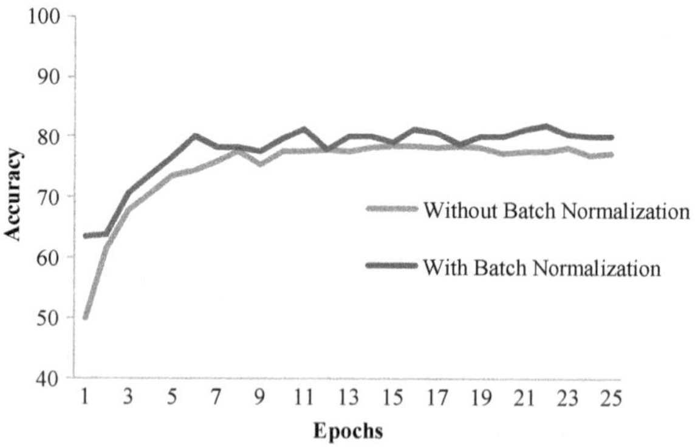

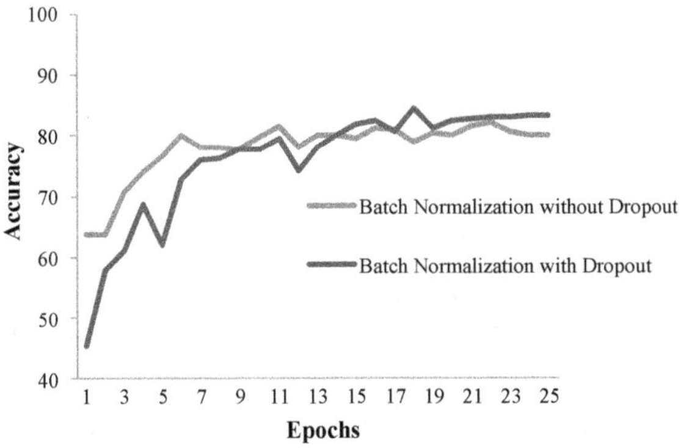

Fig. 10.11 Performance comparison of validation accuracy without dropout and with dropout in batch normalization

filter size, pooling method, number of epochs and layer patterns can improve the accuracy further. The architecture of CNN goes deeper and deeper, the network needs to learn tens of thousands to millions of parameters. A large amount of training data is required to train these parameters properly. Overfitting problem can be caused by poor quality training data and it can be avoided by training the network with noise free training data. Also overfitting due to the small dataset can be reduced with data augmentation, which can increase the training data by performing various transformations.

10.5 Conclusion

A new deep convolution neural network model is proposed in this chapter for image classification in cifar-10 dataset. The proposed model is analyzed with various optimization strategies, the inclusion of dropout and batch normalization. The Adam optimizer reduces the entropy loss over time as compared to other optimizers such as momentum, Adadelta and Rmsprop in this model. The Adam optimizer achieves a maximum validation accuracy of 78% in the first 25 epochs to classify the images in cifar-10 dataset. İntroducing dropout after dense layer prevents the model from learning too detail with the training data and achieves an accuracy of 81.32%. The CNN model with dropout and batch normalization ensures an improved performance in the validation phase with an accuracy of 83.42%. Since CNN model with batch normalization and dropout avoids model deviation, it is recommended to use with higher learning rates. Experimental results show that the proposed model exhibits significant improvement in the performance on classifying the images in the cifar-10 dataset.

References

1. Ionescu, R.T., Popescu, M.: State-of-the-art approaches for image classification. In: Knowledge Transfer between Computer Vision and Text Mining, pp. 41–52, Springer (2016)
2. Hinton, G.E., Osindero, S., Teh, Y.W.: A fast learning algorithm for deep belief nets. Neural Comput. **18**(7), 1527–1554 (2006)
3. LeCun, Y., Bengio, Y., Hinton, G.: Deep learning. Nature **521**, 436–444 (2015)
4. Deng, L., Yu, D.: Deep learning: methods and applications. Found. Trends Signal Process. **7** (3–4), 197–387 (2014)
5. Da, C., Zhang, H., Sang, Y.: Brain CT image classification with deep neural networks. In: Handa, H., Ishibuchi, H., Ong, Y.S., Tan, K. (eds.) Proceedings of the 18th Asia Pacific Symposium on Intelligent and Evolutionary Systems, vol. 1, pp. 653–662. Springer (2015)
6. Paoletti, M.E., Haut, J.M., Plaza, J., Plaza, A.: A new deep convolutional neural network for fast hyperspectral image classification. ISPRS J. Photogramm. Remote Sens. (2017)
7. Sudars, K.: Face recognition Face2vec based on deep learning: small database case. Autom. Control Comput. Sci. **51**(1), 50–54 (2017)
8. LeCun, Y., Bottou, L., Bengio, Y., Haffner, P.: Gradient-based learning applied to document recognition. Proc. IEEE **86**(11), 2278–2324 (1998)
9. Mugahed, A., et. al.: An automatic computer-aided diagnosis system for breast cancer in digital mammograms via deep belief network. J. Med. Biol. Eng., 1–14 (2017)
10. Wang, Y., et. al.: Automatic tumor segmentation with deep convolutional neural networks for radiotherapy applications. Neural Process. Lett., 1–12 (2018)
11. Zhang, Y.D., et. al.: Image based fruit category classification by 13-layer deep convolutional neural network and data augmentation. Multimed. Tools Appl., 1–20 (2017)
12. Chi, J., Walia, E., Babyn, P., Wang, J., Groot, G., Eramian, M.: Thyroid nodule classification in ultrasound images by fine-tuning deep convolutional neural network. J. Digit. Imaging **30** (4), 477–486 (2017)
13. Wang, X., Zhang, W., Wu, X., Xiao, L., Qian, Y., Fang, Z.: Real-time vehicle type classification with deep convolutional neural networks. J. Real-Time Image Process., 1–10 (2017)
14. Krizhevsky, A., Sutskever, I., Hinton, G.: Imagenet classification with deep convolutional neural networks. In: Proceedings of the 25th International Conference on Neural Information Processing Systems, vol. 1, pp. 1097–1105 (2012)
15. Simonyan, K., Zisserman, A.: Very deep convolutional networks for large-scale image recognition. arXiv preprint. arXiv:1409–1556 (2014)
16. Szegedy, C., et. al.: Going deeper with convolutions. In: IEEE Conference on Computer Vision and Pattern Recognition (CVPR) (2015)
17. Zagoruyko, S., Komodakis, N.: Wide residual networks. arXiv:1605.07146 (2017)
18. Huang, G., et. al.: Deep networks with stochastic depth. In: Leibe, B., Matas, J., Sebe, N., Welling, M. (eds.) Computer Vision—ECCV 2016. Lecture Notes in Computer Science, vol. 9908. Springer (2016)
19. Srivastava, N., et. al.: Dropout: a simple way to prevent neural networks from overfitting. J. Mach. Learning Res. **15**, 1929–1958 (2014)
20. Ioffe, S., Szegedy, C.: Batch normalization: accelerating deep network training by reducing internal covariate shift. In: Proceedings of the 32nd International Conference on International Conference on Machine Learning, vol. 37, pp. 448–456 (2015)
21. https://www.cs.toronto.edu/kriz/cifar.html
22. Krizhevsky, A.: Learning multiple layers of features from tiny images, Technical report. University of Toronto (2009)

23. Liu, S., Deng, W.: Very deep convolutional neural network based image classification using small training sample size. In: 3rd IAPR Asian Conference on Pattern Recognition (2015)
24. Bengio, Y., Simard, P.: Frasconi. P.: Learning long-term dependencies with gradient descent is difficult. IEEE Trans. Neural Netw. **5**(2), 157–166 (1994)
25. Ruder, S.: An overview of gradient descent optimization algorithms. arXiv:1609.04747 (2017)

Chapter 11
A Novel Underwater Image Enhancement Approach with Wavelet Transform Supported by Differential Evolution Algorithm

Gur Emre Guraksin, Omer Deperlioglu and Utku Kose

Abstract In this paper, a novel underwater image enhancement approach was proposed. This approach includes use of a method formed by the wavelet transform and the differential evolution algorithm. In the method, the contrast adjustment function was applied to the original underwater image first. Then, the homomorphic filtering technique was used to normalize the brightness in the image. After these steps, the underwater image was separated into its R, G, and B components. Then wavelet transform function was performed on each of the R, G, and B channels with Haar wavelet decomposition. Thus, detailed images were obtained for each of the color channels by wavelet transform low-pass approximation (cA), horizontal (cH), vertical (cV) and diagonal (cD) coefficients. Four parameters of weights (w) of each component cA, cH, cV, and cD situated in the R, G, and B color channels were optimized using differential evolution algorithm. In the proposed method, differential evolution algorithm was employed to find the optimum w parameters for Entropy and PSNR in separate approaches. Finally, unsharp mask filter was used to enhance the edges in the image. As an evaluation approach, performance of the proposed method was tested by using the criteria of entropy, PSNR, and MSE. The obtained results showed that the effectiveness of the proposed method was better than the existing techniques. Likewise, the visual quality of the image was also improved more thanks to the proposed method.

G. E. Guraksin
Department of Biomedical Engineering, Afyon Kocatepe University, Afyonkarahisar, Turkey
e-mail: emreguraksin@aku.edu.tr

O. Deperlioglu
Department of Computer Technologies, Afyon Kocatepe University, Afyonkarahisar, Turkey
e-mail: deperlioglu@aku.edu.tr

U. Kose (✉)
Department of Computer Engineering, Suleyman Demirel University, Isparta, Turkey
e-mail: utkukose@sdu.edu.tr

J. Hemanth and V. E. Balas (eds.), *Nature Inspired Optimization Techniques for Image Processing Applications*, Intelligent Systems Reference Library 150,
https://doi.org/10.1007/978-3-319-96002-9_11

Keywords Underwater image enhancement · Differential evolution algorithm Wavelet transform · Optimization · Artificial intelligence

11.1 Introduction

The physical properties of water have the disruptive effects on taking an underwater image such as the attenuation of light, absorption and scattering of light, and the foggy environment. For this reason, underwater images have poor color quality and low visibility, and also one color dominates on underwater image. To overcome these problems, underwater image enhancement has an important role for underwater scientists. There are several proposed methods or approaches in the literature such as contrast enhancement, the optical priors, fusion, and dehazing etc. for overcoming the problems related to underwater imaging [1–5].

The underwater image processing generally contains two different methods which are the image restoration and the image enhancement methods. The image restoration method aims to recover a row image using a model of reduction and model of original image configuration. These methods are meticulous and very powerful. But, these methods comprise a plurality of variables and parameters [6]. The image enhancement methods use the qualitative subjective criteria of image. But to achieve a more visually pleasing image, they do not need any physical model for image formation. In this context, such methods are usually faster and simpler deconvolution approaches [7]. When we take a look at the related literature, it can be seen that there are many different methods employed in order to achieve enhancement of underwater images [8–15].

Preprocessing techniques have a great importance for the underwater image enhancement methods. Therefore, many researchers developed several preprocessing methods for underwater image enhancement. Sometimes, the filter methods were used alone or with different methods and also sometimes a few filters were used together with other filters [16–19]. For example, Bazeille et al. proposed an automatic method to pre-process underwater images. It reduced underwater perturbations and improved image quality. This method did not require any parameter setting. The method proposed by Bazeille et al. was used as a first process of edge detection. The robustness of the method was analyzed using an edge detection robustness criterion [20].

One of the most supportive improvement activities of the underwater image improvement is artificial intelligence or optimization algorithms such as fuzzy logic, Vortex Optimization Algorithm and Constancy Deskewing Algorithm etc. [3, 4, 21, 22]. Ratna Babu and Sunitha proposed an image enhancement approach based on Cuckoo Search Algorithm with Morphological Operations. First, they selected the best contrast value of an image with Cuckoo Search algorithm. Then, they carried out morphological operations. From the obtained results, they said that "the proposed approach is converted into original color image without noise and adaptive process enhance the quality of images" [23].

Nowadays, the wavelet transform is a very popular method of underwater processing. Usually, it is used in very different ways for image compression and removing noise [2, 24]. Discrete Wavelet Transforms (Haar, Daubechies, etc.) are orthogonal wavelet, and their forward and inverse transforms require only additions and subtractions. Therefore, implementing these functions on the computer are very easy. Today, one of the most favorable techniques is the Discrete Wavelet Transform which uses the Haar functions in image coding, edge extraction and binary logic design [25].

In this paper, a new approach for the underwater images enhancement was proposed. This method contain wavelet transform and differential evolution algorithm. Differential evolution algorithm is one of the most powerful optimization algorithms as having the advantages of evolutionary approaches, requiring less parameter setting, and being efficient on complex optimization problems even though it has a simple structure [26–28]. On the other hand, wavelet transform is a powerful, multidisciplinary technique used widely for solving difficult problems in different fields like mathematics, physics, and engineering by focusing on applications such as signal processing, image processing, data compression, and pattern recognition [29, 30]. Wavelet transform is better than the traditional Fourier methods with its advantages on employing localized basis functions and achieving faster computation [31]. Because of remarkable advantages of these techniques, the authors decided to combine them for the research problem of this study.

Regarding the subject of the paper and the performed research, remaining content was organized as follows: In the second section, wavelet transform, differential evolution algorithm, contrast adjustment, homomorphic filter, and unsharp mask filter were described in details. Following to that, findings obtained from the performed applications on underwater image enhancement and also a general discussion on them were given in the third section. Then, the paper was ended with the last section providing explanations on conclusions and some possible future works.

11.2 Theoretical Background

Just before focusing on the applications on underwater image enhancement, it is important to give brief information about theoretical background regarding the followed enhancement approach. The following subtitled were devoted to that purpose.

11.2.1 Wavelet Transform

When X represents an indexed image, X as well as the output arrays cA, cH, cV, and cD are m-by-n matrices. When X represents a true color image, it is an m-by-n-by-3 array, where each m-by-n matrix represents a red, green, or blue color

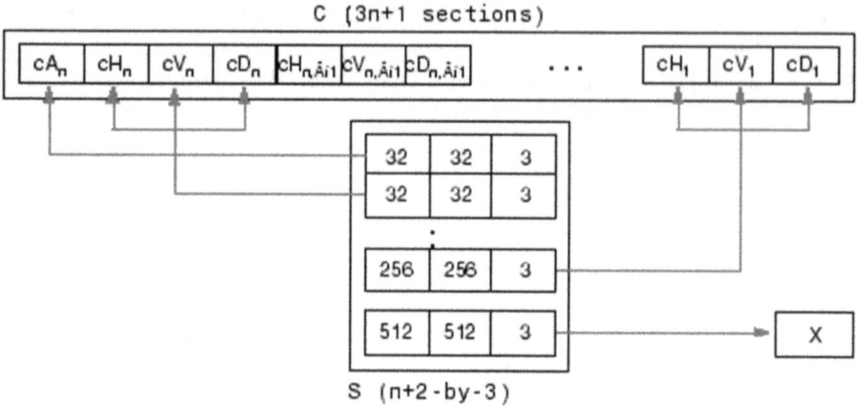

Fig. 11.1 The decomposition vector C and the corresponding bookkeeping matrix S [32]

plane concatenated along the third dimension. The size of vector C and the size of matrix S depend on the type of analyzed image. For a true color image, the decomposition vector C and the corresponding bookkeeping matrix S can be represented as in Fig. 11.1 [32].

For images, the wavelet representation can be computed with a pyramidal algorithm similar to the one-dimensional algorithm for two-dimensional wavelets and scaling functions. A two-dimensional wavelet transform can be computed with a separable extension of the one-dimensional decomposition algorithm. At each step, we decompose $A^d_{2^{j+1}}f$ into $A^d_{2^j}f$, $D^1_{2^j}f$, $D^2_{2^j}f$, and $D^3_{2^j}f$. This algorithm is shown as a block diagram in Fig. 11.2. Firstly, the rows of $A^d_{2^{j+1}}f$ are convolved with a one-dimensional filter, retain every other row. Then the columns of the resulting signals are convolved with another one-dimensional filter and retain every other column. The filters used in this decomposition are the quadrature mirror filters $\hat{H}$ and $\hat{G}$. The structure of application of the filters for computing $A^d_{2^j}$, $D^1_{2^j}$, $D^2_{2^j}$, and $D^3_{2^j}$ are shown in Fig. 11.2 [32]. The wavelet transform of an image is computed $A^d_1 f$ by repeating this process for $-1 \geq j \geq -J$. This corresponds to a separable conjugate mirror filter decomposition [32, 33].

Let us use an orthogonal wavelet for the computation scheme which becomes easy. We start with the two filters of length 2N, denoted h(n) and g(n) and corresponding to the wavelet. Now by induction, let us define the following sequence of functions (Wn (x), n = 0, 1, 2, 3, …) as Eqs. 11.1 and 11.2:

$$W_{2n}(x) = \sqrt{2} \sum_{k=0,\cdots,2N-1} h(k) W_n(2x-k) \tag{11.1}$$

$$W_{2n+1}(x) = \sqrt{2} \sum_{k=0,\cdots,2N-1} g(k) W_n(2x-k) \tag{11.2}$$

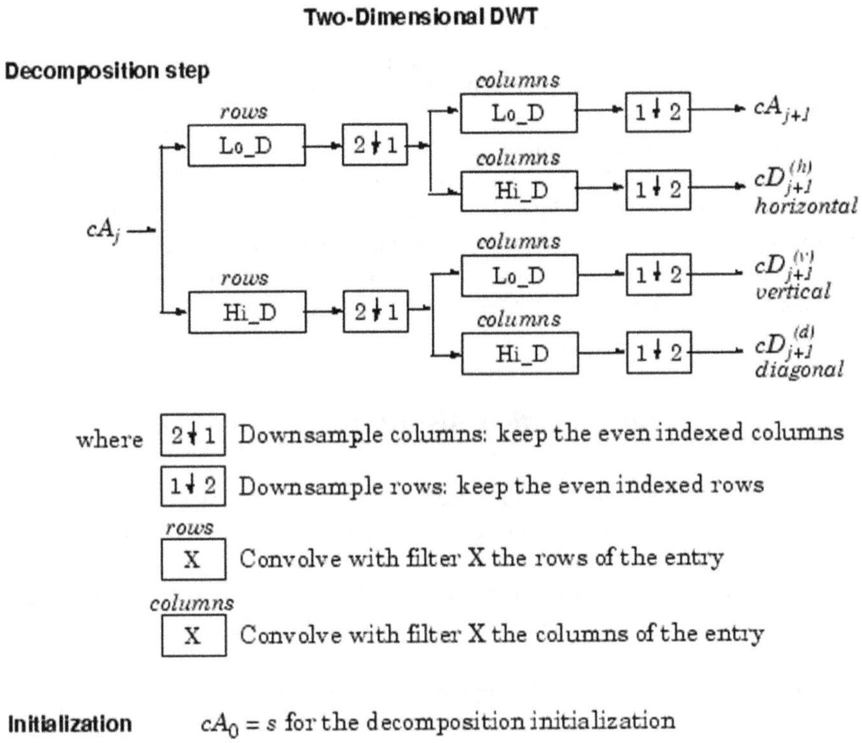

Fig. 11.2 Two-dimensional decomposition of an image [32]

where W0(x) = ϕ(x) is the scaling function and W1(x) = ψ(x) is the wavelet function [32].

The original Haar definition is as Eq. 11.3:

$$haar(0,t) = 1, \quad for\, t \in [0,1); \quad haar(1,t) = \begin{cases} 1, & for\, t \in \left[0, \frac{1}{2}\right), \\ -1, & for\, t \in \left[\frac{1}{2}, 1\right) \end{cases} \tag{11.3}$$

and $haar(k,0) = \lim_{t \to 0} + haar(k,t)$, $haar(k,1) = \lim_{t \to 1} + haar(k,t)$ and at the points of discontinuity within the interior (0, 1) $haar(k,t) = \frac{1}{2}(haar(k,t-0) + haar(k,t+0))$ [25].

For the Haar wavelet, Eqs. 11.1 and 11.2 are transformed into Eqs. 11.6 and 11.7, respectively:

$$N = 1, \quad h(0) = h(1) = \frac{1}{\sqrt{2}} \tag{11.4}$$

$$g(0) = g(1) = \frac{1}{\sqrt{2}} \tag{11.5}$$

$$W_{2n}(x) = W_n(2x) + W_n(2x - 1) \tag{11.6}$$

$$W_{2n+1}(x) = W_n(2x) - W_n(2x - 1) \tag{11.7}$$

$W_0(x) = \phi(x)$ is the haar scaling function and $W_1(x) = \psi(x)$ is the Haar wavelet, both supported in [0, 1]. So W_{2n} can be obtained by adding two 1/2-scaled versions of W_n with distinct support intervals. W_{2n+1} can be obtained by subtracting the same versions of W_n. Starting from more regular original wavelets, using a similar construction, the smoothed version of this system of W-functions is obtained all with support in the interval [0, 2N − 1] [32].

11.2.2 Differential Evolution Algorithm

The Differential Evolution Algorithm (DEA) is an artificial intelligence based on optimization algorithm introduced by Storn and Price in 1997 [28, 34, 35]. The DEA is a stochastic search technique based on population for solving global optimization problems. This algorithm is very simple but powerful, and the algorithm efficiency and effectiveness have been proven in many different applications [26, 36–42].

At first, the DEA starts with the population of Np, and D-dimension vectors with parameter values, which are randomly and uniformly distributed between the pre-specified lower initial parameter bound x_j, low and the upper initial parameter bound x_j, high:

$$x_{j,i,G} = x_{j,low} + rand(0,1).(x_{j,high} - x_{j,low}), \quad j = 1, 2, \ldots, D; \quad i = 1, 2, \ldots, N_p; \quad G = 0 \tag{11.8}$$

In Eq. 11.8, G is the generation index to which the population belongs, index i represents the ith solution of population index, and j indicates the parameter index.

The DEA has three main solution mechanisms as mutation, crossover, and selection. Thanks to these mechanisms, the DEA is known as an evolutionary solution approach (So, it also has the word “evolution” in its name) [43–45]. In detail, Khan et al. expressed pseudo code of the DEA as in Fig. 11.3 [46]. For more examples of pseudo code, readers are referred to [47, 48].

Figure 11.4 provides a brief flow chart of the DAE [49].

With its wide use, the DAE has also had many different modified versions in the related literature and in this way, it has been used along a wide scope of research problems [50–56].

Generate the initial population of individuals
Do
For each individual j in the population
Choose three numbers $n_1, n_2,$ *and* n_3 *that is,* $1 \le n_1, n_2, n_3 \le N$ *with* $n_1 \ne n_2 \ne n_3 \ne j$
Generate a random integer $i_{\text{rand}} \in (1, N)$
For each parameter i

$$y^{i,g} = x^{n_1,g} + F(x^{n_2,g} - x^{n_3,g})$$

$$z_j^{i,g} = \begin{Bmatrix} y_j^{i,g} if \ \text{rand}() \le \text{CR} \ or \ j = j_{\text{rand}} \\ x_j^{i,g} \text{otherwise} \end{Bmatrix}$$

End For
Replace $x^{i,g}$ *with the child* $z^{i,g}$ *if* $z^{i,g}$ *is better*
End For
Until the termination condition is achieved

Fig. 11.3 A pseudo code of the DEA [46]

11.2.3 Contrast Adjustment

The most primary way of image processing is point transform. It means that this transform maps the values at individual pixels in the input image into corresponding pixels in an output image. In the mathematical concept, this is a one-to-one functional mapping from input to output, arithmetic or logical operations on images. These operations can be applied between two images, IA and IB, or can be applied between an image IA and a constant C. For example, Iout = IA + IB or Iout = IA + C.

Contrast adjustment is a method obtained by adding or multiplying a positive constant value C to each pixel location. This operation increases pixel's value and pixel's brightness. In general, it is a multiplier. For the color image, firstly, the color image is separated into its R, G, and B components. Then, contrast adjustment is applied for each R, G, and B components separately [57].

11.2.4 Homomorphic Filter

When the illumination-reflectance model is considered, it is supposed that an image is a function of the product of the illumination and the reflectance as described in Eq. 11.9:

$$f(x, y) = i(x, y) \cdot r(x, y) \tag{11.9}$$

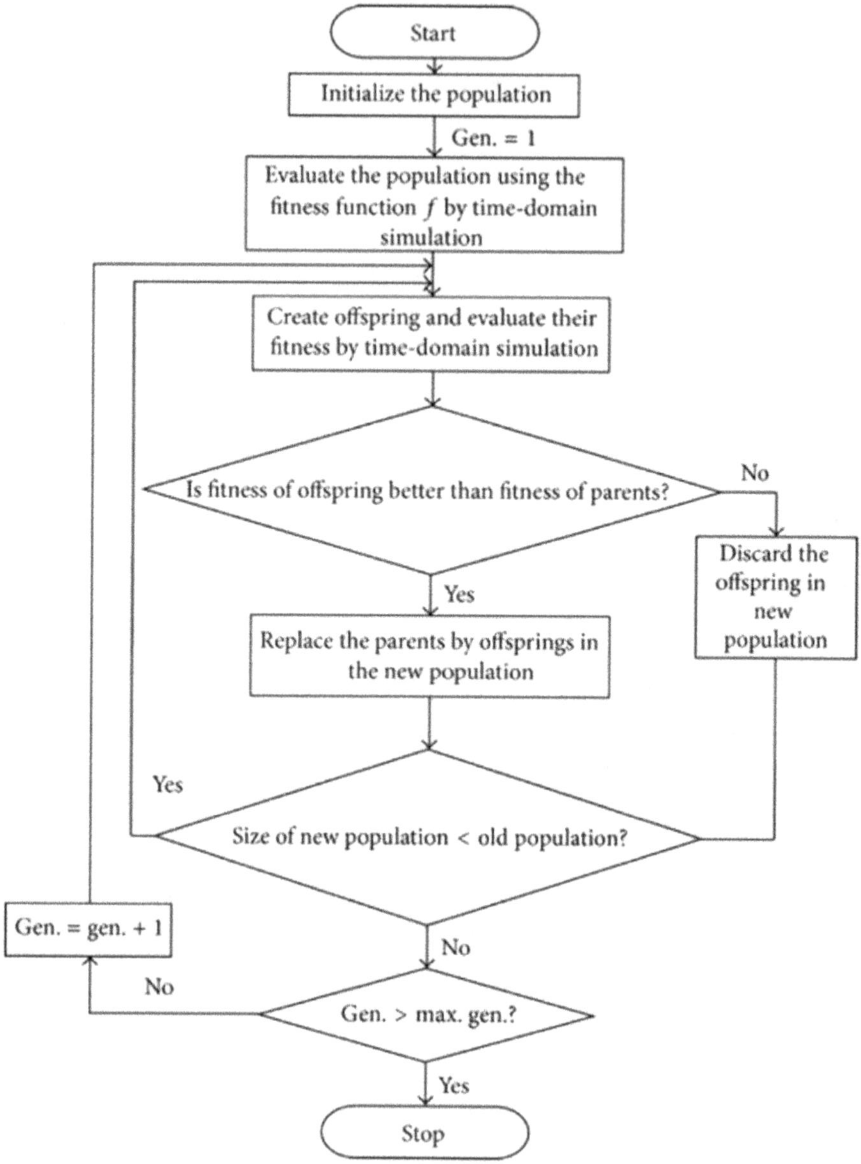

Fig. 11.4 A brief flow chart of the DAE [49]

where f(x, y) is the input image, i(x, y) is the illumination multiplicative factor, and r(x, y) is the reflectance function. In this context, the illumination factor changes slowly through the view field, therefore it represents low frequencies in the Fourier transform of the image. On the contrary, reflectance is associated with high frequency components. The low frequencies can be suppressed with a high-pass filter by multiplying these components. Processes of the algorithm can be discussed step by step as follows:

- As seen in Eq. 11.10, the logarithm is taken to separate components of the illumination and reflectance. Thus, the multiplicative effect is converted into an additive one by the logarithm.

$$g(x,y) = \ln(f(x,y)) = \ln(i(x,y).r(x,y)) + \ln(r(x,y)) \tag{11.10}$$

- Equation 11.11 shows that the computation of the Fourier transform of the log-image.

$$G(w_x, w_y) = I(w_x, w_y) + R(w_x, w_y) \tag{11.11}$$

- Equation 11.12 shows that the high-pass filter applied to the Fourier transform decreases the contribution of illumination and also amplifies the contribution of reflectance. Thus, the edges of the objects in the image is sharpened.

$$\begin{aligned} S(w_x, w_y) &= H(w_x, w_y) \cdot I(w_x, w_y) + H(w_x, w_y) \cdot R(w_x, w_y) \\ with, \quad H(w_x, w_y) &= (r_H - r_L) \cdot (1 - \exp(-(\frac{w_x^2 + w_y^2}{2\delta_w^2}))) + r_L \end{aligned} \tag{11.12}$$

  where $r_H = 2.5$ and $r_L = 0.5$ are the maximum and minimum coefficients values respectively, and δ_w is the factor controlling the cutoff frequency. These parameters are selected empirically [20].

Computation of the inverse Fourier transform is used to come back in the spatial domain and then taking the exponent to obtain the filtered image [20].

11.2.5 Unsharp Masking

The unsharp mask filter is an edge enhancement filter and it is also called as boost filtering. Unsharp filtering works by taking a smoothed version of an image obtained from the original image in order to improve the high-frequency information in the image. Firstly, this function generates an edge image from the original image using Eq. 11.13:

$$I_{edges}(c,r) = I_{original}(c,r) - I_{smoothed}(c,r) \tag{11.13}$$

The smoothed version of the image is generally obtained from filtering the original image with a kernel of a mean or a Gaussian filter. Then, the resulting difference image is added on to the original image to carry out some degree of sharpening as given in Eq. 11.14:

$$I_{enhanced}(c, r) = I_{original}(c, r) + k(I_{edges}(c, r)) \quad (11.14)$$

where the constant scaling factor k is used to ensure the resulting image to be within the proper range. Generally, k can be selected between 0.2 and 0.7 depending on the level of required sharpening. The operation here is called sometimes as boost filtering [57].

11.3 Solution of the Proposed Image Enhancement Approach

In this paper, we proposed a new approach to enhance the underwater images using wavelet transform and differential evolution algorithm. In the first stage of the proposed technique, contrast adjustment procedure was applied to the original underwater image. By the help of this procedure, we adjusted the contrast in the image with the limits of 0.01 and 0.99, so the contrast of the output image was increased. After the contrast adjustment procedure, homomorphic filtering procedure was used to normalize the brightness in the image which is a generalized technique for nonlinear image enhancement and correction. In the next stage the underwater image was separated into its R, G, and B components. Then wavelet transform operation was performed to each R, G, and B channels. For this purpose, the Haar wavelet decomposition was used. Thus, by the help of wavelet transform, lowpass approximation (cA), horizontal (cH), vertical (cV) and diagonal (cD) detailed images were obtained for each color channels. In this part of the algorithm, a weight value for each of the components (totally four weight value) obtained by wavelet transform was assigned by the help of the differential evaluation algorithm. Using the differential evolution algorithm, we optimized these four weight (w) parameters of each component cA, cH, cV and cD residing in the R, G and B color channels.

In the proposed approach, the differential evolution algorithm was employed to find the optimum w parameter maximizing the sum of the entropy of the reconstructed image. Besides, the entropy of the improved image, PSNR, and sum of Entropy and PSNR were also used as fitness function to find the optimum w parameter in the differential evolution algorithm. While using the differential evolution algorithm, there are some parameters like population size (N), length of the chromosome (D), the mutation factor (F), the crossover rate (C), and the maximum generations number (g) in the algorithm which must be initialized at first. These are the main parameters of the differential evolution algorithm. In this study, we supposed that N = 40, F = 0.8, C = 0.8, and g = 1000. Also, as we had four parameters for optimization, the length of a chromosome (D) was four. During the initialization of the population, the w parameters were randomly selected between 0 and 1. The fitness function was employed as the entropy (also we tried PSNR and sum of the entropy and PSNR) of the generated image pointing the richness of information in the image. After the weighting procedure of the wavelet components finished, the new R, G, and B components were obtained by reconstructing the wavelet coefficients. In the next stage of the algorithm, unsharp masking procedure was used to enhance the edges in the image. By the help of unsharp masking, an enhanced version of the image (in which some features such as edges were sharpened) was obtained. A detailed solution flow of the proposed enhancement approach is given in Fig. 11.5.

In Fig. 11.6, an example process for an image by the proposed enhancement approach is given. Briefly, the first image (a) represents the original underwater image. In the second stage: (b), contrast adjustment procedure was applied to the original underwater image. After contrast adjustment procedure, homomorphic filtering operation (c) applied version of the image is given in the third image. In the fourth step of the system (d) filtered image was separated into its R, G and B color components. After that, by the help of wavelet transform lowpass approximation (cA), horizontal (cH), vertical (cV) and diagonal (cD) detailed images were obtained for each color channels and weighting procedure was applied to these color channels. So the new R, G and B color components (e) was obtained. In the sixth image (f), weighting R, G and B components was fused and the new color image was obtained. At the last step (g) unsharp masking procedure was applied to the new colored image and the enhanced version of the underwater image was obtained.

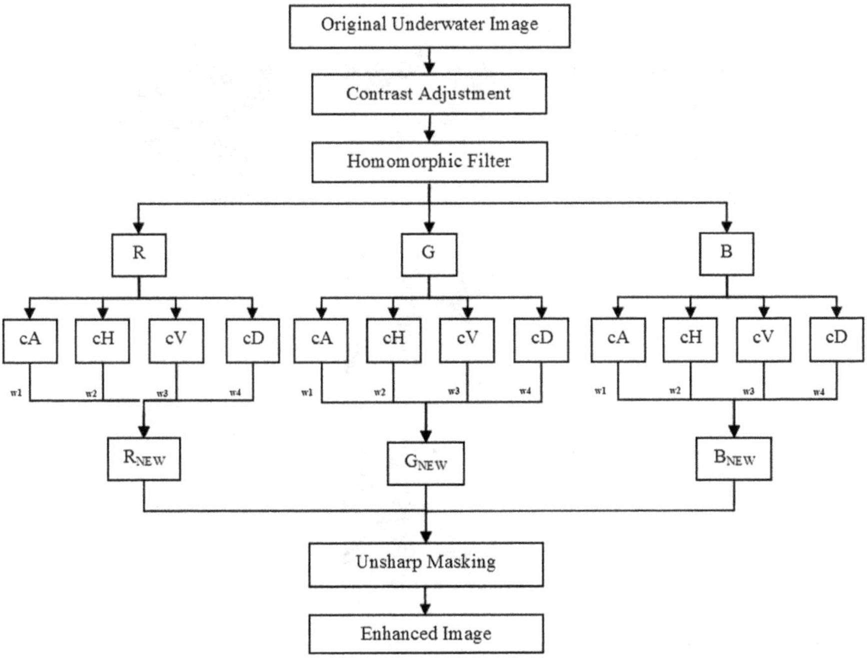

Fig. 11.5 A detailed solution flow of the proposed enhancement approach

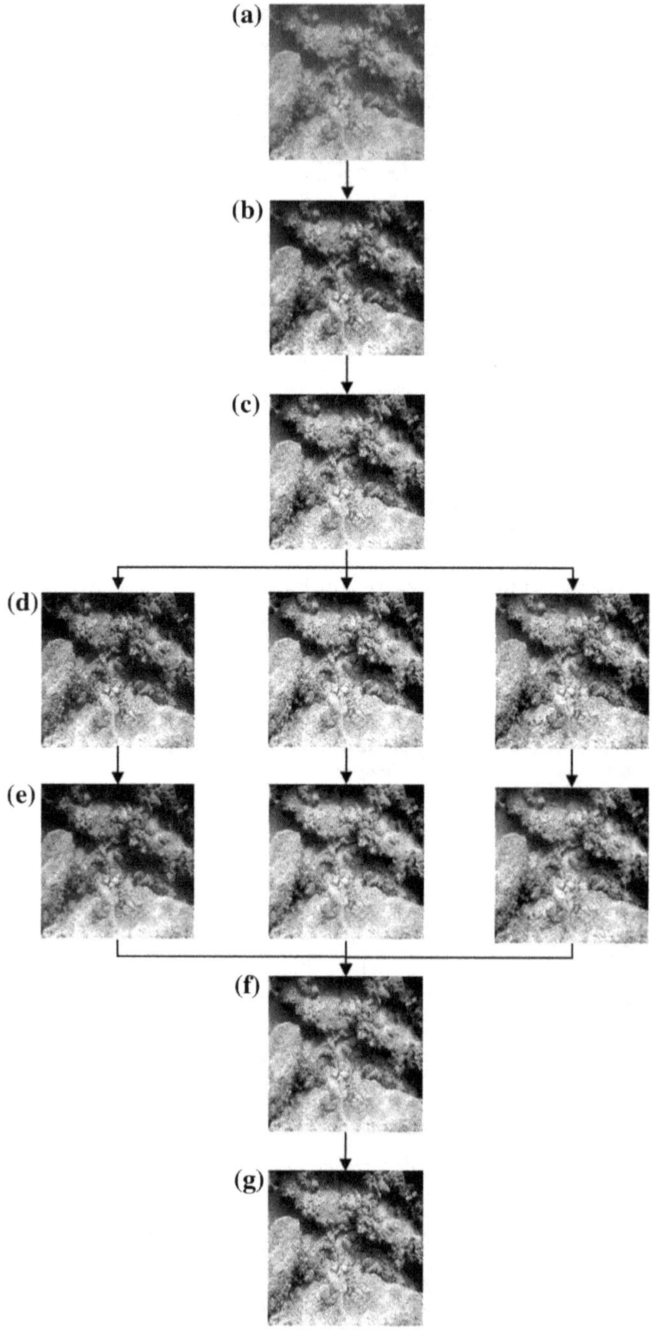

Fig. 11.6 An example process for an image by the proposed enhancement approach

11.4 Applications with the Proposed Approach

In Figs. 11.7, 11.8 and 11.9, there are some examples of the original underwater images and the enhanced underwater images using the proposed methods with the histograms. The underwater images which are image 2, image 6, image 7 and image 8 were collected from Antalya city in Turkey. Image 9 was taken from the publication written by Celebi and Erturk [58]. Image 1, image 3, image 4 and image 5 were taken from the publication written by Ghani and Isa [59].

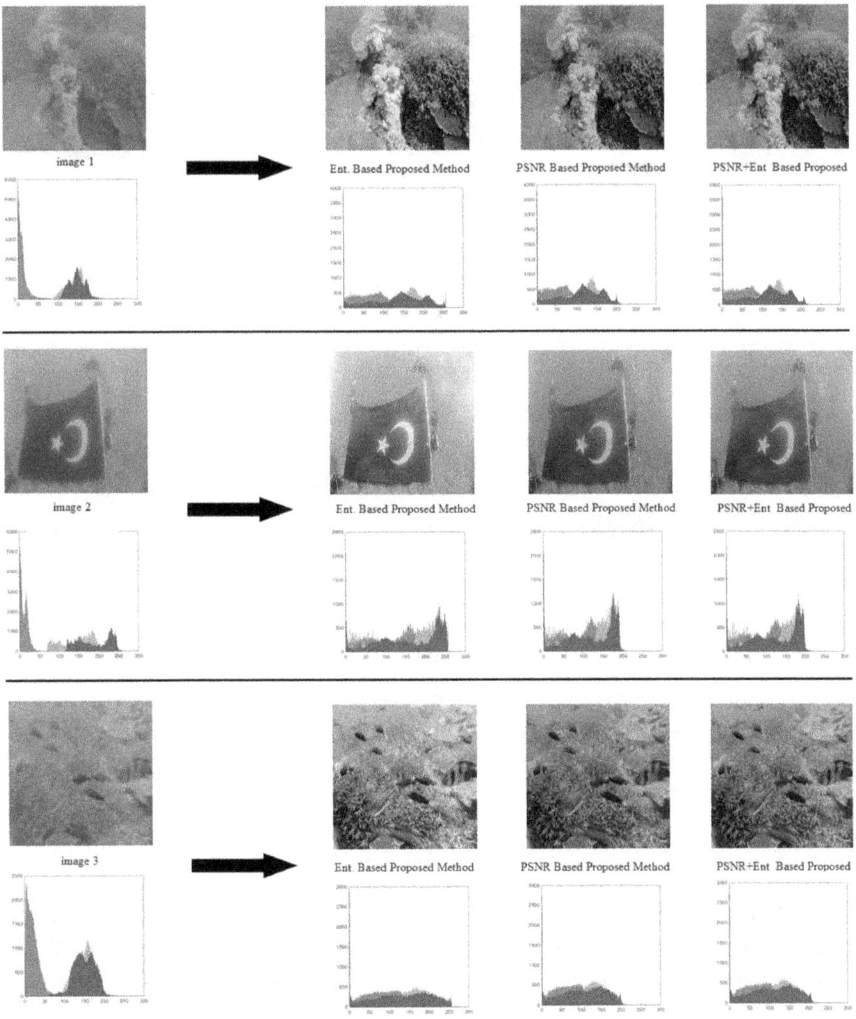

Fig. 11.7 Examples of original underwater images and enhanced underwater images with histograms (image 1, image 2, and image 3)

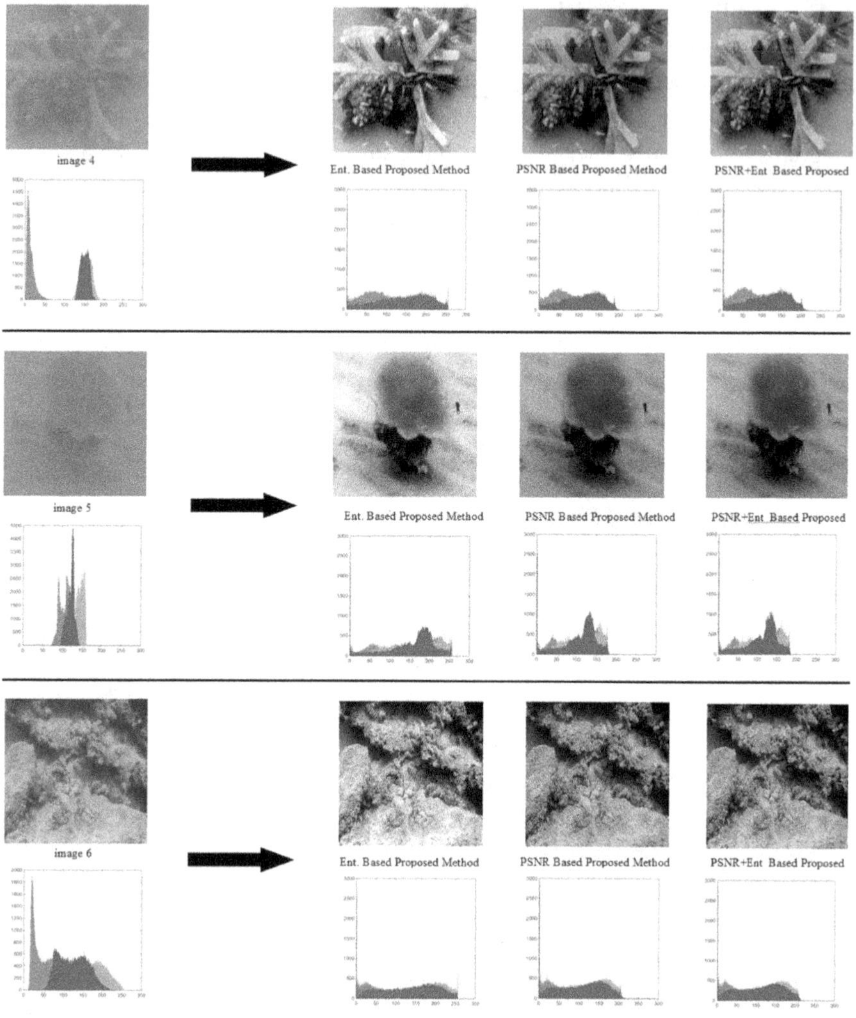

Fig. 11.8 Examples of original underwater images and enhanced underwater images with histograms (image 4, image 5, and image 6)

11.5 Evaluation

As we mentioned before, the proposed method was carried out in three different ways (in terms of comparison with the other studies in the literature) with the entropy value of the reconstructed image, PSNR value of the reconstructed image, and the sum of entropy and the PSNR value of the reconstructed image used for the information about the clarity of the image. The PSNR (Peak Signal to Noise Ratio) is known as the ratio between maximum possible power and corrupting noise that

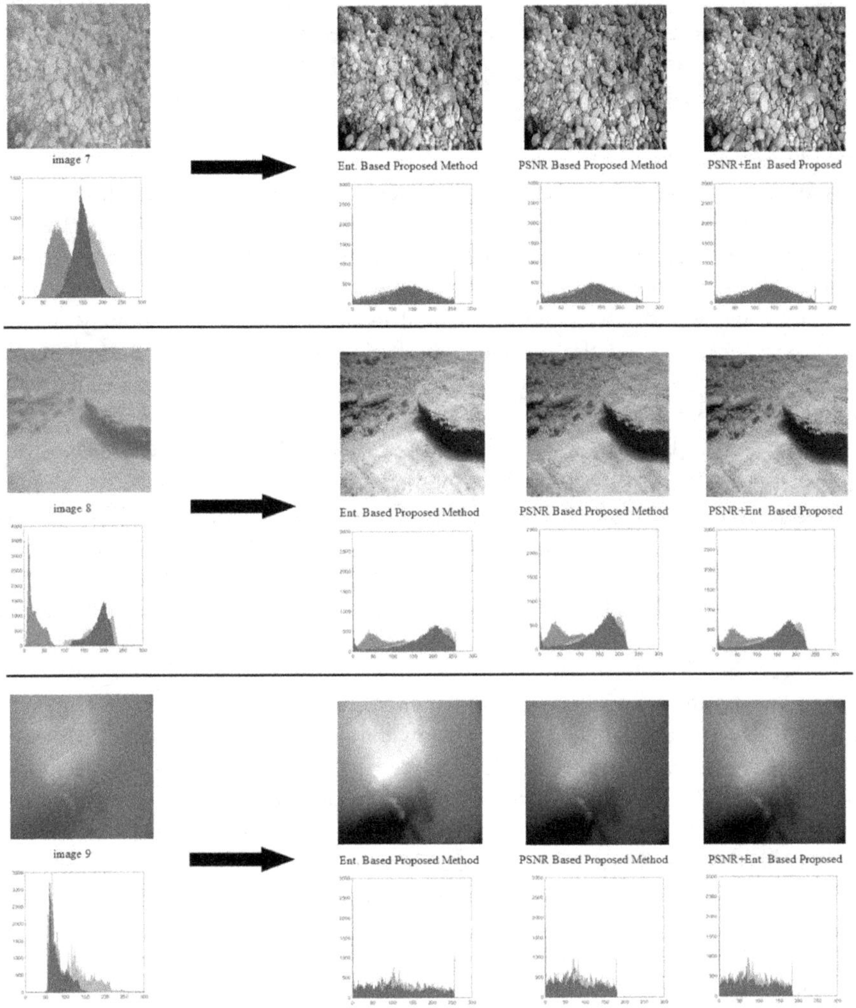

Fig. 11.9 Examples of original underwater images and enhanced underwater images with histograms (image 7, image 8, and image 9)

affect representation of an image [60]. Both PSNR and MSE are commonly used for the reconstructed or enhanced image quality measurement [60–65].

During the evaluation process, six images shown as application examples in Fig. 11.5 were considered. Regarding the evaluation, the results are given in Table 11.1 for proposed methods with PSNR, entropy, and the sum of both of them respectively. As seen in Table 11.1, proposed method with PSNR gives the best results in terms of PSNR and MSE. On the other hand, the proposed method with entropy gives the best results in terms of entropy of the final enhanced images, and the third method with the sum of the entropy and PSNR give the best result in term of the sum of the entropy and PSNR values of the final enhanced images.

For image 6, the entropy, average gradient, and PSNR values for the proposed method and previous study performed by Guraksin et al. [66] are given in Table 11.2. Beside this, the final enhanced images mentioned in Table 11.2 are given in Fig. 11.10.

As shown in Table 11.2, there are three results of the proposed method. In the first approach, we used the proposed method considering only the entropy while calculating the weights with the differential evaluation algorithm. In the second approach, we used the proposed method with considering the PSNR value while calculating the weights with the differential evaluation algorithm. Finally, in the third approach, we used the proposed method with considering the sum of the entropy and the PSNR together while calculating the weights with the differential evaluation algorithm.

As seen from Table 11.2, all the entropy values of proposed approaches are higher than the method performed by Guraksin et al. [66]. Only the average gradient value in the approach of [66] is higher. On the other hand, the PSNR and MSE values of the proposed approaches are higher than the approach of [66]. So it can be said that the proposed approaches are more efficient than the approach performed by Guraksin et al. [66]. The differences can be seen in Fig. 11.10 (final enhanced images for image 11.6).

As seen in Fig. 11.10, all details are more distinguishable in the proposed method besides the original image and the enhanced image performed by Guraksin et al. [66]. If we examine the results of its own among the three methods performed in this study, we can see that the luminosity of the entropy based approach was higher than the PSNR and the sum of the entropy and the PSNR based approaches. Besides this, we can see that the luminosity of the PSNR based approach was lower than the entropy based and the sum of the entropy and the PSNR based approaches. The sum of the entropy and the PSNR based approaches had the luminosity between these two approaches.

It is possible to focus on more comparisons for other images. For image 11.9, the comparison of entropy values for the proposed method and previous studies performed by Celebi and Erturk [58] and Bazeille et al. [20] are given in Table 11.3. As seen from the Table 11.3, the proposed approach provides the best performance while considering the entropy value.

For images 1, 3, 4, and 5, the comparison of Entropy, PSNR, and MSE values for the proposed methods and previous studies performed by Ghani and Isa [59] are given in Table 11.4. As seen from Table 11.4, the proposed approach with entropy provides the best performance while considering the entropy values except image 5. As in image 5, the entropy value of the proposed method is close to the proposed method performed by Ghani and Isa [59]. Again, the proposed approaches with PSNR and the sum of the entropy and the PSNR together provide the best performance while considering the PSNR and MSE values of the enhanced images.

Table 11.1 The entropy, average gradient, PSNR and MSE values for the related underwater images

	Image	Entropy of the original image	Entropy of the enhanced image	Average gradient of the original image	Average gradient of the enhanced image	PSNR	MSE	Entropy of the enhanced image/ entropy of the original image	PSNR + entropy of the enhanced image/ entropy of the original image
Proposed approach with PSNR	Image 1	6.877	7.509	2.490	9.702	13.349	3007.345	1.092	3.033
	Image 2	7.416	7.490	1.481	4.624	11.569	4530.522	1.010	2.570
	Image 3	7.295	7.535	3.645	11.268	12.958	3290.589	1.033	2.809
	Image 4	6.341	7.486	1.589	8.636	11.789	4307.589	1.181	3.040
	Image 5	6.216	7.277	0.647	3.850	14.949	2080.646	1.171	3.576
	Image 6	7.694	7.614	7.236	13.694	16.613	1418.243	0.990	3.149
	Image 7	7.414	7.768	7.524	21.488	14.363	2381.117	1.048	2.985
	Image 8	7.323	7.605	3.224	10.385	12.078	4029.998	1.039	2.688
	Image 9	6.810	7.375	0.713	2.084	16.612	1418.553	1.083	3.522
Proposed approach with entropy	Image 1	6.877	7.813	2.490	12.038	12.324	3808.087	1.136	2.928
	Image 2	7.416	7.861	1.4811	5.968	10.221	6179.940	1.060	2.438
	Image 3	7.295	7.857	3.645	14.070	11.849	4247.781	1.077	2.701
	Image 4	6.341	7.868	1.589	11.329	10.686	5551.847	1.241	2.926
	Image 5	6.216	7.770	0.647	5.442	11.517	4585.775	1.250	3.103
	Image 6	7.694	7.899	7.236	16.109	14.484	2315.596	1.027	2.909
	Image 7	7.414	7.794	7.524	21.612	14.121	2517.627	1.051	2.956
	Image 8	7.323	7.814	3.2243	12.016	11.460	4646.978	1.067	2.632
	Image 9	6.810	7.860	0.713	2.961	13.037	3231.301	1.154	3.069

(continued)

Table 11.1 (continued)

	Image	Entropy of the original image	Entropy of the enhanced image	Average gradient of the original image	Average gradient of the enhanced image	PSNR	MSE	Entropy of the enhanced image/ entropy of the original image	PSNR + entropy of the enhanced image/ entropy of the original image
Proposed approach with entropy + PSNR	Image 1	6.877	7.565	2.490	10.014	13.323	3025.604	1.100	3.037
	Image 2	7.416	7.543	1.481	4.764	11.548	4553.018	1.017	2.574
	Image 3	7.295	7.590	3.645	11.651	12.932	3310.688	1.040	2.813
	Image 4	6.341	7.575	1.589	9.072	11.749	4346.942	1.195	3.047
	Image 5	6.216	7.315	0.647	3.945	14.929	2090.136	1.177	3.579
	Image 6	7.694	7.635	7.236	13.835	16.601	1422.414	0.992	3.150
	Image 7	7.414	7.777	7.524	21.600	14.358	2383.720	1.049	2.986
	Image 8	7.323	7.634	3.224	10.834	12.050	4055.867	1.043	2.688
	Image 9	6.810	7.413	0.713	2.086	16.593	1424.870	1.089	3.525

Table 11.2 The entropy, average gradient, PSNR and MSE values for image 11.6

	Entropy	Average gradient	PSNR	MSE
Original Image	7.694	7.236	–	
Guraksin et al. [66]	6.897	**19.258**	12.569	3599.183
Proposed approach with entropy	**7.899**	16.109	14.484	2315.596
Proposed approach with PSNR	7.614	13.694	**16.613**	**1418.243**
Proposed approach with PSNR + Ent	7.635	13.835	16.601	1422.414

Best values in bold

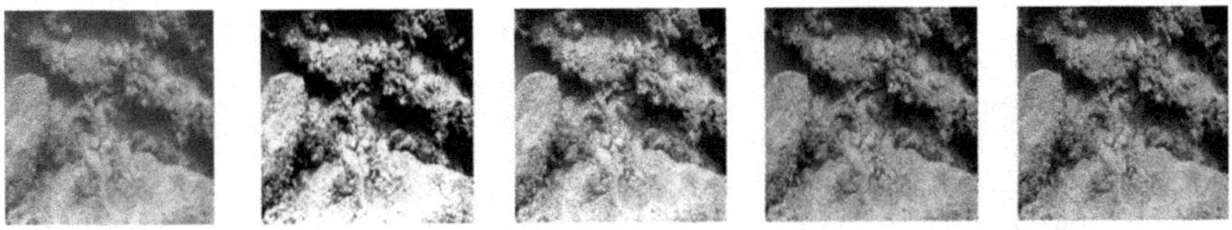

Fig. 11.10 The final enhanced images for image 6

Table 11.3 The entropy and average gradient values of other studies and entropy based proposed approach for image 11.9

Celebi and Erturk [58]	Bazeille et al. [20]	Proposed approach with entropy
Entropy of the original image = 6.19	Entropy of the original image = 6.19	Entropy of the original image = 6.81
Entropy of the enhanced image = 6.66	Entropy of the enhanced image = 6.98	Entropy of the enhanced image = 7.86
Entropy of the enhanced image/entropy of the original image = 1.100	**Entropy of enhanced the image/entropy of the original image = 1.128**	**Entropy of the enhanced image/entropy of the original image = 1.154**
Average grad. of the orig. image = 0.96	Average grad. of the orig. image = 0.96	Average grad. of the orig. image = 0.71
Average gradient of the enhanced image = 5.84	Average gradient of the enhanced image = 1.62	Average gradient of the enhanced image = 2.96
Average gradient of the enhanced image/average gradient of the original image = 6.08	**Average gradient of the enhanced image/average gradient of the original image = 1.69**	**Average gradient of the enhanced image/average gradient of the original image = 4.17**

Table 11.4 Entropy, PSNR, and MSE values for the image 1, image 3, image 4 and image 5

		Entropy (original)	Entropy (enhanced)	PSNR	MSE	Ent. Enh./Ent Org.	PSNR + Ent Enh./Ent. Org.
Image 1	Ghani and Isa [59]	6.902	7.702	12.33	3801	1.116	2.902
	Prop.App (Ent.)	6.877	7.813	12.32	3808	**1.136**	2.928
	Prop.App (PSNR)	6.877	7.509	**13.35**	**3007**	1.092	3.033
	Prop.App (PSNR + Ent)	6.877	7.565	13.32	3026	1.100	**3.037**
Image 3	Ghani and Isa [59]	7.341	7.764	12.79	3424	1.058	2.800
	Prop.App (Ent.)	7.295	7.857	11.85	4248	**1.077**	2.701
	Prop.App (PSNR)	7.295	7.535	**12.96**	**3291**	1.033	2.809
	Prop.App (PSNR + Ent)	7.295	7.590	12.93	3311	1.040	**2.813**
Image 4	Ghani and Isa [59]	6.341	7.866	10.87	5317	1.240	2.955
	Prop.App (Ent.)	6.341	7.868	10.69	5552	**1.241**	2.926
	Prop.App (PSNR)	6.341	7.486	**11.79**	**4308**	1.181	3.040
	Prop.App (PSNR + Ent)	6.341	7.575	11.75	4347	1.195	**3.047**
Image 5	Ghani and Isa [59]	6.218	7.857	13.15	3149	**1.264**	3.378
	Prop.App (Ent.)	6.216	7.770	11.52	4586	1.250	3.103
	Prop.App (PSNR)	6.216	7.277	**14.95**	**2081**	1.171	3.576
	Prop.App (PSNR + Ent)	6.216	7.315	14.93	2090	1.177	**3.579**

Best values in bold

11.6 Conclusions and Future Work

Underwater images have poor contrast and resolution. So enhancement of underwater images has a significant role due to the absorption and scattering of light in underwater environment. In this paper, we proposed a new approach to enhance underwater images using the wavelet transform and differential evolution algorithm. At first, some preprocessing operations like contrast adjustment and homomorphic filtering operations were performed to the raw underwater images. Then image was separated into its R, G, and B color components. After that, by the help of wavelet transform, lowpass approximation (cA), horizontal (cH), vertical (cV) and diagonal (cD) detailed images were obtained for each color channels, and weighting procedure was applied to these color channels obtained using the differential evaluation algorithm. After that, weighted R, G, and B components was fused, and the new color image was obtained. At the last step, unsharp masking procedure was applied to the new color image and the enhanced version of the underwater image was obtained. During the weighting procedure, the proposed method was carried out in three different ways with the entropy value of the reconstructed image, PSNR value of the reconstructed image and the sum of entropy and the PSNR value of the reconstructed image. First of all, the enhanced images were examined using the proposed method with these three attributes used for the information about the clarity of the image.

According to the results obtained, all details were more distinguishable in the proposed method besides the original image and the enhanced image. If we examine the results of its own among the three methods performed in this study, we can see that the luminosity of the entropy based approach was higher than the PSNR and the sum of the entropy and the PSNR based approaches. Therefore, some details in the image could not be seen properly. Besides this, we can see that the luminosity of PSNR based approach was lower than the entropy based and the sum of the entropy and the PSNR based approaches. So this caused the image to be a little darker which was an unwanted situation. The sum of the entropy and the PSNR based approaches had the luminosity between these two approaches. Consequently, in our opinion, the sum of the entropy and the PSNR based approaches was more efficient than the entropy based and PSNR based approaches. In this study, the proposed method was used on different underwater images, and the results were compared with the other studies in the related literature in terms of entropy and PSNR. According to the obtained results, it can be said that the proposed approach effectively improved the visibility of underwater images.

In addition to the performed studies, the authors are also focused on some future researches. In this context, there will be some researches on applying the proposed approach in different underwater conditions and different places to have more and more ideas about success of the approach. On the other hand, different variations of the approach, which are formed with the support of different optimization algorithms, will be applied to the same research problem to see if it is possible to have alternative solution ways for the related literature.

Acknowledgements This paper has been supported by Afyon Kocatepe University Scientific Research and Projects Unit with the Project number 16.KARİYER.46. The authors also would like to thank to Ali Topal for his support to adapt the study content to Springer chapter requirements.

References

1. Chiang, J.Y., Chen, Y.C.: Underwater image enhancement by wavelength compensation and dehazing. IEEE Trans. Image Process. **21**(4), 1756–1769 (2012)
2. Jayasree, M.S., Thavaseelan, G.: Underwater color image enhancement using wavelength compensation and dehazing. Int. J. Comput. Sci. Eng. Commun. **2**(3), 389–393 (2014)
3. Kaur, T., Sidhu, R.K.: Performance evaluation of fuzzy and histogram based color image enhancement. Proc. Comput. Sci. **58**, 470–477 (2015)
4. Lakshmi, R.S., Loganathan, B.: An efficient underwater image enhancement using color constancy deskewing algorithm. Int. J. Innovative Res. Comput. Commun. Eng. **3**(8), 7164–7168 (2015)
5. Lathamani, K.M., Maik, V.: Blur analysis and removal in underwater images using optical priors. Int. J. Emerg. Technol. Adv. Eng. **5**(2), 59–65 (2015)
6. Torres-Méndez, L.A., Dudek, G.: Color correction of underwater images for aquatic robot inspection. In: International Workshop on Energy Minimization Methods in Computer Vision and Pattern Recognition, pp. 60–73. Springer, Berlin (2005)
7. Prabhakar, C.J., Praveen Kumar, P.U.: An image based technique for enhancement of underwater images. Int. J. Mach. Intell. **3**(4), 217–224 (2011)
8. Banerjee, J., Ray, R., Vadali, S.R.K., Shome, S.N., Nandy, S.: Real-time underwater image enhancement: an improved approach for imaging with AUV-150. Sadhana **41**(2), 225–238 (2016)
9. Farhadifard, F., Zhou, Z., von Lukas, U.F.: Learning-based underwater image enhancement with adaptive color mapping. In: 2015 9th International Symposium on Image and Signal Processing and Analysis (ISPA), pp. 48–53. IEEE (2015)
10. Ghani, A.S.A., Aris, R.S.N.A.R., Zain, M.L.M.: Unsupervised contrast correction for underwater image quality enhancement through integrated-intensity stretched-rayleigh histograms. J. Telecommun. Electr. Comput. Eng. **8**(3), 1–7 (2016)
11. Hitam, M.S., Awalludin, E.A., Yussof, W.N.J.H.W., Bachok, Z.: Mixture contrast limited adaptive histogram equalization for underwater image enhancement. In: 2013 International Conference on Computer Applications Technology (ICCAT), pp. 1–5. IEEE
12. Li, X., Yang, Z., Shang, M., Hao, J.: Underwater image enhancement via dark channel prior and luminance adjustment. In: OCEANS 2016-Shanghai, pp. 1–5. IEEE (2016)
13. Lu, H., Li, Y., Xu, X., Li, J., Liu, Z., Li, X., et al.: Underwater image enhancement method using weighted guided trigonometric filtering and artificial light correction. J. Vis. Commun. Image Represent. **38**, 504–516 (2016)
14. Sethi, R., Sreedevi, I., Verma, O.P., Jain, V.: An optimal underwater image enhancement based on fuzzy gray world algorithm and bacterial foraging algorithm. In: 2015 Fifth National Conference on Computer Vision, Pattern Recognition, Image Processing and Graphics (NCVPRIPG), pp. 1–4. IEEE (2015)
15. Sheng, M., Pang, Y., Wan, L., Huang, H.: Underwater images enhancement using multi-wavelet transform and median filter. Indonesian J. Electr. Eng. Comput. Sci. **12**(3), 2306–2313 (2014)
16. Beohar, R., Sahu, P.: Performance analysis of underwater image enhancement with CLAHE 2D median filtering technique on the basis of SNR, RMS error, mean brightness. Int. J. Eng. Innovative Technol. **3** (2013)
17. Eustice, R., Pizarro, O., Singh, H., Howland, J.: UWIT: underwater image toolbox for optical image processing and mosaicking in MATLAB. In: Proceedings of the 2002 International Symposium on Underwater Technology, 2002, pp. 141–145. IEEE (2002)
18. Haile, M.A., Yin, W., Ifju, P.G.: MATLAB® based image preprocessing and digital image correlation of objects in liquid. In: SEM Annual Conference & Exposition on Experimental and Applied Mechanics, pp. 1–11 (2009)
19. Serikawa, S., Lu, H.: Underwater image dehazing using joint trilateral filter. Comput. Electr. Eng. **40**(1), 41–50 (2014)

20. Bazeille, S., Quidu, I., Jaulin, L., Malkasse, J.-P.: Automatic underwater image pre-processing. In: CMM'06, Brest, France (2006).
21. Kose, U., Guraksin, G.E., Deperlioglu, O.: Improving underwater image quality via vortex optimization algorithm. In: International Multidisciplinary Conference IMUCO '16, 21–22 April 2016, Antalya, Turkey, pp. 327–333 (2016)
22. Preethi, S.J., Rajeswari, K.: Membership function modification for image enhancement using fuzzy logic. Int. J. Emerging Trends Technol. Comput. Sci. **2**(2), 115–118 (2013)
23. Babu, R.K., Sunitha, K.V.N.: Enhancing digital images through cuckoo search algorithm in combination with morphological operation. J. Comput. Sci. **11**(1), 7–17 (2015)
24. Li, Q.Z., Wang, W.J.: Low-bit-rate coding of underwater color image using improved wavelet difference reduction. J. Vis. Commun. Image Represent. **21**(7), 762–769 (2010)
25. Porwik, P., Lisowska, A.: The Haar-wavelet transform in digital image processing: its status and achievements. Mach. Graphic. Vis. **13**(1/2), 79–98 (2004)
26. Das, S., Suganthan, P.N.: Differential evolution: a survey of the state-of-the-art. IEEE Trans. Evol. Comput. **15**(1), 4–31 (2011)
27. Neri, F., Tirronen, V.: Recent advances in differential evolution: a survey and experimental analysis. Artif. Intell. Rev. **33**(1–2), 61–106 (2010)
28. Price, K., Storn, R.M., Lampinen, J.A.: Differential Evolution: A Practical Approach to Global Optimization. Springer Science & Business Media, Heidelberg (2006)
29. Addison, P.S.: The Illustrated Wavelet Transform Handbook: Introductory Theory and Applications in Science, Engineering, Medicine and Finance. CRC Press, Boca Raton (2017)
30. Sifuzzaman, M., Islam, M.R., Ali, M.Z.: Application of wavelet transform and its advantages compared to fourier transform. J. Phys. Sci. **13**, 121–134 (2009)
31. Sharma, M., Singh, G., Gupta, R.: Application of wavelet–an advanced approach of transformation. Adv. Res. Electr. Electron. Eng. **1**(1), 28–34 (2014)
32. Misiti, M., Misiti, Y., Oppenheim, G., Poggi, J.-M.: Wavelet Toolbox™ Reference. The MathWorks, Inc., Natick (2016)
33. Mallat, S.G.: A theory for multiresolution signal decomposition: the wavelet representation. IEEE Trans. Pattern Anal. Mach. Intell. **11**(7), 674–693 (1989)
34. Storn, R., Price, K.: Differential Evolution—A Simple and Efficient Adaptive Scheme for Global Optimization Over Continuous Spaces, vol. 3. ICSI, Berkeley (1995)
35. Storn, R., Price, K.: Differential evolution—a simple and efficient heuristic for global optimization over continuous spaces. J. Glob. Optim. **11**(4), 341–359 (1997)
36. Chen, S., Rangaiah, G.P., Srinivas, M.: Differential evolution: method, developments and chemical engineering applications. Diff. Evol. Chem. Eng. Dev. Appl. **6**, 35 (2017)
37. Chakraborty, U.K. (ed.): Advances in Differential Evolution, vol. 143. Springer, Heidelberg (2008)
38. El Ela, A.A., Abido, M.A., Spea, S.R.: Optimal power flow using differential evolution algorithm. Electr. Power Syst. Res. **80**(7), 878–885 (2010)
39. Wang, L., Zeng, Y., Chen, T.: Back propagation neural network with adaptive differential evolution algorithm for time series forecasting. Expert Syst. Appl. **42**(2), 855–863 (2015)
40. Penas, D.R., Banga, J.R., González, P., Doallo, R.: Enhanced parallel differential evolution algorithm for problems in computational systems biology. Appl. Soft Comput. **33**, 86–99 (2015)
41. Augusteen, W.A., Kumari, R., Rengaraj, R.: Economic and various emission dispatch using differential evolution algorithm. In: 2016 3rd International Conference on Electrical Energy Systems (ICEES), pp. 74–78. IEEE (2016)
42. Bas, E.: The training of multiplicative neuron model based artificial neural networks with differential evolution algorithm for forecasting. J. Artif. Intell. Soft Comput. Res. **6**(1), 5–11 (2016)
43. Fan, H.Y., Lampinen, J.: A trigonometric mutation operation to differential evolution. J. Global Optim. **27**(1), 105–129 (2003)
44. Mayer, D.G., Kinghorn, B.P., Archer, A.A.: Differential evolution–an easy and efficient evolutionary algorithm for model optimisation. Agric. Syst. **83**(3), 315–328 (2005)

45. Vesterstrom, J., Thomsen, R.: A comparative study of differential evolution, particle swarm optimization, and evolutionary algorithms on numerical benchmark problems. In: CEC2004. Congress on Evolutionary Computation, 2004, vol. 2, pp. 1980–1987. IEEE (2004)
46. Khan, S.U., Qureshi, I.M., Zaman, F., Shoaib, B., Naveed, A., Basit, A.: Correction of faulty sensors in phased array radars using symmetrical sensor failure technique and cultural algorithm with differential evolution. Sci. World J. (2014)
47. Brownlee, J.: Differential evolution. Clever algorithms: nature-inspired programming recipes. http://www.cleveralgorithms.com/nature-inspired/evolution/differential_evolution.html. Retrieved 25 Sept 2016 (2016)
48. Cortés-Antonio, P., González, J.R., Villa-Vargas, L.A., Ramırez-Salinas, M.A., Molina-Lozano, H., Batyrshin, I.: Design and implementation of differential evolution algorithm on FPGA for double-precision floating-point representation. Acta Polytech. Hung. **11**(4), 139–153 (2014)
49. Sumithra, S., Victoire, T.: Differential evolution algorithm with diversified vicinity operator for optimal routing and clustering of energy efficient wireless sensor networks. Sci. World J. (2015)
50. Liu, J., Lampinen, J.: A fuzzy adaptive differential evolution algorithm. Soft. Comput. **9**(6), 448–462 (2005)
51. Abbass, H.A.: The self-adaptive Pareto differential evolution algorithm. In: Proceedings of the 2002 Congress on Evolutionary Computation, 2002. CEC'02, vol. 1, pp. 831–836. IEEE (2002)
52. Chaturvedi, P., Kumar, P.: Population segmentation-based variant of differential evolution algorithm. In: Proceedings of Fifth International Conference on Soft Computing for Problem Solving, pp. 401–410. Springer, Singapore (2016)
53. Sayah, S., Zehar, K.: Modified differential evolution algorithm for optimal power flow with non-smooth cost functions. Energy Convers. Manag. **49**(11), 3036–3042 (2008)
54. Mallipeddi, R., Lee, M.: An evolving surrogate model-based differential evolution algorithm. Appl. Soft Comput. **34**, 770–787 (2015)
55. Wazir, H., Jan, M.A., Mashwani, W.K., Shah, T.T.: A penalty function based differential evolution algorithm for constrained optimization. Nucleus **53**(2), 155–161 (2016)
56. Zhang, J., Lin, S., Qiu, W.: A modified chaotic differential evolution algorithm for short-term optimal hydrothermal scheduling. Int. J. Electr. Power Energy Syst. **65**, 159–168 (2015)
57. Solomon, C., Breckon, T.: Fundamentals of Digital Image Processing. Wiley-Blackwell, West Sussex (2011)
58. Celebi, A.T., Erturk, S.: Visual enhancement of underwater images using emprical mode decomposition. Expert Syst. Appl. **39**, 800–805 (2012)
59. Ghani, A.S.A., Isa, N.A.M.: Underwater image quality enhancement through composition of dual-intensity images and Rayleigh-stretching. SpringerPlus **3**(1), 1–14 (2014)
60. Kaushik, P., Sharma, Y.: Comparison of different image enhancement techniques based upon PSNR & MSE. Int. J. Appl. Eng. Res. **7**(11), 2010–2014 (2012)
61. Grgic, S., Grgic, M., Mrak, M.: Reliability of objective picture quality measures. J. Electr. Eng. **55**(1–2), 3–10 (2004)
62. Huang, X.Q., Shi, J.S., Yang, J., Yao, J.C.: Study on color image quality evaluation by MSE and PSNR based on color difference. Acta Photonica Sin. **36**(8), 295–298 (2007)
63. Kipli, K., Muhammad, M., Masra, S., Zamhari, N., Lias, K., Azra, D.: Performance of Levenberg-Marquardt backpropagation for full reference hybrid image quality metrics. In: Proceedings of International Conference of Multi-Conference of Engineers and Computer Scientists (IMECS'12) (2012)
64. Wang, Z., Bovik, A.C.: A universal image quality index. IEEE Signal Process. Lett. **9**(3), 81–84 (2002)
65. Wang, Z., Bovik, A.C., Sheikh, H.R., Simoncelli, E.P.: Image quality assessment: from error visibility to structural similarity. IEEE Trans. Image Process. **13**(4), 600–612 (2004)
66. Guraksin, G.E., Kose, U., Deperlioglu, O.: Underwater image enhancement based on contrast adjustment via differential evolution algorithm. In: 2016 International Symposium on INnovations in Intelligent SysTems and Applications (INISTA), pp. 1–5. IEEE (2016)

Chapter 12
Feature Selection in Fetal Biometrics for Abnormality Detection in Ultrasound Images

R. Ramya, K. Srinivasan, B. Sharmila and K. Priya Dharshini

Abstract Feature selection is a processing step that gives a subset of features required for analyzing an image. The process flow of medical image processing includes pre-processing, image segmentation, feature extraction and feature selection. Medical imaging has been developed predominantly nowadays. It assists the physicians to diagnose the diseases through various medical modalities. Fetal defects are the most common congenital abnormality found at birth. Fetal features are selected and extracted to determine fetal biometrics such as Amniotic Fluid Volume, Bi-parietal Diameter, Head Circumference, Abdominal Circumference, Femur Length and Gestational Age. IntraUterine Growth Restriction remains a challenging problem for both the obstetrician and the pediatrician. The vital role of this approach is to detect abnormalities non-invasively and reduce the risk factors in early stages of pregnancy.

Keywords IntraUterine growth restriction · Fetal parameters · Feature selection Ultrasound image

12.1 Introduction

The feature selection technique is an emerging research topic used in machine learning. This helps to remove the unwanted features in the images and also in enhancing the quality of the system. The feature selection is that the information

R. Ramya (✉) · K. Srinivasan · B. Sharmila · K. Priya Dharshini
Department of EIE, Sri Ramakrishna Engineering College, Coimbatore, India
e-mail: ramya.r@srec.ac.in

K. Srinivasan
e-mail: srineekvasan@srec.ac.in

B. Sharmila
e-mail: sharmila.b@srec.ac.in

K. Priya Dharshini
e-mail: priyadharshini.1676004@srec.ac.in

J. Hemanth and V. E. Balas (eds.), *Nature Inspired Optimization Techniques for Image Processing Applications*, Intelligent Systems Reference Library 150,
https://doi.org/10.1007/978-3-319-96002-9_12

contains numerous highlights that are either excess or inconsequential, would thus be able to be evacuated without bringing about much loss of data. Obstetric Ultrasonography is a non-invasive tool used to analyze the growth of the fetus across gestation. Development of a fetus in its growth is an important factor of prenatal medical care. Analysis of fetal development on ultrasound images is widely performed compared to other modalities such as Computed Tomography (CT) or Magnetic Resonance Imaging (MRI).

The common purpose of the ultrasound investigation is to determine the position of the fetus and the placenta, the number of fetuses, Estimated Fetal Weight (EFW), Amniotic Fluid Volume (AFV) Gestational Age (GA) and to detect abnormalities present in fetus. In [1] Support Vector Machine (SVM) is used to detect the outcome of IntraUterine Growth Restriction.

Wang et al. [2] proposed phase based feature selection process exploiting the boundaries of fetal head for the measurement of Bi-Parietal Diameter and Head Circumference. In [3] Haar-like feature is used to extract features from fetal head for the application of Adaboost classifier and the Bi-Parietal Diameter is measured. Yaqub et al. [4] describes Random forest classification framework for segmenting femur and the classification is improved based on selecting strong features. The modified new method gives notable improvement compared to traditional Random Forest. Rahmatullah et al. [5] proposed an automatic method for selecting standard plane for the measurement of fetal biometrics from fetal ultrasound volume. The features are derived from Haar wavelets and feature selection is executed during training process by Adaboost algorithm. Results show that the recall rate is 91.29%.

In [6] innumerable techniques are explained for feature extraction, classification and retrieval and Content-Based Image Retrieval (CBIR). Also analyzed that the features are optimized using Particle Swarm Optimization (PSO). Dorigo et al. proposed Ant colony optimization (ACO) method which can be used for system fault detecting, network load balancing, robotics and other optimization problems [7]. The BPD is measured on a horizontal plane that traverses the thalami and cavum septum pellucidum. An automatic method is implemented that measures Bi-Parietal Diameter of fetal head and length of the femur from ultrasound images. Active contour model is used in measuring the fetal parameters [8]. In [9] morphological watershed segmentation is proposed for estimating the fetal femur length and growth patterns (Fig. 12.1).

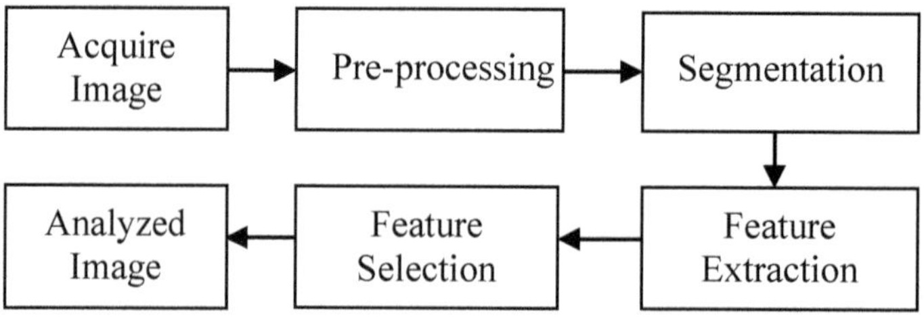

Fig. 12.1 General block diagram of fetal ultrasound images

12.2 Steps Involved in Processing of Fetal Image

The following stages are used for processing a fetal ultrasound image [10].

- Pre-processing
- Segmentation
- Feature Extraction
- Feature selection.

12.2.1 Pre-processing

Image Pre-processing is defined as removing noises from images. The main disadvantage of Obstetric Ultrasonography is poor quality of images due to speckle noise. Speckle is a granular noise that constitutionally exists and degrades the quality of the medical ultrasound images.

Speckle is primarily due to the interference patterns of the returning wave at the transducer aperture. Speckle noise in medical images generally serious, causing difficulties for image interpretation. It is a multiplicative noise that reduces the visual interpretation in fetal ultrasound images. Hence there are many attempts made in researching field to de-noise ultrasound images by developing de-speckling methods. Median filter is widely used to remove speckle noise in fetal ultrasound images.

12.2.2 Segmentation

Segmentation is one of the tedious processes in image processing. Image segmentation is the process that split an image into its constituent parts or segments. The image partitioning process continuous to a level when a particular problem being solved, i.e., the segmentation will end when the objects of interest have been detached in an application. Segmentation algorithms mainly based on discontinuity and similarity. Consequently, the automatic segmentation of anatomical structures in ultrasound imagery is a real challenge due to acoustic interference in these images.

Segmentation in ultrasound images is the process of dividing the structures or organs of fetus such as head, limbs, abdomen, femur, etc. It is carried out by means of many segmentation techniques like thresholding technique, canny segmentation algorithm, etc.

12.2.3 *Feature Extraction*

Features are quantities which distinctively describe a target such as size, shape, composition, location etc. Quantitative measurement of feature extraction and selection provides comprehensive description of an image. Feature selection and extraction plays a significant role in fetal ultrasound images to evaluate parameters such as Bi-Parietal Diameter, Head Circumference, Abdominal Circumference, Femur Length, Amniotic Fluid Volume and Gestational Age.

12.2.3.1 Estimated Fetal Weight

The fetal weight is an essential component that needs to be estimated for recognition of fetal growth. Fetal weight measurement should be carried out to reduce perinatal morbidity and mortality. IntraUterine Growth Restriction or prematurity is allied with low fetal weight. The fetus will be too small due to uteroplacental insufficiency. Fetal macrosomia refers to a fetus with larger weight and it is due to maternal diabetes. Estimated Fetal Weight is determined using parameters such as Bi-Parietal Diameter, Head Circumference, Abdominal Circumference and Femur Length. Hadlock IV formula can be used to calculate fetal weight.

Bi-Parietal Diameter

The Bi-Parietal Diameter is one of the biometric parameters used to estimate fetal weight. Anisotropic diffusion, otherwise called as Perona–Malik diffusion is a technique used to remove noise in ultrasound images. This technique does not affect the significant parts of the image content mainly edges, lines or other details that are important for the analysis of the image. Anisotropic diffusion is a generalization of diffusion process and it produces a family of parameterized images. Each resulting image is a consolidation between the input image and a filter, based on the local content of the input image. As a result, anisotropic diffusion is a non-linear and space-variant transformation of the input image.

Dilation is one of the two basic operations performed in the sector of mathematical morphology, the other being erosion. It is mostly applied to binary images, but there are versions that work on gray scale images. The basic effect of the dilation operator on a binary image is to gradually expand the boundaries of white pixel regions. Thus enlargement of areas of foreground pixels takes place while holes become smaller within the regions.

The Canny edge detector is an edge detection technique containing algorithm to detect edges in images without loss of information. Canny edge detection is a technique to extract significant structural data for further processing of images from different vision objects. The Blob Analysis technique is used to calculate quantities for labelled regions in a binary image. The blob analysis block provides measurements such as the centroid, bounding box, label matrix and blob count.

Algorithm for Bi-Parietal Diameter Calculation

Step 1: Get fetal head image.
Step 2: Smoothing of image is done by Anisotropic diffusion technique and it is defined as,

$$\partial I \partial t = \nabla c \cdot \nabla I + c(x, y, t)\Delta I \tag{12.1}$$

Diffusion coefficient is given by,

$$c(\| \nabla I \|) = e - (\| \nabla I \| / K) \tag{12.2}$$

Step 3: Morphological operation (Dilation) is applied to filtered image and it is defined by,

$$A \oplus B = \cup b \in BAb \tag{12.3}$$

Step 4: Apply canny edge algorithm to the dilated image.
Step 5: Apply Bounding box by means of blob analysis technique.
Step 6: Bi-parietal diameter is calculated by Euclidean distance formula measuring central axis of fetal head.

Head Circumference
The Head Circumference is evaluated from fetal head image. The circumference of head is calculated by fitting an ellipse model. This method gives quantities such as elliptic equation parameter, angle and major and minor axis of an ellipse. From the parameters, head circumference can be calculated. Li et al. [11] developed a learning based framework for the measurement of Head Circumference. A fast ellipse fitting (ElliFit) method is employed and Random Forest classifier is utilized to detect fetal head.
Algorithm for Head Circumference Calculation

Step 1: Acquire fetal head image.
Step 2: Image pre-processing is done by Median filter.
Step 3: Apply Binary thresholding to the filtered image.
Step 4: Head circumference is calculated by Ellipse fitting process.
Eccentricity is given by,

$$e = \mathrm{sqrt}\left(1 - \left(\left(b^2\right)/\left(a^2\right)\right)\right) \tag{12.4}$$

Circumference of ellipse is given by,

$$h = \left((a - b)^2/(a + b)^2\right) \tag{12.5}$$

$$\mathrm{cir} = \left(\mathrm{pi} * (\mathrm{a}+\mathrm{b}) * \left(1 + \left(3 * \frac{\mathrm{h}}{10 + (4 - 3 * \mathrm{h})^{0.5}}\right)\right)\right) \quad (12.6)$$

Abdominal Circumference

Abdominal Circumference is measured from fetal abdomen image. Circle Hough Transform algorithm is used to measure the circumference of abdomen. Circle Hough Transform (CHT) is a feature extraction technique for detecting circles. It is a specialization of Hough Transform. The aim of this technique is to find circles in imperfect input images. The circle parameters are produced in the Hough parameter space and then the local maxima called accumulator matrix is selected. By measuring the centre of the abdomen, concentric circles are made and the circle appropriate to the outer surface of abdomen is selected. The circumference of the circle gives the measurement of abdomen.

Algorithm for Abdominal Circumference Calculation

Step 1: Acquire fetal abdomen image.
Step 2: Image pre-processing is done by Median filter.
Step 3: Morphological operation (Dilation) is applied to filtered image.
Step 4: Canny edge detection algorithm is applied to the dilated image.
Step 5: After edge detection, Circle Hough transform algorithm is used to measure abdominal circumference of fetus.
In a two-dimensional space, a circle can be described by,

$$(\mathrm{x} - \mathrm{a})^2 + (\mathrm{y} - \mathrm{b})^2 = \mathrm{r}^2 \quad (12.7)$$

Coordinates of centre of a circle and radius is determined using Circle Hough peaks.

Femur Length

Femur is the thigh bone of a fetus. The measurement of length of the femur plays an important role in evaluating fetal growth. The femur length can be measured by means of clustering process namely, K-means clustering technique. The K-means algorithm is an iterative technique that is used to partition an image into K clusters. In [12] entropy based segmentation approach is proposed to segment femur for the evaluation of femur length. Based on the density and height-width ratio of the femur, slim and long object selection is designed.

Algorithm for Femur Length Calculation

Step 1: Acquire fetal femur image.
Step 2: Image pre-processing is done by Median filter.
Step 3: Morphological operation (Dilation) is applied to filtered image.
Step 4: Adaptive K means clustering algorithm is applied to the dilated image.
Step 5: Apply Bounding box by means of blob analysis technique.

Step 6: Femur length is calculated by measuring the maximum length of the clusters.

The gender is detected depending on the white intensity values of the processed image and fetal weight is estimated [13]. Cheng et al. [14] estimated the fetal weight using Artificial Neural Network model and accuracy is increased compared to other methods. The Fetal Ultrasonographic parameters are determined and fetal size is classified by K-means algorithm. The BPD is helpful for dating a pregnancy and in estimating intrauterine fetal weight in weight Eqs. (12.1, 12.2, 12.3), other than that its esteem is constrained and can at times be deluding in the assessment of growth in the fetus. Head circumference value is calculated by ellipse fitting algorithm Eq. (12.6). Sharma et al. [15] formulated a method that shows the results of ultrasound and clinical methods. By Dare's Formula, clinical estimation results in low average absolute error for fetal birth weight below 3500 grams and by Johnson's Formula, it gives the least average absolute error for birth weight above 3500 grams. Fetal weight using six formulae (Shepard, Campbell, Hadlock I, II, III, and IV) is estimated [16, 17]. Hadlock IV formula gives the best positive correlation results between actual birth weight and estimated fetal weight. The commonly used formulae to estimate fetal weight are Shepard, Campbell, Hadlock I, II, III and IV. The formula of Hadlock IV had the highest positive correlation with Actual birth weight among six formulae. Hadlock IV formula to manipulate estimated fetal weight is given by,

$$\begin{aligned}\text{Log10}\,(\text{weight}) = {} & 1.3596 - 0.00386 * \text{AC} * \text{FL} + 0.0064 * \text{HC} \\ & + 0.00061 * \text{BPD} * \text{AC} + 0.0424 * \text{AC} + 0.174 * \text{FL}\end{aligned} \tag{12.8}$$

12.2.3.2 Amniotic Fluid Volume

Amniotic fluid is a liquid constituting nutrients, water and biochemical products that surrounds the fetus in a uterus sac. The amniotic fluid is used to control infection, temperature, develop lung and digestive system and support umbilical cord. A low amniotic fluid volume is one of the factors that indicate the presence of IUGR problem. So measurement of Amniotic Fluid Volume becomes an essential part of fetal ultrasound evaluating fetal well-being. Perinatal death and several perinatal outcomes such as abnormal birth weight, premature rupture of membranes, fetal abnormalities and increased risk of obstetric interventions are allied with abnormal AFV. Fetal Weight is associated with Amniotic Fluid Volume during the first half of gestational period. At first, the ratio of amniotic fluid volume to fetal weight increases till 30 weeks of gestation and then appears to drop. The Amniotic Fluid Volume of a fetus is measured from the evaluation of Amniotic

Fluid Index (AFI) and it is one of the components of biophysical profile. When the amniotic fluid index estimation is less than 5 cm, then presence of anomaly is certain. The largest vertical pocket in each of the four quadrants depleted of fetal parts and umbilical cord of a uterus is detected. The sum of all amniotic fluid indices in four quadrants gives Amniotic Fluid Volume. Normal scale of Amniotic Fluid Index values range from 8 to 25 cm. Rashid [18] proposed Single Deepest Pocket (SDP) to determine AFV. SDP is the largest vertical measurement of amniotic fluid which is free from umbilical cord and other fetal parts.

Algorithm for Estimation of Amniotic Fluid Volume

Step 1: Acquire four quadrant images of amniotic fluid.

Step 2: The input image is subjected to morphological operation (Opening). Opening is the dilation of the erosion of a set A by a structuring element B,

$$A \circ B = (A \ominus B) \oplus B \tag{12.9}$$

where,

$\ominus$ and $\oplus$ denote erosion and dilation.

Step 3: Apply binary thresholding to the image.

Step 4: The maximum vertical pocket is segmented by means of black and white boundary feature.

Step 5: Amniotic fluid index is measured for each of 4 quadrants. Total amniotic volume is given by summation of all index value for 4 quadrants.

12.2.3.3 Gestational Age

Gestational age is the common term used to detect the period of pregnancy. It is measured in weeks. For first trimester, Crown Rump Length is the best predictor of gestational age. For second and third trimester, the fetal parameters such as Bi-parietal Diameter, Head Circumference, Abdominal Circumference and Femur Length are used for the calculation of gestational age.

Steps involved in Measurement of Gestational Age

Step 1: Acquire fetal image.

Step 2: Image pre-processing is done by Median filter.

Step 3: Histogram Equalization is applied to pre-processed image. The general histogram equalization formula is given by,

$$h(v) = \mathrm{round}((\mathrm{cdf}(v) - \mathrm{cdfmin} - (M \times N) - 1) \times (L - 1)) \tag{12.10}$$

Step 4: The input image is subjected to morphological operation (Opening). Opening is the dilation of the erosion of a set A by a structuring element B.

$$A \circ B = (A \ominus B) \oplus B \tag{12.11}$$

Step 5: The image is subjected to binary thresholding.

Step 6: The boundary of fetus is segmented by means of black and white boundary feature.

Step 7: Gestational age is calculated by,

$$GA = (CRL \times 1.04)0.5 \times 8.05 + 23.7 \tag{12.12}$$

GA for BPD is given by,

$$GA = ((2 * bpd) + 44.2) \tag{12.13}$$

During routine examinations, the sonographer manually plots minor and major ellipse axes on the ultrasound image and estimates the parameters for automatic detection and measurement of biometric parameters to estimate gestational age. The formula is an approximation that can be used up to 14 weeks of gestational age.

12.2.4 Feature Selection

Image Feature Selection (FS) is a predominant task which affects the performance of image classification and recognition. Feature selection technique increases the classification accuracy by selecting subset of relevant features from the data set. The feature extraction from an image can be accomplished using a numerous image processing techniques. Feature selection aid to minimize the feature space which increases the assessment accuracy and reduces the computation time. This can be attained by removing irrelevant, redundant and noisy features. The feature selection method implemented on three steps which includes screening, ranking and selecting. The screening process removes unwanted predictors having large missing values. Ranking sorts the necessary predictors based on ranks. Selecting process recognize the subset of features by conserving significant predictors [19].

The feature selection techniques for classification process include

- Information Gain
- Relief
- Fisher Score
- Lasso.

Classification is the problem of recognizing to which of a set of group, a new feature belongs based on the training set of data. The images are analysed into a set of features based on a particular model in training phase. The extracted features and label information are utilized by the learning algorithm to learn a classifier. In the testing phase, the classifier will perform on the extracted feature set with the input data and predict the labels.

The classification methods are broadly classified into,

- Linear Classifiers
- Support Vector Machines
- Decision trees and
- Neural Networks (Fig. 12.2).

Features are selected by search algorithms such as Sequential forward Selection, Sequential Backward selection, Genetic Algorithm and Particle Swarm Optimization. For measuring Bi-Parietal Diameter and Head Circumference from fetal head, the outer region of head part should have proper boundaries. This is achieved by phase-based feature selection process. This method is used to detect symmetric (ridge-like) and asymmetric (step edge-like) features. This is achieved by exploiting local phase-based measures computed from a 2D isotropic analytic signal: monogenic signal. Thus the method is used to extract the skeleton of the skull which aids to measure BPD and HC. This can be accomplished with the Multi-Scale Feature Asymmetry (MSFA) and the Multi-Scale Feature Symmetry (MSFS) measure. Good selection of feature will lead to better result of process. For the detection of fetal head in Bi-Parietal diameter measurement, Haar-like feature is used to extract feature from cropping image object and AdaBoost classifier can be used for object detection while Randomized Hough Transform can be applied for biometry measurement [3].

Traditional Random Forest (RF) technique can also be implemented on segmenting femur for the measurement of femur length. The high redundancy of feature selection process is motivated in the traditional RF framework. In the first stage, methods are involved to improve classification by having strong features and

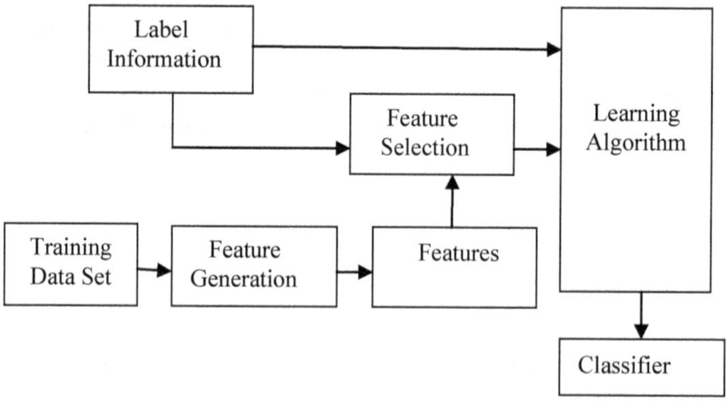

Fig. 12.2 General block diagram of feature selection for classification

neglecting weak ones. Weighting each tree in the forest can be implemented during the testing stage to yield accurate result [4].

Optimization techniques reduce the feature dimensionality by using evolutionary algorithms which includes Genetic Algorithm (GA), Genetic Programming, Evolution Strategy, Evolutionary Programming and Differential Evolution. These algorithms are used for choosing optimal features from the extracted feature set of medical images to diagnose diseases. Genetic algorithm is an optimization technique based on natural selection. It provides solutions to optimization problems depending on operators such as mutation, crossover and selection. GA provides less computational load to classification algorithm. A more important difference between genetic algorithms and other traditional optimization techniques is that population of points at one time is used by GA compared to single point access by traditional optimization methods.

A genetic algorithm needs,

- A genetic depiction of the solution
- A fitness function

Analyse classification process for many diseases with association of GA and SVM techniques for feature selection and classification. Results show that GA-SVM is robust for different medical data set [20]. In prenatal diagnosis application, GA has been used. Fetal macrosomia is one of the complicated fetal features. To differentiate Appropriate Gestational Age (AGA) and Large for Gestational Age (LGA) infants, capillary electrophoresis is used to evaluate amniotic fluid. GA was used for selecting features required for the Bayesian statistics [21]. Fetal weight can also be predicted with the help of GA.

GA is used for reducing the number for features of fetal ultrasound images and these reduced features are further used for classification process that provides better results for abnormality detection.

12.3 Experimental Results and Discussion

The proposed automatic methods were able to measure Amniotic Fluid Volume, Estimated Fetal Weight and Gestational Age. The automatic measurement of fetal biometrics such as Bi-Parietal Diameter, Head Circumference, Abdominal Circumference and Femur Length leads to automatic detection of Estimated Fetal Weight. Automatic detection of Amniotic Fluid Volume shows the fetus suffering with Oligohydramnios or Polyhydramnios. Figure 12.3 shows the processing steps of Amniotic Fluid Volume.

Table 12.1 shows the manual and automatic measurement of Amniotic fluid volume. Presence of abnormality in volume is also shown (Fig. 12.4).

Table 12.2 shows the results of Bi-Parietal Diameter with manual and automatic measurements. Figure 12.5 shows the correlation plot between manual BPD and

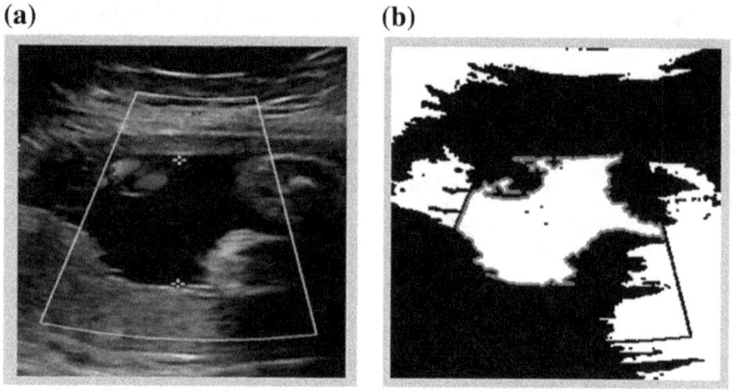

Fig. 12.3 Measurement of amniotic fluid volume **a** input image (I quadrant); **b** boundary segmentation

Table 12.1 Measurement of amniotic fluid volume

Manual measurement (in cm)	Proposed method (in cm)	Status
8.05	8.47	Normal
4.21	4.84	Oligohydramnios
6.75	6.77	Low
24.2	24.7	Polyhydramnios
8.92	8.0	Normal

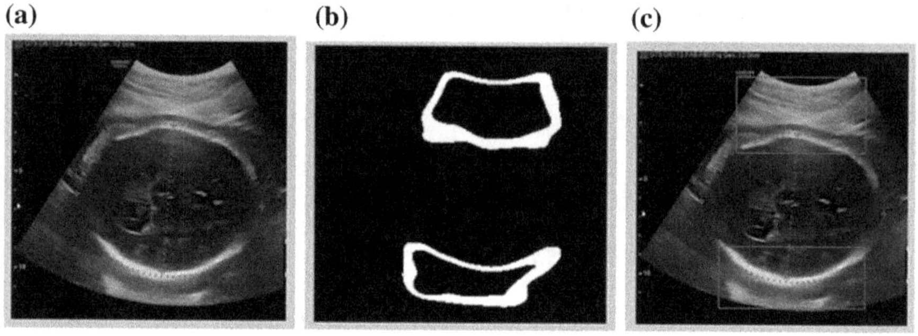

Fig. 12.4 Measurement of Bi-parietal diameter **a** input image; **b** canny edge detection; **c** bounding box image

Automatic BPD. The correlation coefficient is 0.98. Mean value for automatic measurements is 60.67 and error is 1.72 mm and standard error is 60.67 $\pm$ 1.7 mm.

The results of Head Circumference with manual and automatic measurements are shown in Table 12.3. Figure 12.6 shows the processing steps for the measurement of Head Circumference.

Table 12.2 Measurement of Bi-parietal diameter

Patient	Manual measurement (in cm)	Proposed method (in cm)
1	7.2	7.4
2	5.0	5.2
3	6.4	6.4
4	7.8	7.6
5	6.5	6.6

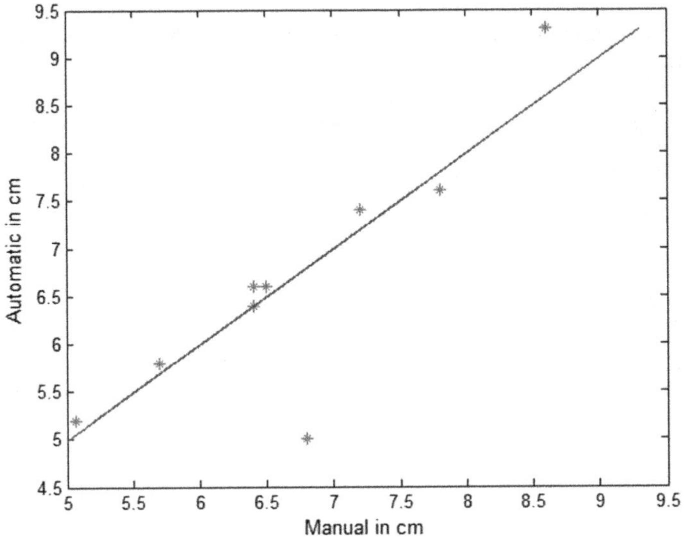

Fig. 12.5 Correlation plot for Bi-parietal diameter

Table 12.3 Measurement of head circumference

Patient	Manual measurement (in cm)	Proposed method (in cm)
1	50.0	54.6
2	24.0	24.5
3	25.0	26.1
4	23.5	23.9
5	24.0	28.6

Figure 12.7 shows the correlation plot between manual HC and Automatic HC. The correlation coefficient is 0.97. Mean value for automatic measurements is 127.87 mm and error is 2.03 mm and standard error is 127.87 ± 2.03 mm (Fig. 12.8).

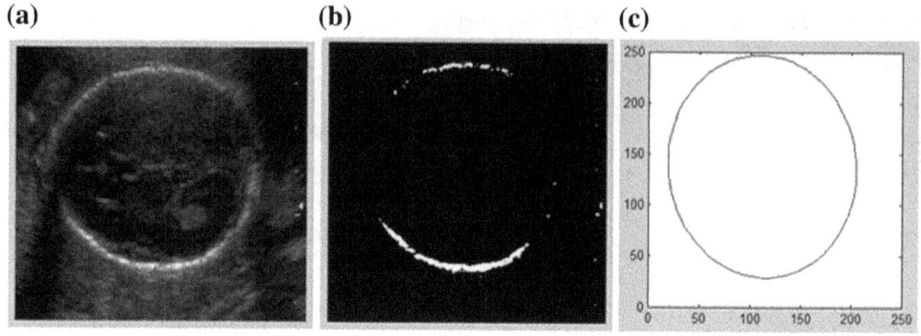

Fig. 12.6 Measurement of head circumference **a** input image; **b** binary thresholding **c** result of ellipse fitting method

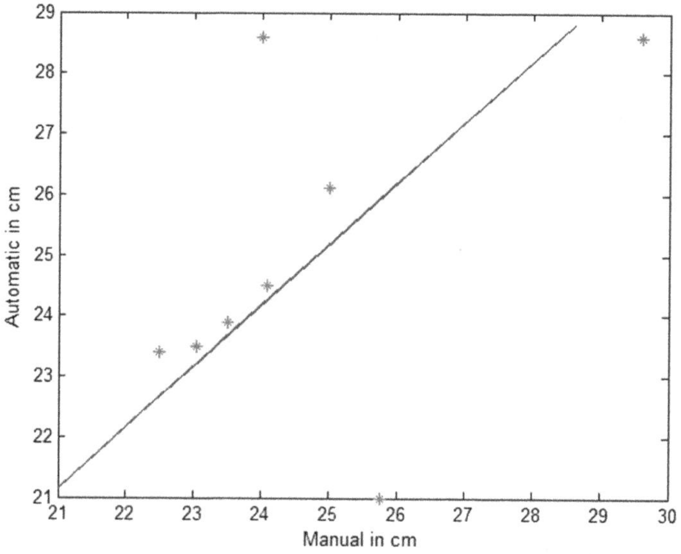

Fig. 12.7 Correlation plot for head circumference

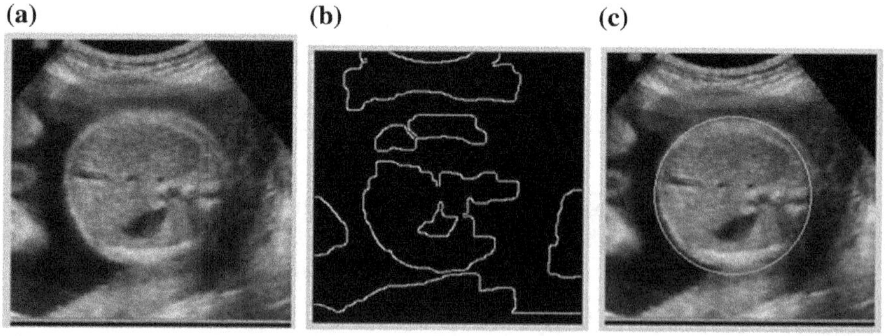

Fig. 12.8 Measurement of abdominal circumference **a** input image; **b** canny edge detection; **c** circle hough transform

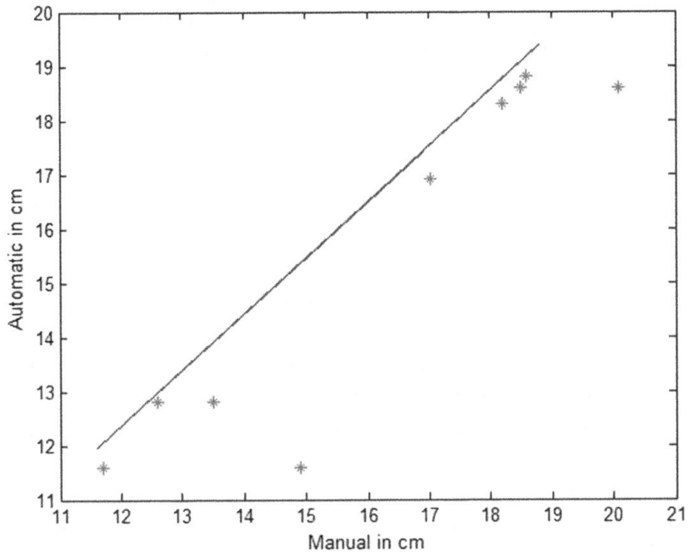

Fig. 12.9 Correlation plot for abdominal circumference

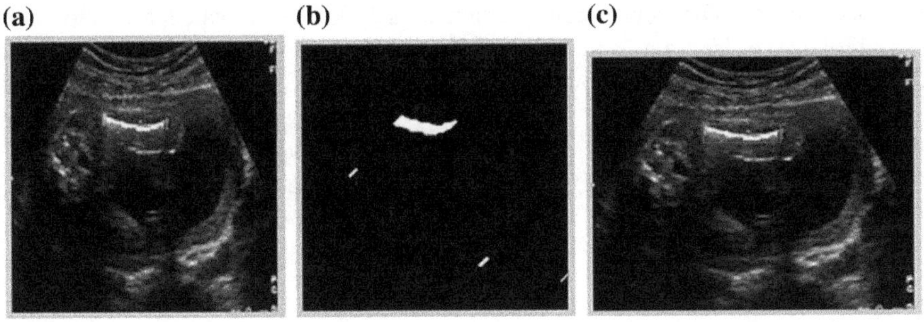

Fig. 12.10 Measurement of femur length **a** input image; **b** adaptive clustering and binary thresholding; **c** bounding box

Table 12.4 Measurement of abdominal circumference

Patient	Manual measurement (in cm)	Proposed method (in cm)
1	18.5	18.6
2	11.7	11.6
3	17.0	16.9
4	12.6	12.8
5	18.6	18.8

Table 12.5 Measurement of femur length

Patient	Manual measurement (in cm)	Proposed method (in cm)
1	2.4	2.6
2	2.2	2.3
3	3.7	3.3
4	3.2	3.3
5	3.8	4.0

Table 12.4 shows the results of Abdominal Circumference with manual and automatic measurements.

Figure 12.9 shows the correlation plot between manual AC and Automatic AC. The correlation coefficient is 0.98. Mean value for automatic measurements is 156.86 mm and error is 2.3 mm and standard deviation error is 156.86 ± 2.03 mm (Fig. 12.10).

Table 12.5 shows the results of Femur Length with manual and automatic measurements. Figure 12.11 shows the correlation plot between manual FL and Automatic FL. The correlation coefficient is 0.94. Mean value for automatic measurements is 31.88 mm and error is 2.2 mm and standard deviation error is 31.88 ± 2.2 mm.

Figure 12.12 shows the correlation plot between manual EFW and Automatic EFW. The correlation coefficient is 0.98. Mean value for automatic measurements is 272 mm and error is 2.8 mm and standard deviation error is 272 ± 2.8 mm. Table 12.6 shows the results of estimated fetal weight with manual and automatic measurements.

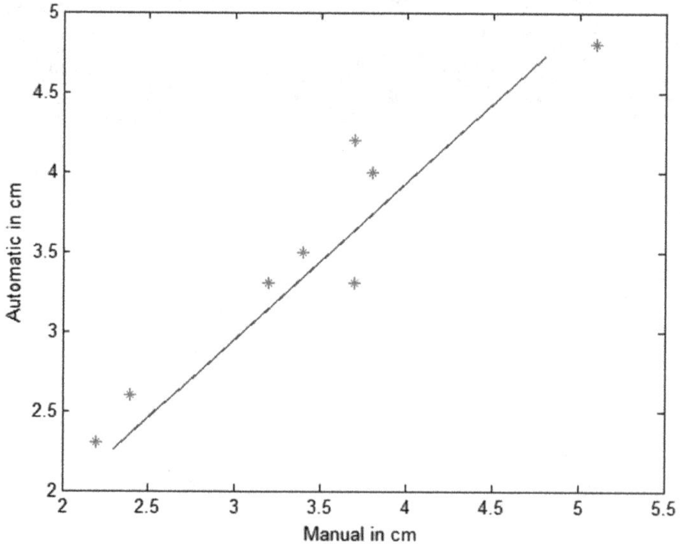

Fig. 12.11 Correlation plot for femur length

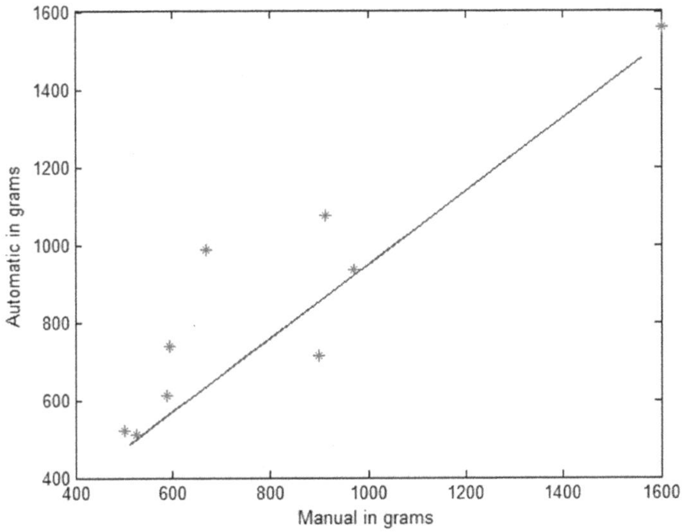

Fig. 12.12 Correlation plot for estimated fetal weight

Table 12.6 Measurement of estimated fetal weight

Gestational age (in days)	Manual measurements (in g)	Proposed method (in g)	Normal/ IUGR
193	590.0	612.0	IUGR
149	525.0	511.0	IUGR
174	594.0	600.0	Normal
197	900.0	875.0	IUGR
177	970.0	935.0	Normal

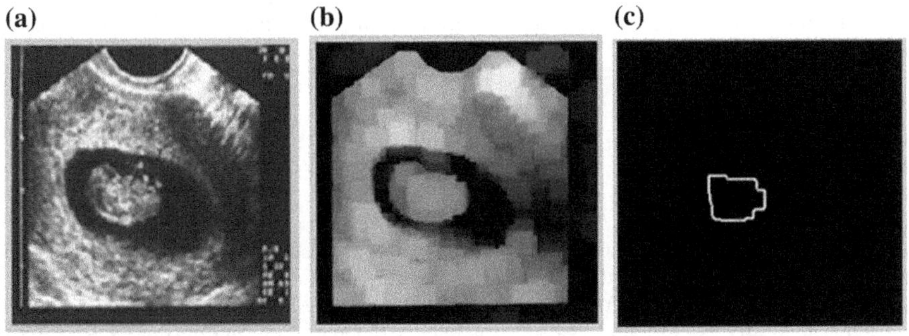

Fig. 12.13 Processing steps of estimating gestational age **a** input CRL image; **b** Histogram equalization and morphological opening; **c** boundary segmentation

Table 12.7 Measurement of gestational age

Manual gestational age	Proposed method for BPD (in mm)	Proposed gestational age
36 weeks 1 day	89.4	36 weeks 5 days
32 weeks 4 days	81.8	32 weeks
37 weeks 6 days	93.2	38 weeks
23 weeks 5 days	60.97	24 weeks
19 weeks 3 days	44.7	19 weeks 6 days

Figure 12.13 shows the processing steps of Gestational Age. The manual and automatic measurement of gestational age is shown in Table 12.7.

12.4 Conclusion

Fetal biometrics for fetal defects were analyzed. A measurement technique for the fetal parameters has been developed and implemented. The evaluation of fetal parameters gives better results for diagnosing the anomalies. Presence of abnormality of Amniotic Fluid Volume such as oligohydramnios and polyhydramnios can also be detected. Prediction of Intra-Uterine growth restricted fetus will be helpful for the obstetricians to take decision regarding the treatment of fetus and this reduces the risk for fetal delivery. This approach can easily identify the child growth restriction, fetal weight, gestational age, whether the fetal is normal or abnormal. For further development of this case having large datasets, optimization techniques can be implemented for optimal solutions using Graphical Processor Unit.

References

1. Gurgen, F., Zengin, Z., Varol, F.: Intrauterine growth restriction risk decision based on support vector machines. Expert Syst. Appl. **39**, 2872–2876 (2012)
2. Wang, W., Zhu, L., Chui, Y.P., Qin, J., Heng, P.A.: Phase based feature detection in fetal ultrasound images. In: 4th IEEE International Conference on Information Science and Technology, pp. 337–340 (2014)
3. Imaduddin, Z., Akbar, M.A., Satwika, I.P., Saroyo, Y.B.: Automatic detection and measurement of fetal biometrics to determine the gestational age. In: 3rd International Conference on Information and Communication Technology, pp. 608–612 (2015)

4. Yaqub, M., Javaid, M.K., Cooper, C., Noble, J.A.: Investigation of the role of feature selection and weighted voting in random forests for 3-D volumetric segmentation. IEEE Trans. Med. Imaging **33**(2), 258–271 (2014)
5. Rahmatullah, B., Papageorghiou, A., Noble, J.A.: Automated selection of standardized planes from ultrasound volume. In: 2nd International Conference on Machine Learning in Medical Imaging, pp. 35–42. Springer, Berlin, Heidelberg (2011)
6. Sasi, K.M., Kumaraswamy, Y.S.: Medical image retrieval system using PSO for feature selection. In: International Conference on Computational Techniques and Mobile Computing, pp. 182–186 (2012)
7. Ling, C., Bolun, C., Yixin, C.: Image feature selection based on ant colony optimization. In: Australasian Joint Conference on Artificial Intelligence, pp. 580–589. Springer, Berlin, Heidelberg (2011)
8. Khan, N.H., Tegnander, E., Dreier, J.M., Eik-Nes, S., Torp, H., Kiss, G.: Automatic detection and measurement of fetal biparietal diameter and femur length–feasibility on a portable ultrasound device. Open J. Obst. Gynecol. **7**(3), 334–350 (2017)
9. Rahmatullah, B., Besar, R.: Comparison of morphological-based segmentation methods for fetal femur length measurements. J. Mech. Med. Biol. **7**(3), 247–263 (2007)
10. Dharshini, K.P., Ramya, R., Srinivasan, K.: Certain investigations on prenatal medical image analysis. In: IEEE International Conference on Electrical, Instrumentation and Communication Engineering, pp. 1–5 (2017)
11. Li, J., Wang, Y., Lei, B., Cheng, J.Z., Qin, J., Wang, T., Li, S., Ni, D.: Automatic fetal head circumference measurement in ultrasound using random forest and fast ellipse fitting. IEEE J. Biomed. Health Inf. **22**(1), 1–9 (2018)
12. Wang, C.W.: Automatic entropy-based femur segmentation and fast length measurement for fetal ultrasound images. In: International Conference on Advanced Robotics and Intelligent Systems, pp. 1–5 (2014)
13. Aditya, Y.N., Abduljabbar, H.N., Pahl, C., Wee, L.K., Supriyanto, E.: Fetal weight and gender estimation using computer based ultrasound image analysis. Int. J. Comput. **7**(1), 11–21 (2013)
14. Cheng, Y.C., Yan, G.L., Chiu, Y.H., Chang, F.M., Chang, C.H., Chung, K.C.: Efficient fetal size classification combined with artificial neural network for estimation of fetal weight. Taiwanese J. Obst. Gynaecol. **51**(4), 545–553 (2012)
15. Sharma, N., Srinivasan, K.J., Sagayaraj, M.B., Lal, D.V.: Foetal weight estimation methods—clinical, sonographic and MRI imaging. Int. J. Scient. Res. Publ. **4**(1), 1–5 (2014)
16. Sowjanya, R., Lavanya, S.: Comparitive study of clinical assessment of fetal weight estimation using johnson's formula and ultrasonic assessment using hadlock's formula at or near term. IOSR J. Dent. Med. Sci. **14**, 20–23 (2015)
17. Kumara, D.M.A., Perera, H.: Evaluation of six commonly used formulae for sonographic estimation of fetal weight in a Srilankan population. J. Obstet. Gynaecol. **31**(1), 20–33 (2009)
18. Rashid, S.Q.: Amniotic fluid volume assessment using the single deepest pocket technique in Bangladesh. J. Med. Ultrasound **21**(4), 202–206 (2013)
19. Pushpalata, P., Jyoti, B.G.: Improving classification accuracy by using feature selection and ensemble model. Int. J. Soft Comput. Eng. **2**(2), 380–386 (2012)
20. Kumar, G.R., Ramachandra, G.A., Nagamani, K.: An efficient feature selection system to integrating svm with genetic algorithm for large medical datasets. Int. J. Adv. Res. Comput. Sci. Soft. Eng. **4**(2), 272–277 (2014)
21. Boisvert, M.R., Koski, K.G., Burns, D.H., Skinner, C.D.: Early prediction of macrosomia based on an analysis of second trimester amniotic fluid by capillary electrophoresis. Biomark. Med. **6**(5), 655–662 (2012)

CPSIA information can be obtained
at www.ICGtesting.com
Printed in the USA
LVHW05*1219300918
591916LV00011B/1260/P

9 783319 960012